SPRINGER HANDBOOK OF AUDITORY RESEARCH

Series Editors: Richard R. Fay and Arthur N. Popper

Springer
New York
Berlin
Heidelberg
Barcelona
Budapest
Hong Kong
London
Milan
Paris
Santa Clara
Singapore
Tokyo

Springer Handbook of Auditory Research

Volume 1: The Mammalian Auditory Pathway: Neuroanatomy
Edited by Douglas B. Webster, Arthur N. Popper, and Richard R. Fay

Volume 2: The Mammalian Auditory Pathway: Neurophysiology
Edited by Arthur N. Popper and Richard R. Fay

Volume 3: Human Psychophysics
Edited by William Yost, Arthur N. Popper, and Richard R. Fay

Volume 4: Comparative Hearing: Mammals
Edited by Richard R. Fay and Arthur N. Popper

Volume 5: Hearing by Bats
Edited by Arthur N. Popper and Richard R. Fay

Volume 6: Auditory Computation
Edited by Harold L. Hawkins, Teresa A. McMullen, Arthur N. Popper, and Richard R. Fay

Volume 7: Clinical Aspects of Hearing
Edited by Thomas R. Van De Water, Arthur N. Popper, and Richard R. Fay

---

Forthcoming volumes (partial list)

Development of the Auditory System
Edited by Edwin Rubel, Arthur N. Popper, and Richard R. Fay

The Cochlea
Edited by Peter Dallos, Arthur N. Popper, and Richard R. Fay

Plasticity in the Auditory System
Edited by Edwin Rubel, Arthur N. Popper, and Richard R. Fay

Thomas R. Van De Water
Arthur N. Popper
Richard R. Fay
Editors

Clinical Aspects of Hearing

With 45 Illustrations

 Springer

Thomas R. Van De Water
Department of Otolaryngology and
Department of Neuroscience
Albert Einstein College of Medicine
Bronx, NY 10461, USA

Arthur N. Popper
Department of Zoology
University of Maryland
College Park, MD 20742-9566, USA

Richard R. Fay
Department of Psychology and
Parmly Hearing Institute
Loyola University of Chicago
Chicago, IL 60626, USA

Series Editors: Richard R. Fay and Arthur N. Popper

Cover illustration: A radial section through the cochlear duct (2nd turn) of a 6-day-old rabbit. From Hans Held. *Trans. Math. Physics Roy. Konigl.,* Vol. 23, No. 5. Sachsischen Association of Science, Leipzig, 1909.

Library of Congress Cataloging in Publication Data
Clinical aspects of hearing / editors, Thomas R. Van De Water, Arthur
  N. Popper, Richard R. Fay.
    p. cm. — (Springer handbook of auditory research : v. 7)
  Includes bibliographical references and index.
  ISBN 0-387-97842-9
  1. Hearing disorders—Pathophysiology. 2. Hearing—Physiological
aspects. I. Van De Water, Thomas R. II. Popper, Arthur N.
III. Fay, Richard R. IV. Series.
  [DNLM: 1. Hearing Disorders—physiopathology. 2. Hearing-
-physiology. W1 SP685F v.7 1995 / WV 270 C6405 1995]
RF291.C56 1995
617.8—dc20
DNLM/DLC
                                                              95-9347

Printed on acid-free paper.

© 1996 Springer-Verlag New York, Inc.
All rights reserved. This work may not be translated or copied in whole or in part without the written permission of the publisher (Springer-Verlag New York, Inc., 175 Fifth Avenue, New York, NY 10010, USA), except for brief excerpts in connection with reviews or scholarly analysis. Use in connection with any form of information storage and retrieval, electronic adaptation, computer software, or by similar or dissimilar methodology now known or hereafter developed is forbidden.
The use of general descriptive names, trade names, trademarks, etc., in this publication, even if the former are not especially identified, is not to be taken as a sign that such names, as understood by the Trade Marks and Merchandise Marks Act, may accordingly be used freely by anyone.
While the advice and information in this book are believed to be true and accurate at the date of going to press, neither the authors nor the editors nor the publisher can accept any legal responsibility for any errors or omissions that may be made. The publisher makes no warranty, express or implied, with respect to the material contained herein.

Production managed by Karina Gershkovich and Terry Kornak; manufacturing supervised by Jeffrey Taub.
Typeset by TechType, Inc., Upper Saddle River, NJ.
Printed and bound by Braun-Brumfield, Ann Arbor, MI.
Printed in the United States of America.

9 8 7 6 5 4 3 2 1

ISBN 0-387-97842-9 Springer-Verlag New York Berlin Heidelberg
ISBN 3-540-97842-9 Springer-Verlag Berlin Heidelberg New York

Jan Wersäll, M.D., Ph.D

This volume on *Clinical Aspects of Hearing* is dedicated to Professor Jan Wersäll, Chairman of the Department of Otolaryngology, Karolinska Institute, Stockholm, Sweden, 1975-1995. On the occasion of his retirement, we honor his long and productive career as both a gifted scientist and a compassionate, skilled clinician.

Series Preface

The *Springer Handbook of Auditory Research* presents a series of comprehensive and synthetic reviews of the fundamental topics in modern auditory research. It is aimed at all individuals with interests in hearing research including advanced graduate students, postdoctoral researchers, and clinical investigators. The volumes will introduce new investigators to important aspects of hearing science and will help established investigators to better understand the fundamental theories and data in fields of hearing that they may not normally follow closely.

Each volume is intended to present a particular topic comprehensively, and each chapter will serve as a synthetic overview and guide to the literature. As such, the chapters present neither exhaustive data reviews nor original research that has not yet appeared in peer-reviewed journals. The series focuses on topics that have developed a solid data and conceptual foundation rather than on those for which a literature is only beginning to develop. New research areas will be covered on a timely basis in the series as they begin to mature.

Each volume in the series consists of five to eight substantial chapters on a particular topic. In some cases, the topics will be ones of traditional interest for which there is a solid body of data and theory, such as auditory neuroanatomy (Vol. 1) and neurophysiology (Vol. 2). Other volumes in the series will deal with topics which have begun to mature more recently, such as development, plasticity, and computational models of neural processing. In many cases, the series editors will be joined by a co-editor having special expertise in the topic of the volume.

Richard R. Fay
Arthur N. Popper

Preface

There is little doubt that one of the motivations for doing research on problems of hearing and communication is to help us better understand, and treat, diseases of the auditory system. While most of the volumes in the Springer Handbook of Auditory Research series are devoted to issues of basic science, we thought it appropriate to include one volume in the series that brings together issues in basic science with issues from the clinical sciences. Our goal is to help the basic scientist understand and appreciate the application of basic research to clinical issues, while, at the same time, to help introduce the clinician to basic science issues directly related to some of the most fundamental clinical issues in the auditory sciences today.

This volume contains eight chapters, seven of which deal with different general clinical issues. Chapter 1, by Robert Ruben, provides an overview and brings together the clinical and basic science issues germane to this volume. The genetic bases of hearing disorders are considered by Steel and Kimberling in Chapter 2. In Chapter 3, Van De Water, Staecker, Apfel, and Lefebvre examine the critical issues of neuronal survival, regeneration of injured neuronal processes, and the neurotrophic factors that might ultimately control such regeneration, repair, and survival. Gravel and Ruben, in Chapter 4, discuss the critical issues associated with auditory deprivation and ask questions about what happens when an animal or human lacks acoustic stimulation during a period formative for the auditory system. A major concern in the clinical literature, and a powerful tool for the basic scientist is the effect of ototoxic drugs on sensory hair cells. Issues of ototoxicity are dealt with in Chapter 5 by Garetz and Schacht. Numerous viral diseases are known to affect the ear, and these are considered by Woolf in Chapter 6. Two specific issues of inner ear function are discussed in the last two chapters. In Chapter 7, Whitehead, Lonsbury-Martin, and McCoy examine otoacoustic emissions, while in Chapter 8, Penner and Jastreboff discuss the biology and etiology of tinnitus, and present a promising animal model for this problem.

Thomas R. Van De Water
Arthur N. Popper
Richard R. Fay

Contents

Series Preface		vii
Preface		ix
Contributors		xiii
Chapter 1	Introduction ROBERT J. RUBEN	1
Chapter 2	Approaches to Understanding the Molecular Genetics of Hearing and Deafness KAREN P. STEEL AND WILLIAM KIMBERLING	10
Chapter 3	Regeneration of the Auditory Nerve: The Role of Neurotrophic Factors THOMAS R. VAN DE WATER, HINRICH STAECKER, STUART C. APFEL, AND PHILIPPE P. LEFEBVRE	41
Chapter 4	Auditory Deprivation and Its Consequences: From Animal Models to Humans JUDITH S. GRAVEL AND ROBERT J. RUBEN	86
Chapter 5	Ototoxicity: Of Mice and Men SUSAN L. GARETZ AND JOCHEN SCHACHT	116
Chapter 6	The Role of Viral Infection in the Development of Otopathology: Labyrinthitis and Autoimmune Disease NIGEL K. WOOLF	155
Chapter 7	Otoacoustic Emissions: Animal Models and Clinical Observations MARTIN L. WHITEHEAD, BRENDA L. LONSBURY-MARTIN, GLEN K. MARTIN, AND MARCY J. MCCOY	199

Chapter 8 Tinnitus: Psychophysical Observations in Humans
and an Animal Model 258
MERRILYNN J. PENNER AND PAWEL J. JASTREBOFF

Index 305

Contributors

Stuart C. Apfel
Department of Neurology, Albert Einstein College of Medicine, Bronx, NY 10461, USA

Susan L. Garetz
Kreesge Hearing Research Institute, Department of Otolaryngology, University of Michigan, Ann Arbor, MI 48109, USA

Judith S. Gravel
Department of Otolaryngology, Albert Einstein College of Medicine, Bronx, NY 10461, USA

Pawel J. Jastreboff
Department of Surgery, University of Maryland School of Medicine, Baltimore, MD 21201-1192, USA

William Kimberling
Center for Hereditary Communication Disorders, Boys Town National Research Hospital, Omaha, NE 68131, USA

Philippe P. Lefebvre
Departments of Audiophonology and Otolaryngology, Department of Human Physiology and Pathophysiology, University of Liege, Liege, Belgium; and Department of Otolaryngology, Albert Einstein College of Medicine, Bronx, NY 10461, USA

Brenda L. Lonsbury-Martin
Department of Otolaryngology, University of Miami Ear Institute, Miami, FL 33101, USA

Glen K. Martin
Department of Otolaryngology, University of Miami Ear Institute, Miami, FL 33101, USA

Marcy J. McCoy
Department of Otolaryngology, University of Miami Ear Institute, Miami, FL 33101, USA

Merrilynn J. Penner
Department of Psychology, University of Maryland, College Park, MD 20742, USA

Robert J. Ruben
Department of Otolaryngology, Montefiore Medical Center, Bronx, NY 10467, USA

Jochen Schacht
Kreesge Hearing Research Institute, Department of Otolaryngology, University of Michigan, Ann Arbor, MI 48109, USA

Hinrich Staecker
Department of Neurosciences, Albert Einstein College of Medicine, Bronx, NY 10461, USA

Karen P. Steel
MRC Institute of Hearing Research, University Park, Nottingham NG7 2RD, UK

Thomas R. Van De Water
Department of Otolaryngology & Department of Neurosciences, Albert Einstein College of Medicine, Bronx, NY 10461, USA; and Department of Human Physiology and Pathophysiology, University of Liege, Liege, Belgium

Martin L. Whitehead
Department of Otolaryngology, University of Miami Ear Institute, Miami, FL 33101, USA

Nigel K. Woolf
Division of Otolaryngology, Department of Surgery, University of California at San Diego, School of Medicine, La Jolla, CA 92093-0666, USA

1
Introduction

Robert J. Ruben

1. Introduction

The last two decades of the twentieth century have witnessed quantal advances in our knowledge of the biological processes that result in hearing. These have come about from the application of contemporary molecular biology, physiology, and imaging techniques that have been made available to all of biology in combination with discoveries in the domain of hearing. This volume, Clinical Aspects of Hearing, presents these pivotal developments as they are and will be applied to the understanding of normal and abnormal hearing in humans. The volume begins with information concerning the molecular genetics underlying hearing and hearing loss (Steel and Kimberling, Chapter 2). The flow of information proceeds to examine the relationship between auditory nerve regeneration and neurotropic factors (Van De Water, Staecker, Apfel, and Lefebvre, Chapter 3); auditory deprivation, which is an example of the effects of central nervous system (CNS) plasticity (Gravel and Ruben, Chapter 4); ototoxicity (Garetz and Schact, Chapter 5); viral infection and immune disease of the inner ear (Woolf, Chapter 6); otoacoustic emissions (Whitehead, Longsbury-Martin and McCoy, Chapter 7), and ends with information concerning tinnitus (Penner and Jastreboff, Chapter 8). The subjects covered in these seven chapters are products of the integration of knowledge of basic science and the morbidity found in clinical otology. Knowledge that will further the prevention, cure, and care of hearing disorders has never been more central to society's success.

In a communication-based culture, a person with a hearing loss, which is manifest through deficient language, is economically and socially disadvantaged, arguably even more so than if he or she were in a wheelchair or had some other physical disability.

It is indeed fortunate that there has been such integration of the two cultures—the laboratory and the clinic—in this pivotal area of human communication so that advances are beginning to be made. There is a substantial probability that even greater, unimagined, ones will be made that will affect the prevention, cure, and care of hearing disorders. The development and understanding of the biological basis of hearing is essential to the continued development of the new information-based society. These seven chapters each contribute information that is essential to the ability of both scientists and clinicians to further their understanding and treatment of hearing disorders.

2. Clinical Aspects of Hearing

2.1 Approaches to Understanding the Molecular Genetics of Hearing and Deafness

Genetic hearing loss accounts for at least 50% and probably more of all hearing disorders. The high prevalence of genetic etiology makes the understanding and application of genetics to the prevention, cure, and care of hearing disorders a high priority for scientific study and application. The knowledge that comes from molecular genetics will be needed to detect, prevent, and possibly cure many of these genetically based disorders. Additionally, the unfolding of the molecular genetic control of the development of the ear will be needed to further our ability to protect, repair, and regenerate cells and organelles of the ear. The strategies for identifying genes that affect the ear are described and include the construction of inner ear cDNA libraries. A major challenge is to determine what information in the cDNA library is most critical to each question concerning normal and abnormal or pathological events occurring in the inner ear. The conservation of genetic information between different species is useful for it allows the scientist to screen many different tissues of the inner ear in quantity, from mouse or other nonprimate species to define a moiety. Once this is defined, then, with the uses of limited human material, it can be assessed to see if it applies to humans.

Another strategy is to work with genes that are known to affect the auditory systems. There are abundant models for this work, especially in the mouse. Identification of human deafness genes has been achieved in a number of disorders that have other associated phenotypic markers, such as the pigmentation abnormalities of Waardenburg's syndrome or the retinal changes in the various types of Usher's disease. A major task is now to identify the gene(s) responsible for "nonsyndromic" hearing loss. This important area of genetics can be enhanced by the study of a number of the mouse mutants for then this will provide genetic models that can be tested in the patients with recessive and/or non syndromic disorders.

2.2 Regeneration of the Auditory Nerve: The Role of Neurotrophic Factors

A major development is the characterization of the repair capabilities of mammalian neural tissue—most notably the growth of neurites. There has been a number of observations that have shown spontaneous attempts of repair — regeneration of neurites in the traumatized inner ears of mammals. The localized application of neurotrophins, neurotrophic growth factors both in vitro and in vivo (Staecker et al. 1995b) has now shown that damaged neuronal processes can be regenerated or repaired. These studies have also shown that the use of these neurotropic growth factors can enable a neuron to be preserved under circumstances that, without the manipulation of the molecular biological environment, would have resulted in their death because of a lack of adequate trophic support.

There are several ways in which these molecular substances can be delivered to the neural structures of the inner ear in vivo. The methods of delivery include the use of osmotic pumps connected to the inner ear, carrier drugs, implantable biological release polymers, transplanted cells programmed to produce the appropriate substances, and mutated viruses that will produce the required substances.

There are abundant data from amphibians and lower vertebrate species that there is a capacity for not only repair of neurites but also regeneration of the proper connections and recovery of auditory function (Corwin 1986; Zakon 1986). Currently this is a primary area for investigation in mammals, including humans. It is theoretically possible for neurons to survive injury and for their neuronal processes to be regenerated if the proper activating factors are known. The information that has been gained from the work of regeneration and pathway control of neurites is a critical step toward this next advance. The information gained in the in vitro and in vivo studies of regeneration of the auditory nerve will have a number of clinical applications. These include the preservation of the neurons and their neural elements so that electrical prostheses can function optimally, and as a necessary part of repair or regeneration of the mammalian cochlear sensory receptor (Staecker et al. 1995a). The description of the molecular control of the growth, repair, regeneration, and maintenance of the neural structures of the auditory system will also be useful for the development of a molecular biological nosology that will describe diseases and indicate possible means for their alleviation. The next phase of this research should show substantial advances in the ability to allow the inner ear to preserve, repair, and possibly regenerate its neural structures so that the cochlea may remain connected to the CNS.

2.3 Auditory Deprivation and Its Consequences from Animal Models to Humans

The anatomical, physiological, and behavioral effects of different types and amounts of auditory deprivation in various species, including humans, have been reviewed and summarized. These external stimulatory environments result in various outcomes, which become part of the acoustic receptive repertory. The organism is affected by many types of sound deprivation, which include unilateral deprivation, conductive loss, and less than severe hearing losses. Many of these deprivations have been studied both in human and in other species and the results, on the whole, are concordant. These cross species data compel the need for effective interventions in humans with these deficiencies. An example is that in the past, unilateral hearing loss in children was not thought to have any linguistic—educational consequence. Studies in this area, as exemplified by the work of Bess (1984) have shown that it in fact entails substantial linguistic and educational morbidity. There is other information concerning the effects of the fluctuating losses that are associated with otitis media with effusion (OME), which are variable between different populations.

The variability of the effect of these different losses may come about from a number of different reasons, some of which have to do with experimental design, controls, and the appropriateness of the measuring instruments. A portion of the difference reported also may be due to the plasticity—resiliency of the CNS. The concept of "a critical period" has been questioned. Are there different critical—sensitive periods for different aspects of auditory communication? Additionally, how malleable or plastic is the CNS when confronted with losses of hearing and/or major changes in the auditory environment?

Optimization of linguistic ability is dependent upon aural communication. Knowledge of the effects of permutations and perversion of these inputs in the CNS and the adaptability of the CNS's response to these changes is essential information in constructing interventions that are directed at the optimization of an aural based language. This is an area of critical importance to the information—communication based society from at least two points of view. The first is the amelioration of the effects of morbid pathological auditory inputs upon humans of all ages. The second would be to devise means to make aural input even more effective in the area of communication and the ability of such stimuli to maximize the linguistic proficiency of each person. Sound—auditory experience has in this age become both a vector of disease/linguistic impairment—and a potential enhance of health—a vehicle to optimize language.

2.4 Ototoxicity

Ototoxic substances have served humankind in perverse ways. Many of them have saved, prolonged life and increased the quality of life in some areas, while at the same time they have caused hearing loss as a price for their effectiveness. A significant aspect is their utility in defining fundamental receptor cell biochemistry and physiology. The characterization of the effects of these diverse substances allows the scientist or the clinician to understand how they act on sensory cells. These substances have been classified both by structure and by use. The detailed information concerning the aminoglycosides, antineoplastic agents, salicylates, diuretics, and other compounds that include quinine, erythromycin, polypeptide antibiotics, and a-difluoromethylornithine is presented systematically and compresses a voluminous literature for effective use. This information serves as the basis for further understanding of the normal and morbid aspects of the biochemistry of the inner ear. From these data, specific interventions will come about for the amelioration of the unwanted effects of these agents and most likely for the interventions other disease states.

2.5 The Role of Viral Infection in the Development of Otopathology: Labyrinthitis and Autoimmune Disease

Hearing loss may be considered as resulting from either intrinsic disorders that are genetically determined or from extrinsic agents of disease such as a bacterium, a virus, or a combination of the two infectious agents. Detailed descriptions, when known, of the mode of action of the mumps, measles (rubeola) cytomegalovirus (CMV), rubella, herpes zoster oticus, influenza types A and B, Parainfluenza serotypes 1, 2, and 3, adenovirus, herpes hominis (simplex types 1 and 2), variola, HIV and AIDS, arenavirus, polio, Columbia SK, yellow fever, and the Epstein-Barr virus are presented. CMV infection may actually be responsible for the largest amount of viral acquired hearing loss. The interaction of CMV with variant host defenses is an area that is now developing. There appears to be a special role of CMV as a purported factor in hearing loss that is associated with HIV—AIDS infection, especially in children. Developing information of this particular relation should be productive in characterizing the mechanisms underlying the host susceptibility. All patients with CMV and/or AIDS do not have a hearing loss. What are the necessary conditions for a morbid process to come about and what are the conditions that serve to prevent the disease? The viral models will play a substantial role in understanding the complexities underlying the variability of susceptibility to infectious agents.

The understanding of the mechanisms of action of viral agents in the inner ear has opened a heretofore unappreciated aspect of this discipline. It is, in retrospect, no surprise that the inner ear has a complex and active immune system. Knowledge of the way in which this immune system interacts with various infectious agents forms a basis for understanding the mechanisms of destruction of the inner ear, and also serves as a guide for their treatment. These immune reactions have been associated with bacterial meningitis and probably underlie the effectiveness of early steroidal therapy in patients with bacterial meningitis (Lebel et al. 1988).

The appreciation of the inner ear as an immunocompetent organ through basic studies has provided a basis for understanding the empirical observations (McCabe 1979) of autoimmune disorders that result in hearing loss. The immunological basis of inner ear disease is considered to account for some of the progressive hearing loss observed in clinical practice. Autoimmune mechanisms may be genetically determined, may be acquired, or may result from an interaction of the two. Knowledge of the immunological properties of the inner ear will enable greater understanding of disease mechanisms of the inner ear and will form the basis for devising effective interventions.

2.6 Otacoustic Emissions: Animal Models and Clinical Observations

Discovery of the motile properties of outer hair cells (Brownell et al. 1985) and the probable associated sound production of the inner ear—otacoustic emissions (OAEs) (Kemp 1978)—has opened new areas for the investigation of basic and pathological aspects of the inner ear. A working nosology groups OAEs into four categories: transiently evoked OAEs (TEOAEs); stimulus-frequency OAEs (SFOAEs); distortion-product OAEs (DPOAEs); and spontaneous OAEs (SOAEs). Each of these is elicited under a different condition and each is used to ascertain different aspects of the functioning of outer hair cells. Although man, monkey, and common non-primate laboratory animals all possess OAEs, there are greater and lesser quantitative and qualitative differences. Delineation of the current knowledge of the singularity of OAEs in various species provides an information resource that should guide the investigator in the interpretation of the results for OAEs and the applicability or generalization of findings to fundamental questions of inner ear and outer hair cell biology.

The OAEs have rapidly been applied to the diagnosis, detection, and the understanding of human diseases. These applications have occurred both in human and animal models. Emphasis has been placed in the area of evaluation of the effects of ototoxicity, including studies of cis-platinum, loop diuretics, aminoglycoside antibiotics, and salicylates. These OAE data are complementary to the information in Chapter 5 by Garetz and Schacht, Ototoxicity: Of Mice and Men. The effect of hypoxia and anoxia, noise trauma, Ménier's syndrome, genetic hearing loss, and idiopathic hearing loss have shown a variety of OAE responses. These studies have implications for the further understanding of the mechanisms that underlie these many different forms of hearing loss. The OAEs have helped to advance and will continue to further the clinical ability for defining a cellular and functional phenotype for these diverse hearing disorders. Each level of evaluation brings a further nosological refinement. Such an example from the rich history of human hearing evaluation would be the introduction of bone conduction to hearing evaluations in the 19th century (Feldman 1970), which led to the differentiation of conductive and nonconductive hearing loss and allowed clinical classification. The classification of the behavioral hearing, bone versus air conduction, was then correlated with the known diseases entities, and this led to the development of effective interventions such as stapedectomy, tympanoplasty, and tympanotomy tubes. It should be assumed that the OAEs coupled with the increased knowledge of molecular biology, genetics, and biochemistry will help to enable similar advances in the care and cure of hearing disorders.

The OAEs are presently being evaluated as screening instruments for the detection of hearing losses in newborn infants (White et al. 1994). OAEs have the advantage of being robust physiological responses and are found in most normal newborns. The disadvantages are the changes in the OAEs in response to conditions of the external and middle ears. There are also the intrinsic problems of any screening program. There are social and economic problems of follow up and tracking and the limitation of testing at one point in time in a spectrum of pathology that has a large component of progressive disorders. These intrinsic limitations must be considered in any screening program and the use of OAEs is no exception. OAEs, even with these limitations, should play a substantial role in newborn screening programs as they provide ready access to a dynamic physiological function of the inner ear—the probable result of the motility of the outer hair cells.

2.7 Tinnitus: Psychophysical Observations in Humans and an Animal Model

Understanding of the basis of an individual's disease is a continuous process. One hundred fifty years ago diseases were characterized by the pattern of the fever. These observations were then related to different disease vectors, the fever became conceptualized as symptom, and the disease was defined by the type of infecting agent, e.g., malaria, tuberculosis, pneumococcus, etc. A similar evolutionary process is now occurring with all of the disorders of the inner ear, tinnitus being but one of many inner ear phenomena for which greater understanding is now being developed. One of the aspects of the nosological development of tinnitus is the use of SOAEs to further define and characterize this troublesome and often serious condition.

The development and exploitation of the salicylate animal model furthers our understanding of what is primarily a sensory phenomenon. The model has further utility as an example of the difficult area of relation between sensation and cellular, biochemical and physiological functions in an animal. An advantage is that invasive and cellular data can be obtained in the animal model that are either difficult or impossible to obtain in humans. A disadvantage is that the manifestation of the sensory experience in the animal is surmised but cannot be proven. The model has been developed so as to allow a reasonable correlation between the animals' acoustic behavior with what can be considered as a form of human tinnitus. These studies are a demanding model for the difficult but necessary investigation of the behavioral aspects of inner ear function.

3. Conclusion

The seven chapters that comprise Clinical Aspects of Hearing contain much of contemporary science that contribute to the acceleration of our understanding of the inner ear and human communication. The advances in our knowledge of the basic molecular regulation of cellular integrity, development, repair, differentiation, and homeostasis, coupled with the knowledge of the biochemistry and immunology of the inner ear is providing a deeper understanding of how we hear. This knowledge is being applied to human diseases with the expectation that effective interventions will be developed for the prevention, cure, and care of the inner ear. The information in this volume serves as a basis for future studies that will further define basic mechanisms and for insights into their application to the treatment of disease. The coming decades will witness the application of this new knowledge. There will be interventions based on the biochemical defects which are caused by genetic disorders. Neural and sensory portions of the inner ear will be repaired or replaced. The use of pharmaceutical and other agents will be potentiated because means will be developed to stop or minimize the adverse effects on the inner ear. The action of infectious agents and the ears' immune reaction will be defined and then utilized to mitigate or prevent disease from destroying it. There will be further refinements of the nosology of hearing loss based on data from OAEs that should enable pathological correlations and lead to the development of effective interventions. The psychophysical models will be applied to peripheral and possibly more central auditory and communication phenomena. Based on these and other basic science developments, the next period should be one that witnesses an enormous increase in our ability to alleviate deafness and its impact on society and the individual.

References

Bess FH, Tharpe AM (1984) Unilateral hearing impairment in children. Pediatrics 74:206-216.

Brownell WE, Bader CR, Bertrand D, de Ribaupierre Y (1985) Evoked mechanical responses of isolated outer hair cells. Science 227:194-196.

Corwin JT (1986) Regeneration and self-repair in hair cell epithelia: Experimental capacities and limitations. In: Ruben RJ, Van De Water TR, Rubel EW (eds). *The Biology of Change*. Amsterdam: Excerpta Medica. pp 291-304.

Feldman H (1970) A history of audiology: A comprehensive report and bibliography from the earliest beginnings to the present. Translation of the Beltone Institute for Hearing Research 22:1-109. Translation of Feldman H (1960) Die geschichtliche Entwicklung der Hörpüfungsmethoden, kurze Darstellung und Bibliographie vonder Anfängen bis zur Gegenwart. Zwanglose Abhandlugen aus dem Gebiet der Hals-Nasen-Ohren-Heilkunde. Leicher H, Mittermaier R, Theissing G. (Eds). Georg Thieme Verlag, Stuttgart.

Kemp DT (1978) Stimulated acoustic emissions from within the human auditory system. J Acoust Soc Am 64:1386–1391.

Lebel MH, et al (1988) Dexamethasone therapy for bacterial meningitis. Results of two double-blind, placebo-controlled trials. N Engl J Med 319:964–971.

McCabe BF (1979) Autoimmune sensorineural hearing loss. Ann Oto Rhino Laryngol 88:585–589.

Staecker H, et al. (1995 A) Technical Notes. Science 261:709–711.

Staecker H, et al. (1995B) The effects of neurotrophins on adult auditory neurons in vitro and in vivo. *Abstracts of the Eighteenth Midwinter Meeting of the Association for Research in Otolaryngology*, 705 p. 177.

White KR, Vohr BR, Maxon, B, Behrens TR, McPherson MG, Mauk GW (1994) Screening all newborns for hearing loss using transient evoked otacoustic emissions. Int J Pediat Otorhinolaryngol 29:203–217.

Zakon H (1986) Regeneration and recovery of function in the amphibian auditory system. In: Ruben RJ, Van De Water TR, Rubel EW (eds) *The Biology of Change*. Amsterdam: Excerpta Medica, pp 305–318.

2

Approaches to Understanding the Molecular Genetics of Hearing and Deafness

KAREN P. STEEL AND WILLIAM KIMBERLING

1. Introduction

Communication provides the means by which the discourses that hold our society together take place. It is communication, or more accurately, the lack of it, that builds barriers among groups of individuals. The fact that a separate culture exists for the deaf underscores the importance of communication in our society. The hearing sense is critical in communication, and further understanding of this process is essential to improving communication skills. This is especially important to those many people who suffer an absence of hearing ability. While lack of hearing is not a life threatening condition in modern society, it is isolating and, if we allow it to be, even dehumanizing. Consequently strong effort must be exerted to understand the causes of hearing loss.

Sixteen percent of all adults in the United Kingdom suffer a significant (>25 dB) hearing impairment (Davis 1989), and this figure is likely to be similar in other countries. The genetic contribution to adult onset hearing loss is poorly understood, but it is reasonable to expect that late onset hearing impairment will have a considerable genetic component. In fact, it is reasonable to believe that those disorders that we may think of as environmental (e.g., otitis media, noise induced hearing loss, drug sensitivity, presbyacusis) each will have some degree of genetic control. In addition, about one in 1000 children is found in the first few years of life to have a hearing impairment that is sufficiently severe that their language development is seriously affected. Of these, around half are usually considered to have a single gene mutation, and in the remaining half the deafness is attributed to primarily environmental causes (e.g., Fraser 1976; Morton 1991). Most reports of the etiology of childhood deafness suggest that two-thirds of the genetic deafness shows autosomal recessive inheritance, one third shows autosomal dominant inheritance, with 1 to 2% of cases being due to X-linked genes. Genetic deafness can also be subdivided into syndromic and nonsyndromic or uncomplicated categories. In the latter group, there are no other features apparent apart from the deafness. In both syndromic and nonsyndromic categories, there are probably tens of gene loci involved. Konigsmark and Gorlin (1976) listed 160 different syndromes including deafness as a feature, and it is likely that many of these represent different gene loci. The nonsyndromic group, particularly those with autosomal recessive inheritance, probably includes a few tens of gene loci, although there may be only a small number of genes commonly involved with a larger number of rare alleles (Stevenson and Cheeseman 1956; Chung et al. 1959; Chung and Brown 1970; Brownstein et al. 1991; Morton 1991).

2. Reasons for Identifying Genes Involved in Hearing and Deafness

There are many reasons for wanting to identify genes that are involved in hearing and deafness. A short-term aim is to provide immediate benefits to families who wish to know more about the deafness in their family through genetic counselling. Once a panel of relevant genes has been identified, it could be used for rapid screening for mutations to provide a diagnosis of hereditary deafness in an individual deaf child when it is otherwise not possible to distinguish genetic from environmental causes. Knowledge about the abnormal gene segregating in a family could be used to identify members of the family who carry one copy of a recessive deafness gene, or who carry a dominant deafness gene that fortuitously is not expressed. Many dominant deafness genes show reduced penetrance or variable expression. Identification of the mutant gene allows accurate prenatal diagnosis when this is requested, for example, in cases where the mutation causes deafness with other severe abnormalities. These benefits will follow rapidly the identification of deafness genes, and in some circumstances even the closely linked markers found before the gene itself is identified will be useful for enhancing the accuracy of genetic counselling.

However, the long-term advantages of identifying genes involved in hearing and deafness are far more wide-ranging. The major long-term benefit will come from using the sequence information and clones to investigate how each gene causes abnormalities of the inner ear. The aim of such research is to understand the whole biological process involved, from how a gene mutation causes protein defects to how the abnormal protein influences cell types that are involved in the development or maintenance of the auditory system in some way. When we understand how the mutations cause deafness, we can start to devise strategies for ameliorating the effects on the ear. This will be particularly relevant to progressive forms of deafness, where it may be possible to halt the deterioration in hearing if we know the biological mechanisms involved. An understanding of the cascade of gene action involved in the differentiation of sensory hair cells, derived from investigations of single gene mutations, should enable us eventually to stimulate the regeneration of hair cells that have degenerated for whatever reason, genetic or environmental, and could prove to be of immense value to aging populations.

It is very difficult, if not impossible at the moment, to gain an understanding of the molecular basis of hearing and deafness from studying the cells and proteins of the inner ear directly. For this reason, a useful approach is to identify the genes involved and to predict the proteins encoded from the gene sequences. Even then, it may not be apparent what the function of the protein product of the gene might be, and so identifying the gene and its protein is the first of many steps in the molecular characterization.

3. Genes Involved in Hearing Versus Genes Involved in Deafness

There are two broad approaches to identifying genes involved in the auditory system. Firstly, genes expressed in the normal inner ear can be identified. One way of doing this is to prepare inner ear cDNA libraries, which are banks of clones of copies of the messenger RNAs from the inner ear tissues, representing the genes that had recently been transcribed from gene to mRNA (e.g., Wilcox and Fex 1992; Beisel et al. 1993; Ryan et al. 1993). Genes of interest can then be identified by selecting from the library cDNA clones that appear to encode relevant proteins. The screening can take the form of searching for base sequence motifs that are common in the types of genes thought likely to be important, such as channel genes, or, alternatively, proteins can be generated from the library for screening with antibodies that have been raised to specific cochlear proteins.

These libraries will contain many housekeeping genes that are expressed in all tissues and so are unlikely to be involved in defining the specific characteristics of the inner ear. The usefulness of such libraries can be enhanced considerably by subtracting them against a cDNA library from another tissue. In this way, cDNA clones common to both tissues will be eliminated from the subtraction library and most housekeeping genes and non-tissue specific genes will be removed in this way. Several cDNA libraries have been prepared in this way (e.g., retina, Swaroop et al. 1991; human fetal cochlea, Morton et al. 1992; guinea pig organ of Corti, Wilcox et al. 1993). A similar approach can also be used to pinpoint genes that are expressed in two tissues that are both known to be affected by certain mutations. An example is the panel of cDNA clones prepared by Beisel et al. (1993) that includes genes found in both a mouse cochlear library and a human retinal library. Any genes found to be expressed in both tissues are likely to be important in that they have been conserved evolutionarily and, in addition, may be necessary for both cochlear and retinal function. This makes them excellent candidates for genes mutated in hearing/visual loss disorders like Usher syndrome.

cDNA libraries from tissues other than the mammalian organ of Corti may be useful in screening for genes that are involved in the process of sensory transduction. Sensory epithelia from the inner ears of non-mammalian vertebrates may be particularly useful because large quantities of hair cell-rich material may be obtained quickly (Sewell and Mroz 1990; Davis et al. 1993; Oberholtzer et al. 1993; Solc et al. 1993), while dissecting enough pure organ of Corti tissue from mammalian cochleas for producing a representative cDNA library is difficult and time consuming. The cochlea contains a large number of different cell types, and even a dissected organ of Corti will consist of a heterogeneous collection of support and sensory cells.

Recent improvements in the techniques for the construction of a cDNA library require less tissue, and it has even been possible to construct a library from as little as a single cell (Eberwine et al. 1992). The technique has only recently been applied to the cochlea on a small population (< 10) of outer hair cells (Beisel, personal communication) using a double round of RNA amplifications with T7 and T3 RNA polymerase. This technique has a great potential for the study of the molecular genetics of the inner ear. The logical next step will be to make subtraction and/or intersection libraries from pairs of different cell populations from the inner ear. For example, the subtraction of an inner hair cell library from an outer hair cell library, and vice versa, would enhance the yield of messages that are important for understanding the specific molecular genetic differences between the inner and outer hair cells. Similarly, one can more precisely examine the regulation of gene expression during development by constructing a series of libraries from a differentiating cell line. Subtraction could be used to identify which genes are turned on and off during development.

The main problem with the construction of human cochlear libraries at this time is the paucity and limited quality of human cochlear tissue. The surgeon is seldom justified in removing a human cochlea. When it is done, the material is variably contaminated by surrounding bone and connective tissue. It may be impossible to produce a quality library from "whole" human cochlear tissue. In fact, only one human cochlear library has been made so far and this was from human fetal material (Morton et al. 1992). This situation should change in the future since the technique of Eberwine (1992) can be used to construct high quality and cell type specific human libraries.

Identifying genes expressed in the normal adult cochlea in this way will be extremely useful in elucidating the molecular basis of auditory transduction. However, there are some limitations in this approach that should be appreciated. Key elements in transduction may be represented by very small numbers of mRNAs, so a large number of cDNA clones may have to be screened in order to detect the relevant gene. The library will only contain copies of mRNAs that were present in the tissue at the time of isolation, so any message that has a rapid turnover may be underrepresented. Similarly, only genes active at the time that the tissue was taken will be represented in the library, so genes that are vital for the development of the inner ear rather than for its maintenance in the adult are unlikely to be identified using a library from a mature cochlea; cDNA libraries from developing inner ears will be needed to detect these developmentally important genes. It will be necessary to confirm that the gene identified is expressed in the expected sites by techniques like in situ hybridization of tissue sections using probes that hybridize specifically to the relevant mRNA. Many genes that are vital for normal development of the inner ear are not expressed in the otocyst or the neuroepithelium of the developing sensory regions at all, but are expressed in adjacent tissues such as the mesenchyme or the neural tube (e.g., *c-kit* and *steel*, Steel et al. 1992; *Pax-3*, Goulding et al. 1991; *int-2*, Represa et al. 1991; Mansour et al. 1993; *Hox-1.6*, Lufkin et al. 1991; Chisaka et al. 1992). Therefore, screening inner ear cDNA libraries is not likely to generate candidates for all of the genes that are important for the development of the sensory epithelium. Finally, identifying genes as being of possible importance in the function of the normal inner ear by the methods outlined above is just the first step in the investigation. It will still be necessary to discover the effects of removing the gene by producing transgenic mice before it can be established that the gene is necessary for normal cochlear function. Many transgenic mice produced with interruptions of other genes, unrelated to the auditory system, have been surprising in their lack of the predicted effects on the phenotype, suggesting a degree of redundancy in gene function that was not previously expected. However, transgenic mice with specific cochlear proteins eliminated will be extremely useful tools for investigating normal sensory transduction.

The second approach to identifying genes involved in the auditory system is to identify genes with mutations already known to affect the inner ear. Some of the strategies for achieving this are described in the remainder of the chapter. The advantages of this approach are that it is known beforehand that the genes identified will have a vital role in auditory function; if the genes are identified in humans first, then it is clear that the identified gene will be involved in causing human deafness and thus will be of direct clinical relevance. One disadvantage of this approach is that genes will only be identified if mutations in them allow viable offspring to be produced. Genes that are important in auditory function may also be vital for other purposes in development such that mutations in them are invariably lethal. Identifying genes on the basis of viable mutations in them will be unlikely to lead to a rational unfolding of the molecular basis of normal cochlear function and may fail to uncover the gene for specific functions like the transducer channel, for example, if no viable mutations exist in the populations presently available. However, mutations causing deafness will be very useful in pinpointing genes that otherwise would not be expected to be involved in cochlear function.

The choice of approach — genes involved in hearing or genes involved in deafness — will obviously depend upon the questions to be addressed, but both approaches will undoubtedly contribute to our knowledge of the molecular basis of hearing. The rest of this chapter will focus on the identification of genes already known to cause deafness.

4. Requirements for Identifying Genes

There are three basic requirements for identifying a deafness gene in the absence of any clear candidate genes that have been described previously. These three requirements are (a) a large number of DNA markers that can be used to localise the gene to particular regions of the genome, (b) a large number of meioses (100-200) available for analysis of recombination rates between DNA markers and the deafness gene, and (c) a homogeneous group of families, each with members carrying the same deafness gene.

Meeting the first requirement has been a priority area for human genome mapping programs in the last few years, and there is now a large and increasing number of suitable markers covering much of the human and mouse genomes. Many of the markers that are the easiest to use depend upon using polymerase chain reaction (PCR). PCR is a biochemical reaction that can amplify specific small regions of DNA. It is very sensitive, requiring only small amounts of original DNA for detection. Regions in the genome, known as dinucleotide repeats (DNRs), have been found to be highly polymorphic in both humans and mouse. The most common DNR is (CA)n, where n refers to the number of repeats and C and A are two of the bases in DNA. For any particular DNR, the n often varies from one person to the next and is inherited as a straightforward genetic marker. Pairs of small oligonucleotides of 20-30 base pairs in length that flank a known dinucleotide can be used in combination with PCR to amplify the intervening DNA containing the DNR. Fragments from different chromosomes will vary in size and the different sized fragments can be separated by electrophoresis and visualized by an appropriate method (e.g., incorporation of radioactive ^{32}P into the amplification reaction). The sizes are scored as different alleles and their inheritance followed. The PCR technique coupled with the discovery of the highly polymorphic nature of the DNRs has revolutionized the approach to mapping disorders and traits in humans as well as mice. In fact, a very complete framework genetic map for both species using such markers has recently been published (Dietrich et al. 1994; Gyapay et al. 1994). In addition to PCR of dinucleotide repeats, restriction fragment length polymorphisms (RFLPs) using labelled probes of known genes for detection will continue to be important in mapping because the use of probes for genes rather than for sequences of unknown function (anonymous DNA probes) facilitates cross-species comparisons. The RFLP depends upon a base change occurring within a restriction site, the site at which a specific sequence of bases permit a restriction enzyme to cleave the DNA. The RFLP was the mainstay of gene localization studies until the advent of PCR allowed DNRs to be detected easily. Further markers for a specific region can be generated by microdissection of the relevant part of the chromosome or flow sorting of Robertsonian chromosomes followed by random cloning of the retrieved DNA (e.g., Greenfield and Brown 1987; Bahary et al. 1992).

The requirement for a large number of meioses available for study is often a severely limiting factor in human genetics. Human families are usually relatively small, and it is often not practical to group families together for analysis because each family may have a mutation of a different gene in a different part of the genome. Mice do not have this limitation, as they can be bred in large numbers using a breeding strategy designed to maximize the information provided.

The third requirement, for a homogeneous group, is equally problematic in the human population. There are many different genes causing deafness in humans, and sometimes it is not possible to assume that the deaf members of one family are all carrying the same gene causing deafness. For example, in dominantly inherited deafness, strong assortative mating is present: the deaf tend to marry other deaf individuals. This produces pedigrees like the one in Figure 2.1. Here, there is no clear pattern of segregation, nor is there any assurance that only one gene is responsible. Indeed, because there are nine unrelated individuals, at least nine different etiologies could be involved. The pedigree given here is typical of deaf pedigrees in the USA and Europe. Such families are completely unusable for linkage because of the inability to score the correct genotype of the affected individuals.

Recessively inherited deafness is also highly heterogeneous within the population. The existence of deaf by deaf marriages with both parents having presumed recessive deafness yet producing normal offspring demonstrates this. Only about 18% of the children of such deaf by deaf matings are deaf. This can occur only because of heterogeneity, since two individuals homozygous for the mutant allele at the same locus should produce all affected offspring. Chung and Brown (1970) have estimated the number of genes responsible for nonsyndromic hearing loss and calculated a value ranging between five and 108 genes (see also Stevenson and Cheeseman 1956; Chung et al. 1959). The methods of estimating the number of recessive genes depend upon a number of assumptions about the nature of the data set. It is reasonable to conclude, as do Chung and Brown, that four or five genes may be responsible for most of the recessive hearing loss in the population, but that several, anywhere from 10 to 100, loci may each be responsible for many different rare types. A similar conclusion was reached more recently by Brownstein et al. (1991), who reported estimates of 6.7

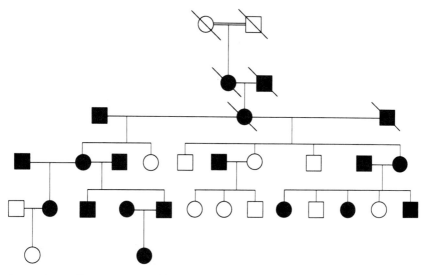

FIGURE 2.1 Pedigree of a family with congenital profound deafness. There is a strong propensity for the deaf to marry other deaf people. Because of this fact, it is difficult to obtain uncomplicated dominant deaf pedigrees for genetic analysis. This pedigree illustrates this phenomenon and is fairly typical of pedigrees of deaf families. It is obvious that there is no segregation of a single, identifiable gene causing the deafness. Consequently, no meioses can be scored for linkage analysis and the family, despite its size, is useless for linkage analysis. Squares represent males and circles females; filled symbols represent the deaf and open symbols the normally-hearing individuals; hash marks indicate dead individuals; double line represents a consanguinous union.

and 22.0 loci for intra- and inter-ethnic matings, respectively, in the Israeli Jewish population.

The experience with Usher Syndrome may parallel that with nonsyndromic recessive hearing loss. Usher syndrome causes hearing impairment and progressive retinitis pigmentosa. The more severe type, type I, shows congenital severe or profound sensorineural hearing impairment, early onset retinitis pigmentosa, and vestibular dysfunction, while type II involves a broader range of hearing impairments and later onset of retinal degeneration with no vestibular defects detectable (Smith et al. 1994). The pathology is not well understood, but the ear defects are most likely neuroepithelial rather than morphogenetic in type. For Usher Syndrome, a minimum of five genes are involved, of which four have been localized. Three of these loci have mutations that cause type I Usher syndrome, and two cause type II (Kimberling et al. 1990; Lewis et al. 1990; Pieke Dahl et al. 1991; Kaplan et al. 1992; Kimberling et al. 1992; Smith et al. 1992). The discovery of more loci is possible as more unusual phenotypes are examined. There does seem to be one common gene for each of type I and II; however, and the other variants are rarer. The existence of this heterogeneity would have made the localization of the Usher genes very difficult if the mix of different genetic types had been more even. A similar but more extensive scenario should be expected for nonsyndromic deafness. If, for the sake of argument, there are only four loci that account for 80% of the cases of recessive deafness, and if these are all equally prevalent, then a sample of 20 multiplex (i.e., more than one affected individual present) families would almost never provide sufficient evidence in favor of linkage even if linkage existed for one of the main subtypes. Nor is there a test of homogeneity that would detect the presence of such extensive heterogeneity (see Ott 1991 for further discussion of heterogeneity). Over half of the families showing recessive deafness will have only one affected child, while most of the remaining sibships would include only two affected children and an exceptionally large family would be one with four affected children. The statistical power of using groups of such families is severely limited.

Mice again have a real advantage over humans in avoiding the problems of heterogeneity. Large numbers of mice can be bred for linkage analysis that are all known to have deafness due to the same mutant gene. The confounding effects of other genes in the genetic background that may modify the effects of the primary mutant gene can also be controlled in mice by using inbred strains that show very little heterozygosity at any loci throughout the genome.

There are some disadvantages in using mice as a means of identifying the genes causing deafness in humans. For example, the populations of deaf mice and deaf humans are likely to have a different mixture of mutant genes involved because the populations have been selected in very different ways. Mice with syndromal deafness predominate among deaf mouse mutants because the additional abnormalities are more easily detected by animal handlers than deafness. Uncomplicated autosomal recessive deafness does occur in mice, however (e.g., *deafness*, Deol and Kocher 1958; Steel and Bock 1980), and there are cases in which a locus has one allele causing uncomplicated deafness and a second allele causing syndromal deafness (e.g., *waltzer v* and *deaf v^{df}*, Deol 1956). Nonetheless, it is possible that some of the major genes involved in causing uncomplicated autosomal recessive deafness in humans may not be represented among the deaf mouse mutants.

5. Strategies for Identifying Genes in Humans

There are strategies that can be employed to circumvent the problems of heterogeneity described above. These include the study of genetic isolates, study of large families segregating for deafness, and the development of better clinical tools (and their use). In the future, it may be feasible to use a candidate gene approach to identifying mutations in families of average size segregating for deafness by screening them for known deafness genes that have previously been identified using other techniques.

5.1 Genetic Isolates

One way to minimize the effects of heterogeneity when looking for genes causing uncomplicated autosomal recessive deafness is to target a closed community, such as a religious or geographic isolate. This will minimize the difficulties in linkage analysis caused by assortative mating between the deaf bringing additional deafness genes into a pedigree. One can reasonably expect that recessive deaf sibships drawn from an isolate will all have the same mutation at a single locus. By targeting an isolate, one improves the chances of identifying the responsible gene. Once the location of a gene is known, linkage studies can be done in a more complex mixture of families to estimate the proportion of the linked type. Unfortunately, one limitation of this approach is that genetic isolates are by definition relatively small, so that a large number of linked families will be difficult to identify. Consequently, the jump from gene location to gene itself by positional cloning will be compromised by the fact that the location of the gene will not be very precisely defined.

5.2 Large Families

A second strategy for minimizing the heterogeneity within a group is to study a large, extended family including many deaf and normally hearing individuals, preferably through several generations. Such families do exist for all patterns of inheritance of deafness, but are more common for sex-linked and dominantly inherited deafness. If twenty or more informative individuals are available, it is highly likely that a significant linkage can be obtained by using only one family. Thus, pooling families is not necessary, and the problems of heterogeneity can be avoided. This has been a successful approach for the localization of a gene for dominantly inherited progressive hearing loss occurring in Costa Rica (Monge's deafness, Leon et al. 1992). One large family descended from a single ancestor was identified. The family was large enough that when markers on chromosome 5q were typed, significant evidence of linkage (known as a LOD score — the higher the LOD score the greater the probability of linkage between the marker and the deafness gene) was obtained. It was important that this analysis was limited to a single family, since subsequently, in three USA families with dominant progressive hearing loss, linkage to 5q markers was not found, demonstrating genetic heterogeneity for this disorder.

5.3 Clinical Tools

The development and implementation of better clinical tools will be an important aid to gene localization. Any clinical description that subdivides nonsyndromic deafness into different groups lessens the impact of heterogeneity. Audiometric configuration has been used in this way to subdivide different families. For example, Konigsmark and Gorlin (1976) defined several subtypes based upon the audiometric profiles: (1) congenital severe loss, (2) progressive low frequency loss, (3) midfrequency loss, (4) high frequency loss, and (5) early onset progressive hearing loss. This was a tentative attempt to classify the types of hearing losses and did not take advantage of other characteristics associated with hearing losses. For example, vestibular function has proven essential in the division of type I Usher Syndrome from type II. Vestibular defects are also relatively common in other types of genetic deafness; for example, Konigsmark and Gorlin (1976) list a total of 160 syndromes with deafness, of which 28 have vestibular dysfunction in some or all cases, 25 have normal vestibular responses, and the remainder have not had vestibular testing. In addition, many cases of uncomplicated deafness in humans with no obvious balance defects do show vestibular dysfunction when this is tested directly (Arnvig 1955; Everberg 1960; Rosenblut et al. 1960; Kaga et al. 1981; Horak et al. 1988). Thus, the division of families into groups with and those without vestibular dysfunction would be a step toward minimizing heterogeneity in a study population. Further subdivisions may be possible based upon the presence or absence of inner ear malformation and the type of malformation present. Imaging techniques are improving and physical attributes of the inner ear may be useful in subdividing causes of deafness (e.g., X-linked deafness, Phelps et al. 1991). Features of residual hearing other than the pure tone threshold (e.g., frequency selectivity, evoked acoustic emissions, temporal resolution, etc.) may also prove to be useful discriminators.

6. Use of Mouse Models

The study of mouse mutants that may be models for human genetic diseases has proven to be a very useful aid in identifying relevant genes in humans. The similar structure of the inner ears of humans and mice, the similarities in the types of inner ear pathology in genetic deafness seen in the two species, and the similar range of associated defects seen in syndromal deafness in mice and humans all suggest that similar sets of genes are involved in directing the development of the auditory system (Deol 1968; Steel 1991; Brown and Steel 1994). Furthermore, many genes unrelated to hearing that have been identified and sequenced in one species have later been shown to have a closely related homologue in the other species. Thus, deafness genes identified in mice are highly likely to have a homologue in humans. A practical strategy for finding human deafness genes would therefore be to identify such genes in mice, which, as described earlier, have a number of significant practical advantages, and then to use the sequence information to clone the equivalent human gene. The human clone can then be used to test for linkage in appropriate human families. Dozens of mouse mutants have been described that are known to be or are likely to be hearing-impaired, and many cause autosomal recessive deafness and so are good candidates for that group of human deafnesses that is most difficult to study directly.

Investigation of the arrangement of groups of homologous genes on chromosomes in the two species suggests that there has been a considerable amount of conservation in the organization of the genome (e.g., Nadeau et al. 1991). Sets of genes, called syntenic groups, are found in identical order along parts of mouse and human chromosomes (as well as in other species, e.g., O'Brien et al. 1993). This observation has been valuable in pinpointing potential mouse models of particular human genetic diseases, as well as suggesting candidate genes for involvement in a disease. A recent example of the advantages in this approach is the identification of mutations in the *PAX 3* gene in Waardenburg syndrome, which involves deafness associated with a pigmentation defect and widely-spaced eyes. This led directly from the finding of *Pax-3* mutations in the mouse homologue, *splotch*. *Splotch* was localized to mouse chromosome 1 many years ago, and in 1990 a group of families with Waardenburg syndrome type 1 (WS1) was shown to exhibit linkage of the disease gene with markers on human chromosome 2q37 (Foy et al. 1990). They proposed *splotch* as a potential mouse homologue because it too showed a pigmentation defect in the form of a belly spot, and it was located in a region of conserved synteny between mouse chromosome 1 and human 2q. Around this time, a clone called HuP2 from human DNA was described and characterized as a paired box-type gene (Burri et al. 1989), and *Pax-3* was identified in the mouse as a paired box gene with similarities to HuP2 (Goulding et al. 1991). *Pax-3* was located on mouse chromosome 1 in a region showing conservation with human chromosome 2q. No recombinations were found between *Pax-3* and *splotch* in 117 backcross mice, supporting the idea that *splotch* may be a mutation in the *Pax-3* gene (Fleming and Steel, unpublished observations). This was indeed the case, as a report in late 1991 demonstrated a small deletion in *Pax-3* from the Sp^{2H} allele of *splotch* (Epstein et al. 1991). These findings in the mouse pointed to *PAX3* in humans as a strong candidate for involvement in WS1, and just three months later, mutations in *PAX3* were described in human WS1 DNA (Tassabehji et al. 1992, 1993; Baldwin et al. 1992). Without the observations in the mouse, the gene for WS1 would still be a long way from being found. Ironically, the mouse mutant *splotch* escapes the auditory effects of the gene, having normal hearing (Steel and Smith 1992).

7. Candidate Genes

A candidate gene is a gene that by virtue of its known structure and function, in combination with its location in the appropriate chromosomal region, makes it a likely possibility to be the gene responsible for a specific disorder. The existence of candidate genes improves the prospects of jumping from gross localization to the gene itself, as described above for WS1 when *PAX3* emerged as a candidate gene. A set of candidate genes for hearing impairment disorders will make the search for deafness genes much easier. In fact, with a set of such candidates and a set of cases of deaf individuals, one could begin screening for mutations and gradually identify cases corresponding to specific mutations.

There are several methods of developing a list of candidate genes. Genes identified from cochlea-specific cDNA libraries will be included, along with any genes already identified as being involved in deafness in other species such as the mouse. Other genes close to the deafness gene should be considered as candidates until they are shown to be separated by meiotic recombination. Any families of genes that might a priori be expected to be involved in the development and function of the inner ear should also be included in the list of candidates. These might include genes known to be important in head development such as *Hox* and *Pax* genes, genes for actins and myosins and other structural proteins that are known to occur in the cochlea, channel genes, and other genes previously shown to be involved in sensory receptors and the peripheral nervous system.

Genes for syndromal deafness should be included as candidate genes even if the target is uncomplicated deafness, because it is always possible that different mutations of the same gene may have different effects. Examples of this phenomenon in the mouse auditory system were given earlier, and it is now clear that different mutations of a single gene in humans can lead to what were thought to be clinically-distinct disorders (e.g., the variety of eye defects caused by mutations of the peripherin gene *RDS*, see Travis and Hepler 1993 for references). Theoretically, it is possible that many of the genes that cause syndromic hearing losses also produce isolated hearing loss. For example, it is not unreasonable to expect that alleles of Usher Syndrome produce only hearing loss or only retinitis pigmentosa. Indeed, some of the syndromic hearing loss genes may actually be mutations such as deletions that involve more than one adjacent locus (a contiguous gene syndrome). If so, then once the Usher Syndrome gene is identified and sequenced, one will want to ask the question whether there are nonsyndromic deafness cases with mutations in the Usher gene. Similar questions occur for Waardenburg Syndrome, namely, is there a form of WS that presents with hearing loss alone. Any syndromic hearing losses must be considered as possible candidates for the nonsyndromic hearing losses.

The investigation of candidate genes for their possible involvement in the particular disorder being studied is a vital part of any strategy for identifying a disease gene, and can sometimes lead very quickly to the mutation. A good example of this was provided by the finding of mutations in the rhodopsin gene in a number of individuals with retinitis pigmentosa (Dryja et al. 1990). However, if the mutation occurs in a gene encoding a previously undescribed type of protein, as in the case of cystic fibrosis, then the candidate gene approach alone is not likely to be fruitful and a positional cloning strategy is the main remaining option. In most cases, a combined approach is adopted, involving localizing the abnormal gene, looking at candidate genes in the appropriate chromosomal region, and building up a detailed genetic map of the area as a prelude to positional cloning (Collins 1992).

8. Positional Cloning

The first step in positional cloning of a disease gene is to localize the gene to a specific region of a chromosome by linkage analysis. Two genes or DNA markers that are close together on a chromosome will be separated only rarely from each other by recombination during meiosis, and are therefore closely linked. Conversely, genes that are widely spaced on a chromosome or are on different chromosomes will be separated frequently during meiotic division, giving recombination rates up to around 50%. The greater the amount of recombination between two markers, the greater the genetic distance will be between them. In positional cloning, a panel of genes and DNA markers that are linked to the disease gene is assembled and the order of the markers established. This may have to be done by assessing genetic distances between genes and inferring their order from the distances, or may be achieved more accurately in suitably informative families by haplotype analysis—charting the parental source of each marker on an individual chromosome and pinpointing the exact position of any recombination breakpoint. Haplotype analysis can easily be done in mice by prior determination of the most informative way of setting up the cross (e.g., Brown et al. 1992) (Fig. 2.2). Haplotype analysis depends upon the assumption that recombination is not very frequent and that two or three recombinations close together on a chromosome is an unlikely event. Therefore, a set of data from individual chromosomes with recombinations can be used to predict the most likely order of markers. This recombinant panel can be used rapidly to position new markers in the relevant region of the chromosome. The more recombination breakpoints that are available for ordering markers, the more detailed the genetic map of markers will be, hence the requirement listed earlier for a large number of offspring to facilitate positional cloning.

A genetic map reflects the likelihood of meiotic recombination occurring between two sites rather than the physical map that reflects the distances between markers in terms of true physical distance (i.e., number of bases); the two distances are related in a nonlinear and highly variable manner, although marker order is the same. The relationship between the physical and genetic maps varies with position in the genome, but averages in human and mouse around 1 centimorgan per 1 to 2 megabases. Since a centimorgan

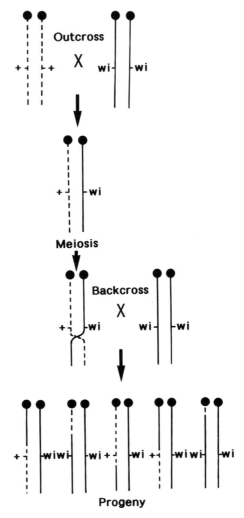

FIGURE 2.2 The plan for an informative cross in the mouse, using the autosomal recessive mutation *whirler* (wi) as an example. Only the relevant chromosome is shown. Solid lines represent chromosomes originally from the whirler stock, and dashed lines represent a different stock carrying the wild type copy of the *whirler* gene (+) and chosen to maximise the genetic difference between it and the *whirler* parental stock. The F1 offspring from the initial outcross carry one complete chromosome from each parent, until crossing over occurs during meiosis. The F1 mice are backcrossed to the parental whirler stock and the progeny from this cross analyzed. Five examples of the progeny are shown, but in practise anything from 100 to 1000 progeny are collected and analyzed. Each of the progeny have one complete chromosome from the parental *whirler* stock, and so the analysis focuses upon mapping the parental origin of markers along the other chromosome. The origin of the gene at the *whirler* locus is determined by recording the behavioral phenotype of each mouse. This type of backcross allows easy haplotype analysis because only one chromosome is analyzed in each mouse.

is defined as 1% recombination, one can predict that 100 or more meioses would be required to map a gene to a region of one million base pairs (1 Mb). Such a region should contain about 80 genes, and the identification of the specific gene being mapped requires either a lot of luck or a great deal of work. In humans, it is difficult to map closer than 0.5 to 1 centimorgan. In mouse, it is possible to map to within 0.1 centimorgan (e.g., Brown et al. 1992), and hence to narrow the physical mapping step to 1 to 2kb or about eight genes. This is a decided advantage and should lead to more rapid identification of deafness genes in the mouse.

An approach that complements recombination analysis is deletion mapping or the use of other chromosomal rearrangements that disrupt or otherwise uncover a gene causing hearing impairment. This approach depends upon cytogenetically visible chromosomal rearrangements being detected in individuals with features that include deafness. For example, the first clue that a gene for Waardenburg syndrome mapped to 2q came from a report of an inversion of part of 2q in a child with the syndrome (Ishikiriyama et al. 1989). Deletion mapping has been especially useful in localizing genes on the X chromosome, where overlapping deletions extending to different parts of the chromosome in males can be correlated with the associated phenotype to localize genes for each component (Brunner et al. 1991; Bach et al. 1992) (Fig. 2.3). A similar approach using deletion panels has been useful in localizing deafness genes in the mouse (Rinchik et al. 1991).

Having constructed a detailed genetic map of the region around the disease gene, the nearest markers on either side of the gene are selected for library screening. Any genomic DNA library of the relevant species could be used, but a YAC (Yeast Artificial Chromosome) library is often chosen because the fragments of DNA in each clone are much larger generally than in phage or cosmid libraries (Burke et al. 1991; Chartier et al. 1992). Any YAC clones identified that contain either flanking marker are then examined, and physical maps are compiled, including the site of the gene used to screen as well as rare restriction enzyme cutter sites. The cloned genomic DNA in each YAC is then used to generate additional markers in a variety of ways, and the new DNA markers that map near to the ends of the YAC clone are used for a further round of screening. Each further YAC clone picked out in this way is then subjected to the cycle of physical mapping, generation of new markers, selection of markers near the ends of the YAC, and use of these new markers to screen the library once more until an overlapping array of YACs (a YAC contig) is built up that crosses the site of the disease gene and reaches the nearest marker on the other side of the gene (Fig. 2.4). The new DNA markers that are generated to extend the YAC contig are in addition used to refine the genetic map, using the recombinant panel of chromosomes. In this way, the position of the recombination breakpoints can be determined more accurately and the smallest region that shows no recombination with the disease gene can be defined and compared with the physical map. This whole process results in

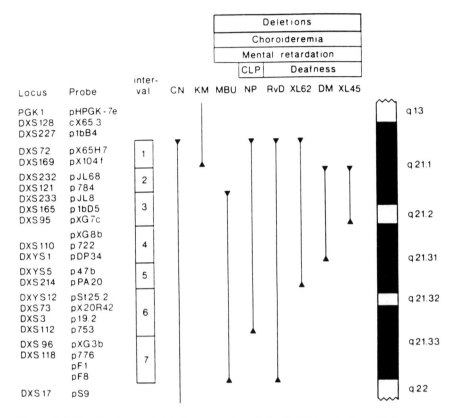

FIGURE 2.3 Duplication and deletion mapping of the Xq21 band of the X chromosome. Duplicated (CN and KM) and deleted segments are indicated by vertical lines and break points by arrowheads. Interval 2 could include a deafness gene because it is deleted in all patients with deafness but is present in patient MBU who is not deaf. Locus is the term used for the position on a chromosome, and probe is the name of the small stretch of DNA used to identify each locus in a specific manner. Interval in this context refers to lengths of chromosome defined at each end by the deletion breakpoints illustrated by vertical lines. Each vertical line represents the deleted segment in an individual person. The panel at the top right indicates the symptoms shown by individuals represented below. The idiogram on the right represents parts of the X chromosome that can be recognized cytologically. From Cremers et al. (1989) with permission.

a set of YAC clones of genomic DNA that span the entire non-recombinant region and so must include the gene that is being sought.

The next step is to examine the genomic DNA within the non-recombinant region, obtained from the relevant YAC clones, for indications of sequences that may be candidates for the disease gene. There are a number of ways of doing this. Exon trapping is one useful approach (Duyk et al. 1990; Buckler et al. 1991). This technique exploits the fact that most

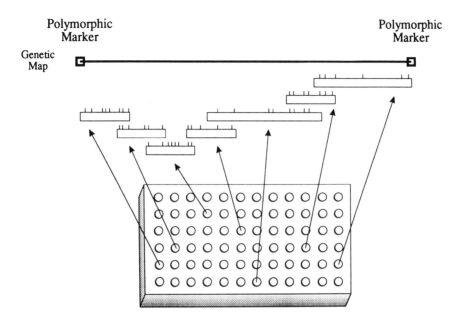

FIGURE 2.4 Construction of a YAC contig between known genetic markers. The first step in correlating a particular gene with a specific disorder is to map the gene to a region between a set of polymorphic flanking markers. Once that distance has been reduced to as small as possible, the next step is to clone all of the DNA between the intervening markers. This is done through the construction of a YAC contig, which is accomplished by repeatedly finding YACs that contain sequences of DNA for the marker genes and to use the ends of those YACs to create new sites that can be used for further screening of the YAC library to extend the cloned region internally. Once the YAC contig is constructed, and as long as there are no occult gaps, the investigator can be assured that the gene for the disorder lies in one of the contiguous clones. The YAC clones can then either be used directly or be subcloned in order to search for CpG islands or to identify exons.

genes are represented in genomic DNA by a series of short stretches of coding sequence, the exons, separated by regions of DNA that do not encode the protein product of the gene, and these are called introns. During the transcription process, the RNA copies of the exons are spliced together, excluding the introns. This splicing requires a specific recognition site that occurs in the intron. The splice site recognition sequences that are found in genomic DNA are relatively conserved and are functional in mammalian expression systems, so they can be used to identify exons within the non-recombinant region by exon trapping. Trapped exons are amplified by polymerase chain reaction and can then be examined as possible components of the disease gene. As this technique depends upon the use of splice sites, any genes that have no introns, consisting of a single exon, will have no splice sites and will not be isolated in this way. Most genes, however, do have multiple exons.

A second method used to identify potential candidate sequences in the nonrecombinant region is to use frequent-cutting restriction enzymes to reduce each YAC clone to smaller fragments, and to use these smaller fragments of DNA to make probes to hybridize to zoo blots. Zoo blots are Southern blots of digested and separated DNA from a number of different species. If a fragment hybridizes with DNA from several species, giving a small number of discrete bands, this suggests that part of the fragment used as a probe contains a sequence that has been conserved during evolution and so is likely to include part of a gene. Many genes are conserved and so will be detected by this method.

A third feature used to pinpoint possible genes in the nonrecombinant region is to look for CpG islands. These are regions of genomic DNA that include a cluster of hypomethylated CpG nucleotide residues, and these islands often occur at the 5' ends of genes (Bird 1986). Some rare-cutting restriction enzymes will identify and cut the genomic DNA at the CpG island, and so when the YAC clones are digested with these enzymes and the restriction sites mapped, a cluster of such restriction sites will indicate the presence of a gene. The presence of hypomethylation, a feature of CpG islands, can be established by using pairs of restriction enzymes that recognize identical base sequences but are differentially sensitive to the degree of methylation of the DNA. Many, but not all, genes will be located using this approach.

Having established some parts of the nonrecombinant region as potential coding regions by exon trapping, conservation across species or the presence of a CpG island, these parts need to be sequenced to establish whether they include an open reading frame. This is simply a stretch of bases that should encode a reasonably sized string of amino acids. As sequencing becomes faster, it will become more practical to sequence entire nonrecombinant regions, provided these are not too large, and to use computer programs to scan for open reading frames. Sequences of any open reading frames identified can be used to compare with published sequences of known genes using computer databases. Similarities with sequences of previously described genes that encode the predicted type of protein (e.g., a channel or transmitter receptor) will be very useful in targetting the most likely candidate sequences for the disease gene.

The next step in positional cloning is to discover whether the candidate sequences identified are expressed at the relevant sites in the body and at appropriate stages of development. Again, there are several ways of testing for appropriate expression, and ideally a combination of techniques will be used. Firstly, Northern blots of mRNA from specific tissues can be hybridized with the candidate sequence. If all tissues appear to contain the complementary mRNA, or if the predicted site of expression (e.g., inner ear, neural tube) does not show any sign of hybridization with the candidate sequence, the probability of the sequence representing the gene sought is reduced. Secondly, the candidate sequence can be used to screen a cDNA library prepared from the relevant tissue (see earlier discussion of cDNA libraries in this chapter). The isolation of cDNA clones, including the candidate sequence from such a library, not only suggests that the candidate is expressed in the predicted location, but also the clone is likely to provide the remaining exonic sequences. As mentioned earlier, even cDNA libraries prepared from the organ of Corti alone will include many messenger RNAs from supporting cells, and so may not necessarily be sufficiently selective. The third approach to investigating expression is to use in situ tissue hybridization of relevant stages of development. This is most easily done using the mouse, as tissue specimens are rarely available from suitable human material. Tissue sections hybridized with the candidate sequence should enable precise definition of the cell types that are expressing the gene at the time. All of these expression studies are best carried out with DNA from normal individuals in the first instance, but once a strong candidate emerges, expression studies should also be carried out using material from abnormal specimens. The mutation may alter the amount of mRNA produced or affect its turnover rate, resulting in a reduced level of detectable mRNA in tissues. Such mutations may occur in noncoding regions of the gene, such as promoters or enhancers.

When a small number of candidate sequences are defined, the homologous regions of genomic DNA from affected individuals are examined for evidence of mutations. It is a great advantage at this stage to have access to a number of different mutations affecting the same gene, because some mutations will be difficult to locate, particularly if they occur in noncoding parts of the gene. Mutations are usually located by sequencing the mutant gene and comparing the sequence with the normal sequence. If differences are found in two or more different mutant animals or affected humans, this is good evidence that the candidate gene is in fact the disease gene. However, caution is needed if only one mutation is available, because any sequence differences found may be nondeleterious polymorphisms. A good example of the importance of considering this possibility was provided by the sequencing of the *Trp-1* gene, at the *brown* locus in the mouse (Zdarsky et al. 1990). In this case, four separate nucleotide differences were found between the *brown* allele and its wild-type homologue. Two of these were predicted to lead to different amino acids being incorporated into the protein chain, and the critical mutation of these two was identified by sequencing a revertant to wild type; the other amino acid substitution did not appear to affect the activity of the protein in melanogenesis. In addition to sequencing, new mutation screening techniques are increasingly being devised to use particularly with human DNA, where large numbers of samples may need to be screened. Some of these screening techniques depend upon using altered electrophoretic mobility to detect conformational changes in single stranded DNA (SSCP, Single Strand Conformational Polymorphism) or double stranded DNA formed from mutant and normal strands (heteroduplex detection). Some of the Waardenburg syndrome mutations in *PAX3* were identified using heteroduplex detection (Tassabehji et al. 1992).

Ultimately, for some genes it may prove necessary to use transgenic techniques to confirm that a candidate sequence is the disease gene, by testing its involvement in the phenotype. Transgenic mice are used as models for human genetic diseases to investigate the mode of action of the gene, but transgenesis may also be needed to identify disease genes that occur in mice and humans if there is not enough evidence from sequence or expression analysis to confirm the genomic cause of the disease. Transgenesis involves introducing a foreign piece of DNA into the genome of an animal. One very useful form of transgenesis involves introducing a modified copy of a normal gene into a cell where it replaces the wild type gene by homologous recombination. The cell is then used to generate a new stock of mice carrying only the modified version of the gene. Usually, the modified version of the gene has a mutation that effectively stops the production of the protein, and these are called "knock-out" experiments. The resulting mice, with no effective protein product of the gene being tested, are examined for signs of the predicted phenotype. Another use of transgenesis is to do the reverse experiment, and replace a mutant gene with a normal, wild-type gene, to see if the phenotype of the mutant can thereby be rescued. Either approach will contribute to establishing whether a candidate gene is the disease gene, but a successful rescue experiment can be regarded as definitive proof. Knock-out experiments can sometimes give equivocal findings when the mouse homologue of a human gene is being studied, as mentioned earlier in this chapter. Nonetheless, the potential value of transgenic mice makes the technology a vital part of the repertoire of techniques for investigating how mutations cause genetic diseases. Once a gene has been identified, targetted mutations interrupting specific parts of the protein product can be introduced by homologous recombination as part of a long term strategy to investigate the function of the protein.

9. Which Human Genes for Deafness Have Been Identified?

A small number of human genes causing deafness have been identified, and these have been reviewed recently (Duyk et al. 1992; Steel and Brown 1994; see Table 2.1). All of the genes identified so far have caused deafness as part of a syndrome, and most of these involve late-onset hearing loss rather than congenital deafness. It is only in the last few months that any genes have even been localized for the most common type of genetic deafness,

TABLE 2.1 Identified genes in human and mouse causing hearing impairment.

Human syndrome	Gene	Mouse mutant	Gene
Waardenburg syndrome type 1, WS1	PAX3	Splotch, Sp	Pax-3
Grieg cephalopolysyndactyly syndrome, GCPS	GLI3	Extra toes, Xt	Gli3
		Histidinaemia, Hal	his
		int-2 transgenic	Fgf-3
		Hox-1.6 transgenic	Hox-1.6, Hoxa-1
Piebald trait	KIT	Dominant spotting, W	c-kit
		Steel, Sl	kit ligand, mgf
Waardenburg syndrome type 2, WS2	MITF	Microphthalmia, mi	mi
		Light, B^{lt}	Trp-1
X-linked Alport syndrome	COL4A5		
Stickler syndrome	COL1A1		
Osteogenesis imperfecta	COL1A1/COL1A2	Mov-13 transgenic and other transgenics	Col a1
Osteopetrosis		Osteopetrosis transgenic	c-src
		Osteopetrosis, op	Csfm
Charcot-Marie-Tooth disease type 1A	PMP-22	Trembler, Tr	pmp-22
Charcot-Marie-Tooth disease type 1B	P_o		
X-linked Charcot-Marie-Tooth disease	Cx32		
Cockayne syndrome	ERCC		
Hunter syndrome	IDS, SIDS		
Hurler syndrome	IDUA		
Neurofibromatosis 2, NF2	NF2		
Norrie disease	NDP		
Autosomal recessive generalised resistance to thyroid hormone, arGRTH	hTRβ		
Aminoglycoside-induced deafness	Mitochondrial 12S rRNA		
Diabetes mellitus with hearing loss	Mitochondrial tRNA$^{Leu(UUR)}$		
Wolfram syndrome	7.6 kb mitochondrial deletion		
Diabetes mellitus and deafness	10.4 kb mitochondrial deletion		

nonsyndromal autosomal recessive deafness, and none have been identified (Guilford et al. 1994a,b). There are good reasons for the bias toward localizing and identifying genes for syndromal deafness. Firstly, if a syndrome is well characterized clinically, it is more likely that a homogeneous group can be assembled that will enable linkage analysis to be successful. For example, the careful clinical characterization of Waardenburg syndrome and Usher syndrome into distinct sub-types was a prerequisite for the localization of some of the genes involved. Secondly, the presence of other features in addition to the hearing impairment allows more informed guesses to be made about possible candidate genes, because the other features provide additional clues to the underlying pathology. An example of this is Alport syndrome, where the kidney and inner ear pathologies had in common a breakdown of the basal lamina that suggested a basal lamina component such as a collagen gene might be a good candidate. Thirdly, the other features of the syndrome may suggest potential homologies with mouse mutants, as occurred in Waardenburg syndrome where the white forelock was reminiscent of the white spotting of the coat in *splotch* mice. When the *splotch* gene was identified, the similarity in phenotype suggested that the same gene would be a good candidate for WS1.

It is interesting to examine exactly how genes causing deafness have been localized and/or identified in humans. X-linkage is often straightforward to determine from examination of pedigrees, and several genes involving deafness have been localized to the X-chromosome as a consequence (e.g., Albinism-deafness syndrome, Alport's syndrome, Hunter syndrome, Norrie's disease, Oto-palato-digital syndrome and X-linked progressive mixed deafness with perilymphatic gusher; see Duyk et al. 1992, Brunner et al. 1991 for references). Mitochondrial inheritance is another pattern that often can be determined by study of the pedigree, as the defect will show maternal inheritance. Several examples of hearing loss associated with mitochondrial inheritance have been reported, two of which show deletions of mitochondrial DNA (e.g., Ballinger et al. 1992; Rötig et al. 1993). Cytogenetically visible chromosomal rearrangements and deletions in isolated cases have been extremely useful in localizing genes. For example, a chromosome 2 inversion in a Japanese boy with Waardenburg syndrome type 1 gave the first clue that the gene for WS1 might be on 2q (Ishikiriyama et al. 1989). Analysis of panels of deletions has enabled more precise localizations of genes causing deafness on the X chromosome; comparison of the extent of each deletion with the spectrum of phenotypic features enables genes causing each feature to be separated (e.g., Brunner et al. 1991; see Fig. 2.3). In the case of Monge's deafness, a very large family with many hearing impaired members in several generations was available for study, allowing linkage to 5q to be determined (Leon et al. 1992). The use of a genetic isolate, the Acadian population in Louisiana, allowed the localization of one of the genes for Usher syndrome to chromosome 11p (Smith et al. 1992). Other genetic isolates in Tunisia were studied to localize two genes causing autosomal recessive nonsyndromic deafness (Guilford et al. 1994a,b).

Identified genes causing hearing impairment are listed in Table 2.1. Of the few genes that have been identified and include deafness as a feature, several are collagen genes. These include Stickler syndrome, which can be caused by a defect in COL2A1, osteogenesis imperfecta caused by mutations of COL1A1 or COL1A2, and Alport syndrome with mutations in the COL4A5 gene. The phenotypic features of these diseases all suggested that a structural gene might be involved, and the collagen genes were a priori good candidates. The accumulations of mucopolysaccharides seen in Hunter and Hurler syndromes, leading ultimately to hearing loss, suggested an inborn error of carbohydrate metabolism, and this was indeed the case for both syndromes. A further example of a gene causing deafness that was identified because of a phenotypic clue is Cockayne syndrome, in which the premature aging suggested a defect in DNA repair mechanisms. The gene affected in Norrie syndrome was identified by a classical positional cloning strategy: firstly the region of the X chromosome containing the gene was defined by deletion mapping and recombination studies, then physical maps using cosmid and YAC clones were constructed and gene sequences identified within the nonrecombinant region were used to screen retinal libraries, and finally a series of microdeletions and missense mutations were found in DNA from Norrie syndrome patients (Berger et al. 1992; Chen et al. 1992a,b; Meindl et al. 1992). Lastly, three genes causing deafness in humans have been identified after their mouse homologues had been characterized. These are WS1, where *PAX3* mutations were found after *Pax-3* had been found to be mutated in *splotch* mutant mice as described earlier in the chapter, Charcot-Marie-Tooth disease in which duplications of the *PMP-22* gene were found after a mutation in the homologous gene in mice was described in *trembler* mutants (Matsunami et al. 1992; Patel et al. 1992; Suter et al. 1992; Timmerman et al. 1992; Valentijn et al. 1992), and Piebaldism, which was shown to be due to mutations in the *KIT* gene after the mouse version of the gene, *c-kit*, was shown to be mutated in several alleles at the *W* locus (Chabot et al. 1988; Geissler et al. 1988; Giebel and Spritz 1991; Spritz et al. 1992). Only one presumed homozygote for piebald trait has been described, and he was deaf (Hultén et al. 1987), so this gene is probably not a major cause of deafness in the human population.

In summary, most of the strategies described earlier in this chapter have contributed toward the identification or localization of some of the deafness genes in humans. It is obvious that a great deal remains to be done in identifying both syndromal deafness genes and, particularly, nonsyndromic deafness genes that have such an important role in the etiology of deafness in human populations.

10. Summary

Mutations in many genes cause hearing impairment in both humans and mice. This heterogeneity makes localizing and identifying the genes involved particularly difficult in humans, and special strategies need to be adopted to minimize the difficulties. These include using very large families for linkage analysis rather than a group of small families, choosing a group from a genetic isolate, and taking care that the clinical characterization of the type of deafness is as uniform as possible within the study group. These approaches have been useful in localizing a handful of genes causing syndromic deafness, a gene for dominant late-onset nonsyndromal deafness, and two genes causing uncomplicated autosomal recessive deafness. For this last type of deafness, a very useful strategy appears to be to identify deafness genes in mice and then to find the homologous human gene. Mice have significant advantages for the identification of genes, because very large groups of 1000 or more offspring can be generated, all known to be carrying the same mutation, providing a powerful resource for positional cloning. Positional cloning is likely to be very important in the identification of deafness genes, and it involves identifying a gene on the basis of its position on the chromosome. An outline of a strategy for positional cloning is given in the chapter. The investigation of candidate genes is another important approach to identifying deafness genes. Candidates include genes that encode a protein product that seems likely a priori to be involved in the auditory system, and genes known to be expressed in the inner ear or other relevant structures such as the developing neural tube will be included. Inner ear cDNA libraries will be useful therefore in suggesting suitable candidate genes. In addition, genes known to be located in the relevant region of the genome will be considered candidates even if their expression in the inner ear has not been investigated. The identification of deafness genes is important because it will aid our understanding of normal ear development and function as well as providing a scientific basis for devising strategies for medical intervention to ameliorate the effects of mutations on hearing.

References

Arnvig J (1955) Vestibular function in deafness and severe hardness of hearing. Acta Otolaryngol 45:283–288.

Bach I, Brunner HG, Beighton P, Ruvalcaba RHA, Reardon W, Pembrey ME, van der Velde-Visser SD, Bruns GAP, Cremers CWRJ, Cremers FPM, Ropers H-H (1992) Microdeletions in patients with gusher-associated, X-linked mixed deafness (DFN3). Am J Hum Genet 50:38–44.

Bahary N, Pachter JE, Felman R, Leibel RL, Albright K, Cram S, Friedman JM

(1992) Molecular mapping of mouse chromosomes 4 and 6: Use of a flow-sorted Robertsonian chromosome. Genomics 13:761-769.

Baldwin CT, Hoth CF, Amos JA, da-Silva EO, Milunsky A (1992) An exonic mutation in the H*u*P2 paired domain gene causes Waardenburg's syndrome. Nature 355:637-638.

Ballinger SW, Shoffner JM, Hedaya EV, Trounce I, Polak MA, Koontz DA, Wallace DC (1992) Maternally transmitted diabetes and deafness associated with a 10.4 kb mitochondrial DNA deletion. Nature Genet 1:11-15.

Beisel KW, Kennedy JE, Swaroop A (1993) Identification of cDNAs shared between the inner ear and the retina. Abstracts of the Sixteenth Midwinter Meeting of the Association for Research in Otolaryngology, p138.

Berger W, Meindl A, van de Pol TJR, Cremers FPM, Ropers HH, Döerner C, Monaco A, Bergen AAB, Lebo R, Warburg M, Zergollern L, Lorenz B, Gal A, Bleeker-Wagemakers EM, Meitinger T (1992) Isolation of a candidate gene for Norrie disease by positional cloning. Nature Genet 1:199-203.

Bird, AP (1986) CpG-rich islands and the function of DNA methylation. Nature 321:209-213.

Brown KA, Sutcliffe M, Steel KP, Brown SDM (1992) Close linkage of the olfactory marker protein gene to the mouse deafness mutation *shaker-1*. Genomics 13:189-193.

Brown SDM, Steel KP (1994) Genetic deafness—progress with mouse models. Hum Molec Genet 3:1453-1456.

Brownstein Z, Friedlander Y, Peritz E, Cohen T (1991) Estimated number of loci for autosomal recessive severe nerve deafness within the Israeli Jewish population, with implications for genetic counseling. Am J Med Genet 41:306-312.

Brunner HG, Smeets B, Smeets D, Nelen M, Cremers CWRJ, Ropers H-H (1991) Molecular genetics of X-linked hearing impairment. In: Ruben RJ, Van De Water TR and Steel KP (eds) Genetics of Hearing Impairment. Ann N Y Acad Sci 630:176-190.

Buckler AJ, Chang DD, Graw SL, Brook JD, Haber DA, Sharp PA, Housman DE (1991) Exon Amplification: A strategy to isolate mammalian genes based on RNA splicing. Proc Natl Acad Sci, USA 88:4005-4009.

Burke DT, Rossi JM, Leung J, Koos DS, Tilghman S (1991) A mouse genomic library of yeast artificial chromosome clones. Mammal Genome 1:65.

Burri M, Tromvoukis Y, Bopp D, Frigerio G, Noll M (1989) Conservation of the paired domain in metazoans and its structure in three isolated human genes. EMBO J 8:1183-1190.

Chabot B, Stephenson DA, Chapman VM, Besmer P, Bernstein A (1988) The proto-oncogene *c-kit* encoding a transmembrane tyrosine kinase receptor maps to the mouse *W* locus. Nature 335:88-89.

Chartier FL, Keer JT, Sutcliffe MJ, Henriques DA, Mileham P, Brown SDM (1992) Construction of a mouse yeast artificial chromosome library in a recombination-deficient strain of yeast. Nature Genet. 1:132-136.

Chen Z-Y, Hendriks RW, Jobling MA, Powell JF, Breakefield XO, Sims KB, Craig IW (1992) Isolation and characterisation of a candidate gene for Norrie disease. Nature Genet 1:204-208.

Chen Z-Y, Sims KB, Coleman M, Donnai D, Monaco A, Breakefield XO, Davies

KE, Craig IW (1992) Characterisation of a YAC containing part or all of the Norrie disease locus. Hum Molec Genet 1:161-164.

Chisaka O, Musci TS, Capecchi MR (1992) Developmental defects of the ear, cranial nerves and hindbrain resulting from targeted disruption of the mouse homeobox gene *Hox-1.6*. Nature 355:516-520.

Chung CS, Brown KS (1970) Family studies of early childhood deafness ascertained through the Clarke school for the deaf. Am J Hum Genet 22:630-644.

Chung CS, Robison OW, Morton NE (1959) A note on deaf mutism. Ann Hum Genet 23:357-366.

Collins FS (1992) Positional cloning: Let's not call it reverse anymore. Nature Genet 1:3-6.

Cremers FPM, van de Pol TJR, Diergaarde PJ, Wieringa B, Nussbaum RL, Schwartz M, Ropers H-H (1989) Physical fine mapping of the choroideremia locus using Xq21 deletions associated with complex syndromes. Genomics 4:41-46.

Davis AC (1989) The prevalence of hearing impairment and reported hearing disability among adults in Great Britain. Int J Epidemiol 18:911-917.

Davis JG, Burns F, Eberwine JH, Greene MI, Oberholtzer JC (1993) Isolation and analysis of candidate auditory epithelium-specific cDNAs from a *Lepomis macrochirus* saccular macula cDNA library. Abstracts of the Sixteenth Midwinter Meeting of the Association for Research in Otolaryngology, p89.

Deol MS (1956) A gene for uncomplicated deafness in the mouse. J Embryol Exp Morphol 4:190-195.

Deol MS (1968) Inherited diseases of the inner ear in man in the light of studies on the mouse. J Med Genet 5:137-158.

Deol MS, Kocher W (1958) A new gene for deafness in the mouse. Heredity 12:463-466.

Dietrich WF, Miller JC, Steen RG, Merchant M, Damron D, Nahf R, Cross A, Joyce DC, Wessel M, Dredge RD, Marquis A, Stein LD, Goodman N, Page DC, Lander ES (1994) A genetic map of the mouse with 4006 simple sequence length polymorphisms. Nature Genet 7:220-245.

Dryja TP, McGee TL, Reichel E, Hahn LB, Cowley GS, Yandell DW, Sandberg MA, Berson EL (1990) A point mutation of the rhodopsin gene in one form of retinitis pigmentosa. Nature 343:364-366.

Duyk GM, Kim S, Myers RM, Cox DR (1990) Exon trapping: A genetic screen to identify candidate transcribed sequences in cloned mammalian genomic DNA. Proc Natl Acad Sci USA 87:8995-8999.

Duyk G, Gastier JM, Mueller RF (1992) Traces of her workings. Nature Genet 2:5-8.

Eberwine J, Yeh H, Miyashiro K, Cao Y, Nair S, Finnel M, Zettel M, Coleman P (1992) Analysis of gene expression in live neurons. Proc Natl Acad Sci USA 87:3010-3014.

Epstein DJ, Vekemans M, Gros P (1991) *Splotch* (Sp^{2H}), a mutation affecting development of the mouse neural tube, shows a deletion within the paired homeodomain of *Pax-3*. Cell 67:767-774.

Everberg G (1960) Unilateral total deafness in children. Clinical problems with a special view to vestibular function. Acta Otolaryngol 52:253-269.

Foy C, Newton V, Wellesley D, Harris R, Read AP (1990) Assignment of the locus for Waardenburg syndrome type 1 to human chromosome 2q37 and possible homology to the *Splotch* mouse. Am J Hum Genet 46:1017-1023.

Fraser, GR (1976) The Causes of Profound Deafness in Childhood. Baltimore, MD: Johns Hopkins University Press.

Geissler EN, Ryan MA, Housman DE (1988) The dominant-white spotting (*W*) locus of the mouse encodes the *c-kit* proto-oncogene. Cell 55:185-192.

Giebel LB, Spritz RA (1991) Mutation of the *KIT* (mast/stem cell growth factor receptor) protooncogene in human piebaldism. Proc Natl Acad Sci USA 88:8696-8699.

Goulding MD, Chalepakis G, Deutsch U, Erselius JR, Gruss P (1991) Pax-3, a novel murine DNA binding protein expressed during early neurogenesis. EMBO J 10:1135-1147.

Greenfield AJ, Brown SDM (1987) Microdissection and microcloning from the proximal region of mouse chromosome 7: Isolation of clones genetically linked to the pudgy locus. Genomics 1:153-158.

Guilford P, Ben Arab S, Blanchard S, Levilliers J, Weissenbach J, Belkahia A, Petit C (1994) A non-syndromic form of neurosensory, recessive deafness maps to the pericentromeric region of chromosome 13q. Nature Genet 6:24-28.

Guilford P, Ayadi H, Blanchard S, Chaib H, Le Pasier D, Weissenbach J, Drira M, Petit C (1994) A human gene responsible for neurosensory, non-syndromic recessive deafness is a candidate homologue of the mouse *sh-1* gene. Hum Molec Genet 3:989-993.

Gyapay G, Morissette J, Vignal A, Dib C, Fizames D, Millasseau P, Marc S, Bernardi G, Lathrop M, Weissenbach J (1994) The 1993-94 Genethon human genetic linkage map. Nature Genet 7:246-339.

Horak FB, Shumway-Cook A, Crowe TK, Black FO (1988) Vestibular function and motor proficiency of children with impaired hearing, or with learning disability and motor impairment. Dev Med Child Neurol 30:64-79.

Hultén MA, Honeyman MM, Mayne AJ, Tarlow MJ (1987) Homozygosity in piebald trait. J Med Genet 24:568-571.

Ishikiriyama S, Tonoki H, Shibuya Y, Chin S, Harada N, Abe K, Nikawa N (1989) Waardenburg syndrome type I in a child with de novo inversion (2) (q35q37.3). Am J Med Genet 33:506-507.

Kaga K, Suzuki JI, Marsh RR, Tanaka Y (1981) Influence of labyrinthine hypoactivity on gross motor development of infants. Ann N Y Acad Sci 374:412-420.

Kaplan J, Gerber S, Bonneau D, Rozet JM, Delrieu O, Briard ML, Dollfus H, Ghazi I, Dufier JL, Frézal J, Munnich A (1992) A gene for Usher Syndrome Type I (USH1A) maps to chromosome 14q. Genomics 14:979-987.

Kimberling WJ, Weston MD, Möller C, Davenport SLH, Shugart YY, Priluck IA, Martini A, Milani M, Smith RJH (1990) Localization of Usher Syndrome Type II to chromosome 1q. Genomics 7:245-249.

Kimberling WJ, Möller CG, Davenport S, Priluck IA, Beighton PH, Greenberg J, Reardon W, Weston MD, Kenyon JB, Grunkemeyer JA, Pieke Dahl S, Overbeck LD, Blackwood DJ, Brower AM, Hoover DM, Rowland P, Smith RJH (1992) Linkage of Usher Syndrome Type I gene (USH1B) to the long arm of chromsome 11. Genomics 14:988-994.

Konigsmark BW, Gorlin RJ (1976) Genetic and Metabolic Deafness. Philadelphia, PA: W B Saunders and Co.

Leon PE, Raventos H, Lynch E, Morrow J, King M-C (1992) The gene for an inherited form of deafness maps to chromosome 5q31. Proc Natl Acad Sci USA 89:5181-5184.

Lewis RA, Otterud B, Stauffer D, Lalouel J-M, Leppert M (1990) Mapping recessive ophthalmic diseases: Linkage of the locus for Usher Syndrome Type II to a DNA marker on chromosome 1q. Genomics 7:250-256.

Lufkin T, Dierich A, LeMeur M, Mark M, Chambon P (1991) Disruption of the *Hox-1.6* homeobox gene results in defects in a region corresponding to its rostral domain of expression. Cell 66:1105-1119.

Mansour S, Goddard JM, Capecchi MR (1993) Mice homozygous for a targeted disruption of the proto-oncogene *int-2* have developmental defects in the tail and inner ear. Development 117:13-28.

Matsunami N, Smith B, Ballard L, Lensch MW, Robertson M, Albertsen H, Hanemann CO, Müller HW, Bird TD, White R, Chance PF (1992) Peripheral myelin protein-22 gene maps in the duplication in chromosome 17p11.2 associated with Charcot-Marie-Tooth 1A. Nature Genet. 1:176-179.

Meindl A, Berger W, Meitinger T, van de Pol D, Achatz H, Dorner C, Haasemann M, Hellebrand H, Gal A, Cremers F, Ropers H-H (1992) Norrie disease is caused by mutations in an extracellular protein resembling C-terminal globular domain of mucins. Nature Genet 2:139-143.

Morton CC, Bieber FR, Gutierrez-Espeleta GA, Khetarpal U, Robertson NG (1992) Cloning genes involved in hearing from a human fetal cochlear cDNA library. Proceedings of the Meeting on The Molecular Biology of Hearing and Deafness, 86.

Morton NE (1991) Genetic epidemiology of hearing impairment. In: Ruben RJ, Van De Water T R and Steel K P (eds) Genetics of Hearing Impairment. Ann N Y Acad Sci 630:16-31.

Nadeau JH, Kosowsky M, Steel KP (1991) Comparative gene mapping, genome duplication, and the genetics of hearing. In: Ruben RJ, Van De Water TR and Steel KP (eds) Genetics of Hearing Impairment. Ann N Y Acad Sci 630:49-67.

Oberholtzer JC, Cohen EL, Davis JG, Greene MI (1993) Use of a chick cochlea cDNA library for the identification and characterization of peripheral auditory tissue transcripts. Abstracts of the Sixteenth Midwinter Meeting of the Association for Research in Otolaryngology, p88.

O'Brien SJ, Womack JE, Lyons LA, Moore KJ, Jenkins NA, Copeland NG (1993) Anchored reference loci for comparative genome mapping in mammals. Nature Genet 3:103-112.

Ott J (1991) Analysis of Human Genetic Linkage. Baltimore, MD: Johns Hopkins University Press.

Patel PI, Roa BB, Welcher AA, Schoener-Scott R, Trask BJ, Pentao L, Snipes GJ, Garcia CA, Francke U, Shooter EM, Lupski JR, Suter U (1992) The gene for the peripheral myelin protein PMP-22 is a candidate for Charcot-Marie-Tooth disease type 1A. Nature Genet 1:159-165.

Pieke Dahl S, Weston MD, Kimberling WJ, Gorin MB, Shugart YY, Kenyon JB (1991) Possible genetic heterogeneity of Usher syndrome type 2: a family unlinked to chromosome 1q markers. Am J Hum Genet (suppl) 49:A1077.

Represa J, León Y, Miner C, Giraldez F (1991) The *int-2* proto-oncogene is responsible for induction of the inner ear. Nature 353:561-563.

Rinchik EM, Johnson DK, Margolis FL, Jackson IJ, Russell LB, Carpenter DA (1991) Reverse genetics in the mouse and its application to the study of deafness. In: Ruben RJ, Van De Water TR, Steel KP (eds) Genetics of Hearing Impairment. Ann N Y Acad Sci 630:80-92.

Rosenblut BR, Goldstein R, Landau WM (1960) Vestibular responses of some deaf and aphasic children. Ann Otol Rhinol Laryngol 69:747-755.

Rötig A, Cormier V, Chatelain P, Francois R, Saudubray JM, Rustin P, Munnich A (1993) Deletion of mitochondrial DNA in a case of early-onset diabetes mellitus, optic atrophy, and deafness (Wolfram syndrome, MIM 222300). J Clin Invest 91:1095-1098.

Ryan AF, Housley GD, Harris JP (1993) Isolation of gene sequences from the inner ear using PCR and cDNA library screening. Abstracts of the Sixteenth Midwinter Research Meeting of the Association for Research in Otolaryngology, p88.

Sewell WF, Mroz EA (1990) Purification of a low-molecular-weight excitatory substance from the inner ears of goldfish. Hearing Res 50:127-138.

Smith RJH, Lee EC, Kimberling WJ, Daiger SP, Pelias MZ, Keats BJB, Jay M, Bird A, Reardon W, Guest M, Ayyagari R, Hejtmancik JF (1992) Localization of two genes for Usher Syndrome Type I to chromosome 11. Genomics 14:995-1002.

Smith RJH, Berlin CI, Hejtmancik JF, Keats BJB, Kimberling WJ, Lewis RA, Möller CG, Pelias MZ, Tranebjaerg L (1994) Clinical diagnosis of the Usher syndromes. Am J Med Genet 50:32-38.

Solc CK, Meyers V, Duyk GM, Corey DP (1993) Cloning and molecular characterization of myosins from bullfrog macula sensory epithelium. Abstracts of the Sixteenth Midwinter Research Meeting of the Association for Research in Otolaryngology, p21.

Spritz RA, Giebel LB, Holmes SA (1992) Dominant negative and loss of function mutations of the *c-kit* (mast/stem cell growth factor receptor) proto-oncogene in human piebaldism. Am J Hum Genet 50:261-269.

Steel KP (1991) Similarities between mice and humans with hereditary deafness. In: Ruben RJ, Van De Water TR and Steel KP (eds) Genetics of Hearing Impairment. Ann N Y Acad Sci 630:68-79.

Steel KP, Bock GR (1980) The nature of inherited deafness in *deafness* mice. Nature 288:159-161.

Steel KP, Brown SDM (1994) Genes for deafness. Trends in Genet 10:428-435.

Steel KP, Smith RJH (1992) Normal hearing in *Splotch* (Sp/+), the mouse homologue of Waardenburg syndrome type 1. Nature Genet 2:75-79.

Steel KP, Davidson DR, Jackson IJ (1992) TRP-2/DT, a new early melanoblast marker, shows that steel growth factor (c-kit ligand) is a survival factor. Development 115:1111-1119.

Stevenson AC, Cheeseman EA (1956) Hereditary deaf mutism, with particular reference to Northern Ireland. Ann Hum Genet 20:177-231.

Suter U, Welcher AA, Özcelik T, Snipes GJ, Kosaras B, Francke U, Billings-Gagliardi S, Sidman RL, Shooter EM (1992) *Trembler* mouse carries a point mutation in a myelin gene. Nature 356:241-244.

Swaroop A, Xu J, Agarwal N, Weissman SM (1991) A simple and efficient cDNA library subtraction procedure: isolation of human retina-specific cDNA clones. Nucl Acids Res 19:1954.

Tassabehji M, Read AP, Newton VE, Harris R, Balling R, Gruss P, Strachan T (1992) Waardenburg's syndrome patients have mutations in the human homologue of the *Pax-3* paired box gene. Nature 355:635-636.

Tassabehji M, Read AP, Newton VE, Patton M, Gruss P, Harris R, Strachan T (1993) Mutations in the *PAX3* gene causing Waardenburg syndrome type 1 and type 2. Nature Genet 3:26-30.

Timmerman V, Nelis E, Van Hul W, Nieuwenhuijsen BW, Chen KL, Wang S, Ben Othman K, Cullen B, Leach RJ, Hanemann CO, De Jonghe P, Raeymaekers P, van Ommen G-JB, Martin J-J, Müller HW, Vance JM, Fischbeck KH, Van Broeckhoven C (1992) The peripheral myelin protein gene *PMP-22* is contained within the Charcot-Marie-Tooth disease type 1A duplication. Nature Genet 1:171-175.

Travis GH, Hepler JE (1993) A medley of retinal dystrophies. Nature Genet 3:191-192.

Valentijn LJ, Bolhuis PA, Zorn I, Hoogendijk JE, van den Bosch N, Hensels GW, Stanton VP, Houseman DE, Fischbeck KH, Ross DA, Nicholson GA, Meershoek EJ, Dauwerse HG, van Ommen G-JB, Baas F (1992) The peripheral myelin gene *PMP-22/GAS-3* is duplicated in Charcot-Marie-Tooth disease type 1A. Nature Genet 1:166-170.

Wilcox ER, Fex J (1992) Construction of a cDNA library from microdissected guinea pig organ of Corti. Hearing Res 62:124-126.

Wilcox ER, Gruber C, Fex J (1993) Construction of a guinea pig ampulla cDNA library and cloning of cDNA sequences common to the ampulla and organ of Corti. Abstracts of the Sixteenth Midwinter Meeting of the Association for Research in Otolaryngology, p146.

Zdarsky E, Favor J, Jackson IJ (1990) The molecular basis of *brown*, an old mouse mutation, and of an induced revertant to wild type. Genetics 126:443-449.

3
Regeneration of the Auditory Nerve: The Role of Neurotrophic Factors

THOMAS R. VAN DE WATER, HINRICH STAECKER,
STUART C. APFEL, AND PHILIPPE P. LEFEBVRE

1. Introduction

This review is intended to introduce the reader to the field of regeneration–repair processes and the trophic and tropic factors that can affect these processes in the developing and mature nervous system, and then to relate these neurotrophic factors to regeneration–repair processes within the inner ear and more specifically to the auditory system. The experiments reviewed in this chapter demonstrate that the auditory system has the potential to support the survival of auditory neurons and to regenerate or repair neuronal processes at least to a limited extent. It appears likely that several members of the neurotrophin family and other neurotrophic factors are intimately involved with both the development and maintenance of auditory neurons and their pattern of innervation. Once the basic molecular mechanisms of these neurotrophic factors in the auditory system is understood, therapy to prevent degeneration of neurons after damage or even as therapy to establish functional reinnervation may become a reality. To assist the reader with all of the shorthand used in identifying the neurotrophic factors and their receptors a table of identification has been prepared (Table 3.1).

The dogma that neurons are post mitotic terminally differentiated cells is no longer considered accurate. Although at present there is little evidence for differentiation of neurons from neuronal "stem cells" after neuronal injury, neurons in the peripheral nervous system have been found to repair themselves after axonotomy and in some cases reestablish functional patterns of innervation. Neuronal repair has been studied since the late nineteenth century. Ramón y Cajal's classic *Degeneration and Regeneration of the Nervous System* was the first major work to examine this process using at that time newly developed neuron-specific histologic methods (Ramón y Cajal 1928). Continuing research in this area has focused largely on skeletal muscle and motor neurons or the visual system. However, significant contributions to this field have also been made by studying the auditory–vestibular system.

Functional regeneration can occur in amphibians (Sperry 1945) and

TABLE 3.1 Abbreviations for neurotrophic factors and their receptors.

Ligands
Nerve growth factor–NGF
Brain derived neurotrophic factor–BDNF
Neurotrophin three–NT-3
Neurotrophin four and five–NT-4/5
Neurotrophin six–NT-6
Ciliary neurotrophic factor–CNTF
Insulin-like growth factor 1–IGF-I
Insulin-like growth factor 2–IGF-II
Acidic fibroblast growth factor–aFGF, FGF-1
Basic fibroblast growth factor–bFGF, FGF-2
Int-2 proto-oncogene product–int-2, FGF-3
Transforming growth factor beta 1 to beta 3–TGFβ_1; TGFβ_2; TGFβ_3

Receptors
Low-affinity NGF (pan neurotrophin) receptor–p75^{LNGFR}
High-affinity NGF receptor–trk-A
High-affinity BDNF and NT-4 receptor–trk-B
High-affinity NT-3 receptor–trk-C
High-affinity CNTF receptor–CNTF$R\alpha$
Leukemia inhibitory factor receptor–LIFRβ
Oncostatin M receptor–gp130

histological evidence of some nerve regeneration has been shown in mammals (Spoendlin and Suter 1976; Spoendlin 1988; Bohne and Harding 1992). In light of the recent interest in the function of neurotrophic factors in neuronal development and injury–repair processes, a series of eighth cranial nerve (VIIIn) regeneration experiments carried out over the last fifty years takes on increasing significance. Recent experiments testing for neurotrophic effects of various growth factors on cultures and explants of auditory neurons suggest that many of the regenerative events previously observed may depend upon, or be enhanced by, growth factors such as basic fibroblast growth factor (bFGF, FGF-2), transforming growth factor beta one (TGFβ_1), insulin, nerve growth factor (NGF), brain derived neurotrophic factor (BDNF) and neurotrophin 3 (NT-3).

2. Regeneration VIIIth Cranial Nerve (VIIIn)

2.1 Amphibians

One of the earliest studies of nerve regeneration in the inner ear was performed in the vestibular system of an amphibian (Sperry 1945). Using an adult tree frog (*Hyla squirelia*) Sperry sectioned the eight cranial nerve

(VIIIn) at its entry point to the brainstem. Given that various divisions of the VIIIn innervate the cristae of the three semicircular ducts, the maculae of the utricle, the lagena, the pars neglecta and the pars basileus, random reinnervation should cause perturbation of reflexes, whereas, the presence of normal reflexes after nerve sectioning would indicate regeneration with functional reinnervation of the correct targets. Five days after sectioning, Sperry began testing the frog's response to angular acceleration. After three weeks of recovery, a reflex response to horizontal acceleration was noted and the frogs progressively recovered reflexes over the next three weeks, leading him to conclude that even after a complete axonotomy, regenerating nerve fibers could find and form functional synapses with their original targets in the brainstem. Sperry also tested recovery of auditory function by observing the frog's response to an auditory stimulus, finding that 3/5ths of the frogs responded to the sound of paper being crumpled. Although mentioning the role of putative growth factors while discussing regeneration of the optic nerve, Sperry stopped short of drawing similar conclusions in these experiments.

Gleisner and Wersäll (1975) in a different species of frog (*Rana temporaria*) undertook a detailed histological study of the VIIIn after sectioning. In a simple VIIIn transection, it was found that the gaps between the central and peripheral stumps were sealed after three days, and new nerve fibers were reestablished within the nerve sheath after about seven days. When nerve stumps were transposed (intentionally switched), repair was found to occur on a similar schedule with an important exception: several nerve fibers were found to reinnervate their original correct target as well as the transposed target. Ultrastructural observations defined a degenerative phase that started about six hours posttransection, and within 24 hours showed increased electron density of the axoplasm and a loss of synaptic vesicles. By day three, increased Schwann cell phagocytic activity was noted with the overall process of neuronal degeneration affecting the afferent nerve fibers more rapidly and severely than efferent ones. The regenerative phase was described as starting at day 7, when the first new nerve fibers could be distinguished in the basal sensory epithelium, but with no identifiable synapses between the regenerating nerves and the sensory cells. Histological evidence of complete reinnervation was observed within two weeks of sectioning complete with nerve-hair cell synaptic complexes. Conversely to the degenerative process, afferent neuronal processes were reestablished before efferent ones. It was also noted that the afferent synaptic structures within the hair cells were fairly well preserved during the period of denervation (Gleisner and Wersäll 1975).

In a similar series of experiments, Zakon and Capranica studied the electrophysiological properties of the auditory portion of the VIIIn in the leopard frog (*Rana pipiens*). After sectioning the nerve as it entered the medulla, animals were allowed a minimum of three months recovery time. Binaural cells in the animals' superior olive reliably have identical, characteristic frequencies for stimulation of the right and left ears (Zakon and Capranica 1981a,b; Zakon 1986). In frogs, these cells receive neural input from the dorsal medullary nucleus (the equivalent of the cochlear nucleus in mammals). It was found that the binaural cells of the regenerated side have a similar frequency response as the nonsectioned side, showing that during regeneration of the auditory nerve, central connections are functionally reestablished (Zakon and Capranica 1981a,b; Zakon 1986). The tuning curves, latency, and threshold measurements of the regenerated fibers were also found to be similar to those of the nonsectioned side (Zakon and Capranica 1981a,b; Zakon 1986). Establishment of functional reinnervation by electrophysiological criteria was first seen five weeks postsectioning and proceeded through about twelve weeks postsectioning. The one major difference noted in regenerated nerves was that the tuning curves of 10% of the superior olivary nucleus cells tested were abnormal, assuming an uncharacteristic "W" shape never seen in normal animals (Zakon and Capranica 1981a,b; Zakon 1986).

Similar VIIIn regeneration studies were also undertaken by Newman and colleagues (Newman et al. 1986) to examine the regeneration of central projections, first described by Sperry in 1945. Using the bullfrog, *Rana catesbiana*, as a model, the VIIIn was transected at its entry point into the brainstem. Regenerated nerve fibers could be traced after five weeks using horseradish peroxidase injection. Both thick and thin fibers were found within the regenerated VIIIn. Regenerated fibers projected to the vestibular nuclei in a pattern similar to that observed in normal frogs, the only difference being the inclusion of thick and thin fibers projecting to the medial nucleus in the regenerated specimens. There was no evidence of aberrrant projections of the regenerated VIIIn fibers. However, paths of the regenerating nerve fibers were at times tortuous. Some nerve fibers even left the nerve tract and grew along the periphery of the brain stem before making the appropriate synaptic connections. Newmann et al. (1986) discuss this in terms of the theory of "chemoaffinity", proposed by Sperry (1945), which in light of recent research on neurotrophins is strongly supported. Physiological tests verified a return of function after VIIIn nerve regeneration (Newmann, Honrubia, and Bell 1987).

2.2 Mammals

Given this successful series of nerve regeneration studies in amphibians, we again encounter the question that is frequently asked about neural regeneration: Why doesn't it work in mammals? Unfortunately, few studies on mammals have been undertaken. The work of Spoendlin (1976), Terrayama et al. (1977) and Bohne and colleagues (Bohne and Harding 1992; Sun, Bohne, and Harding 1995) showed that the ability of the nerve to initiate self-repair and to begin to form new projections is not the problem, but rather the targeting of the regenerated neuronal processes. Spoendlin and

Suter (1976) reported that after transecting the VIIIn in the inner meatus of cats (*Felis domesticus*), regeneration of efferent and afferent nerve fibers could be observed. After the initial sectioning of the VIIIn, 95% of the auditory neurons were seen to degenerate over two months (Fig. 3.1). After twelve months of recovery, an increase of unmyelinated fibers was described in the first turn of the cochlea. Myelinated fibers and type II ganglion cells (i.e., the unmyelinated neurons that compose less than 10% of the spiral ganglion neurons and supply afferent innervation to the outer hair cells) were noted to be retained at a fairly constant number. In animals that were followed for a long period of time, large myelinated fibers were seen to pass the spiral ganglion entirely. Spoendlin believed that these large myelinated fibers were regenerating efferents that transversed the spiral ganglion in a radial fashion, following the most direct route into the cochlea. However, Spoendlin was unable to locate their points of termination, and thus summarized that they were "unable to reach their final destination." The increase of unmyelinated fibers was found to correspond to an increase in the remaining branches of the afferent nerves. Spoendlin suggested that the increase in unmyelinated fibers was a compensation by numbers for the lost sensory input from the sectioned afferents of the VIIIn.

In a slightly different approach, Terrayama et al. (1977) brought about the degeneration of guinea pig (*Cavia porcellus*) auditory neurons by perfusing the organ of Corti with streptomycin. After an initial degenerative phase that started about four days after perfusion, the organ of Corti collapsed, and infiltration of the nerves by macrophages occurred about ten days after perfusion. Electron microscopic evidence of nerve degeneration was clearly identifiable. The first signs of nerve regeneration were evident shortly afterwards. By twenty days postperfusion, Schwann cells could be seen encircling regenerated neuronal processes. Despite the observation that neuronal repair was taking place, this group observed an overall decline in the total number of cochlear nerve fibers. Animals examined 2 and 4 months post injury showed a great decrease of fibers as well as degeneration of the efferent olivo-cochlear bundle. Apparently, the new nerve fibers degenerated after a brief period of regeneration (Terrayama et al. 1977).

Other researchers have also found evidence of neuronal regeneration in mammals. Bohne and colleagues described the regeneration of auditory nerve fibers that grew along abnormal pathways in the cochleae of noise damaged chinchilla (*Chinchilla laniger*) inner ears. These fibers failed to stain for acetylcholinesterase activity, leading to the conclusion that these regenerated fibers were afferents (Sun, Bohne, and Harding 1995).

The VIIIn fibers both in amphibians and mammals appear to be capable of regeneration to varying degrees. The auditory and vestibular nerve fibers of anurans not only regenerate both peripheral and central axons but also are capable of restoring functional hearing and balance responses. In contrast, the nerve fibers of mammalian cochleae appear to be capable of

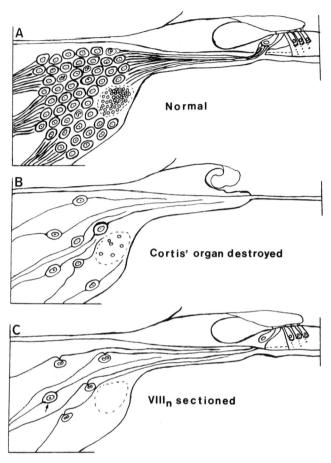

FIGURE 3.1 A schematic representation of two different patterns of secondary retrograde degeneration of spiral ganglion neurons that occur following (B) destruction of the organ of Corti and (C) sectioning of the VIII_n as compared to (A) the normal complement of auditory neurons and innervation of the auditory hair cells. After destruction of the organ of Corti, about 90 percent of the type I and type II auditory neurons degenerate (compare B to A). After section of the VIII_n, only the greater majority of the type I auditory neurons associated with the inner hair cells degenerate, whereas the type II auditory neurons and the entire afferent nerve supply of the outer hair cells remain normal (compare C to A and to B). A few type I neurons of the inner hair cell system always remain as unmyelinated type three neurons (see C; arrow). Modified from Spoendlin H (1988) Biology of the vestibulocochlear nerve in Alberti PW, Ruben RJ (eds) Otologic Medicine and Surgery Volume I, New York: Churchill Livingstone.

only a very limited amount of regrowth, as shown by Spoendlin (1976), and seem unable to find their targets after a period of postinjury neuritogenesis. In light of the concept of "chemoaffinity" advanced in several of the amphibian regeneration papers and in recent work on the neurotrophins, it seems plausible that there is a fundamental difference between amphibians and mammals in neurotrophin production by neuronal targets or in the ability of regenerating nerve fibers to react to trophic factors. Recent studies have shown that VIIIn sectioning in chinchillas results in the up-regulation of the p75 low affinity neurotrophin receptor (Fina, Popper, and Honrubia 1994), indicating that neurotrophins are most probably involved in the VIIIn response to injury and the neural regeneration process.

3. Development of Innervation of the Auditory System

3.1 Temporal Patterns

To understand the mechanics of regeneration it is important to first understand the sequence of events that define auditory innervation during normal development (Sher 1971; Van De Water 1986, 1988; Sobkowicz 1992; Van De Water et al. 1991, 1992; see Figs. 3.2 and 3.3). For simplicity, we will review primarily the development of afferent innervation. Unless specified, all research was carried out in laboratory mice (*Mus musculus*). At embryonic day eight (E8), the otic placode, destined to become the neurosensory epithelium of both the cochlear vestibular organs, begins to invaginate to form the otic vesicle. Shortly before the completion of this step (E9), a group of cells is identifiable at the basal end of the otic flask that is destined to become the ganglion cells that will form both the spiral and vestibular ganglia (Ruben 1967). These differentiating neurons can be identified by day E10.5 by immunostaining with anti 66 kD neurofilament (Galinovic-Schwartz et al. 1991). Within the next 24 h, ingrowth into the areas of presumptive sensory epithelium of the otocyst begins. Proliferation of neurons continues to day E13 (Ruben 1967). Hair cells develop and begin to differentiate between days E13 and 14, and afferent synaptogenesis occurs between E18 and the fifth day postpartum (5PP) (Sobkowicz 1992). During the postnatal period, several important events conclude the maturation of the cochlea. Schwann cells continue to undergo mitosis and complete their terminal division on 3 to 5 days PP (Ruben 1967). Myelination of neuronal fibers occurs 4 to 5 days postnatally (Romand and Romand 1982) and is extended to the neuronal somas by about 7 days PP. In the mouse, ultrastructural differentiation of spiral ganglion neurons into type I (innervating inner hair cells) and type II (innervating outer hair cells) occurs postnatally (Hafidi and Romand 1989). Furthermore, during the postnatal period, spiral ganglion neurons undergo programmed physiological cell death (i.e., apoptosis). At day 5 PP, the rat (*Rattus norvegicus*) loses 22% of its spiral ganglion auditory neurons (Rueda et al. 1987). This change corresponds to the completion of synaptogenesis. Completion and optimization of neuronal tuning is not completed until 4 weeks PP.

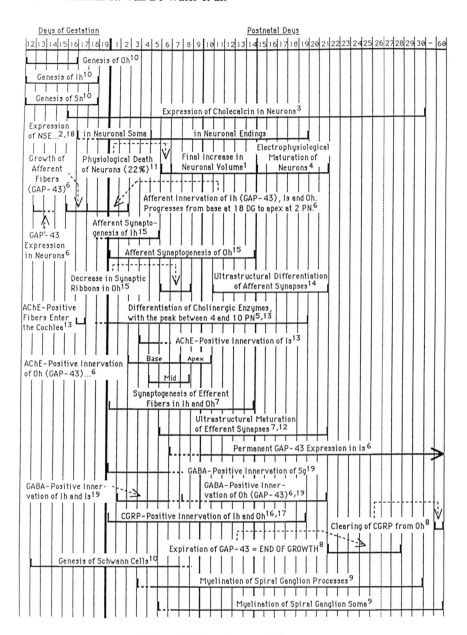

FIGURE 3.2. Caption on facing page.

Observations made by Sobkowicz et al. (1980) in postnatal organotypic cochlear cultures suggest that these postnatal changes are driven by the hair cells. In these cultures, the opening of the cochlear duct results in growth of the hair cells away from the spiral ganglion. The neuronal processes subsequently elongate to maintain synaptic contact with the displaced hair cells (Sobkowicz et al. 1980). As discussed later in this chapter, trophic factors, both soluble and matrix-bound, may provide a potential mechanism for both the development and later maintenance of spiral ganglion neurons. Formation of efferent synapses occurs from days 6 to 16 PP (Lenoir, Shnerson and Pujol 1980). If hair cells are destroyed by ototoxic or noise trauma, degeneration of auditory neurons results (Spoendlin 1971; Bichler, Spoendlin, and Rauchegge 1983; Spoendlin 1988), suggesting a trophic link between the sensory cells and the auditory neurons. Finally, staining for various structural proteins and neurotransmitters has shown that maturation of neurons is not completed until about 60 days PP. A timetable summarizing the important developmental events occurring from the onset (i.e., E12) to the completion (i.e., PP 60) of innervation of the cochlea in the mouse is presented in Figure 3.2 (Sobkowicz 1992).

←—————————————————————————————

FIGURE 3.2 Developmental timetable: maturational events as suggested by the current state of knowledge. The data concern the mouse, with the exception of refs. 11 and 17 (rat), and ref. 4 (gerbil), both species developing postnatally. The gestation time of the mouse ranges from 18 to 21 days (Rugh 1968); in this timetable, 19 days represents the average gestation time. The bold black line between 19 DG (gestation day) and 1 PN (postnatal day) represents the time of birth. The speckled lines at 10 and 14 PN represent the onset of hearing (Shnerson, Devigne, and Pujol 1982). The superscripted numbers on the events in this timetable represent the references listed below, the full references can be found listed alphabetically in the bibliography of this chapter: 1. Anniko 1988; 1989 (gerbil); 5. Emmerling and Sobkowicz 1988; 6. Emmerling and Sobkowicz 1990; 7. Emmerling et al. 1990; 8. Emmerling unpublished; 9. Romand and Romand 1990; 10. Ruben 1967; 11. Rueda et al. 1987 (rat); 12. Shnerson, Devigne, and Pujol 1982; 13. Sobkowicz and Emmerling 1989; 14. Sobkowicz 1982; 15. Sobkowicz et al. 1986; 16. Sobkowicz et al. 1988; 17. Tohyama et al. 1989 (rat); 18. Whitlon and Sobkowicz 1988; 19. Whitlon and Sobkowicz 1989. Symbols used in this timetable: |___ onset of differentiation; |___ endpoint of differentiation; |ᵃ|ᵇ|ᶜ|ᵈ| starting point is from a up to d; _ _ _ _ _ starting point unknown; _____ _ _ _ endpoint unknown; → ongoing synthesis of GAP-43. *Abbreviations:* acetylcholine esterase (AChE); calcitonin gene related peptide (CGRP); gamma aminofutyric acid (GABA); growth associated peptide forty-three (GAP43); inner hair cell (Ih); outer hair cell (Oh); neuron specific enolase (NSE); spiral ganglion neurons (Sn). Reproduced from Sobkowicz H (1992) The development of innervation in the organ of Corti in Romand R (ed) Development of the Auditory and Vestibular Systems 2, Amsterdam: Elsevier.

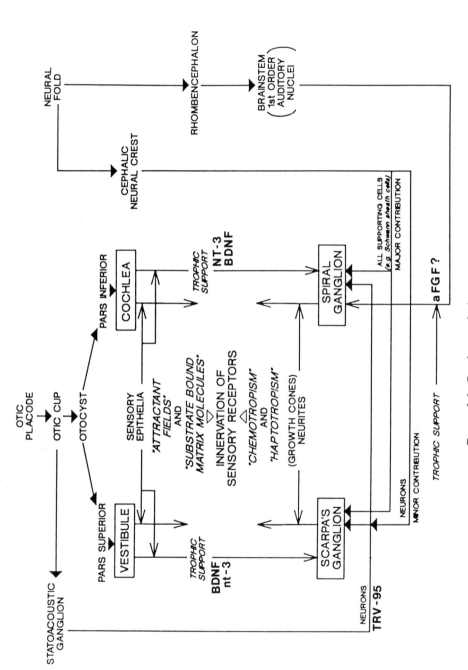

FIGURE 3.3. Caption on facing page.

3.2 Chemoattractive Fields

The mechanics of trophic fields in early embryonic development was extensively investigated by Van De Water and Ruben (Van De Water and Ruben 1984; Van De Water, Galinovic-Schwartz, and Ruben 1989) based on observations that denervated lateral line target fields of frog tadpoles (*Rana clamitans*) attracted nearby nerves to the denervated neuromasts (Speidel 1948). By coculturing two E12.5 mouse otocysts with a single statoacoustic ganglion complex, it was shown that neuronal ingrowth occurred into the developing sensory structures of both otocysts. This observation suggested that chemoattractant fields were being produced by the developing sensory epithelium (Van De Water 1986, 1988; Van De Water et al. 1991). Ard and colleagues observed in the chick (*Gallus domesticus*) that the peripheral target tissue (i.e., otocyst) of the statoacoustic ganglion supported neuronal survival in this ganglion (Ard, Morest, and Haugher 1985). In a further set of experiments in a mammal (*Mus musculus*), explanted statoacoustic ganglia were cultured alone or in the presence of peripheral (otocyst) or central (rhombencephalon) target tissues. Neuronal survival was found to depend on the presence of either central or peripheral target tissue. If both central and peripheral tissues were present, neuronal survival was even further augmented (Zhou and Van De Water, 1987). These observations (Ard, Morest, and Haugher 1985; Zhou and Van De Water 1987) and others mentioned above led to the theory that a soluble neurotrophic factor(s) (e.g., NT-3) could be acting in a trophic manner to support auditory nerve development and perhaps acting to establish trophic fields that attract the neuritic processes of the developing auditory neurons to its peripheral target (i.e., the neuroepithelium of the differentiating organ of Corti (Fig. 3.3; Van De Water 1988).

FIGURE 3.3 Flow chart depicting the major developmental interactions that influence and control the differentiation of inner ear sensory structures and the establishment of their innervation. Otic components in boxes represent final inner ear structures. Lines with solid arrowheads trace the sequence of changes that occur during the development of inner ear structures, while lines with open arrowheads indicate influences that operate at each stage of otic development. Influences that as yet lack definitive proof but are strongly suggested by indirect evidence are indicated by open arrowheads without lines. Neurotrophic factors in capitals indicate a major trophic effect while lower case suggests a minor trophic effect. The presence of a ? mark indicates suggestive data exist but a lack of definitive proof for trophic activity in-situ. *Abbreviations:* acidic fibroblast growth factor (aFGF); brain derived neurotrophic factor (BDNF); neurotrophin three (nt-3). Modified from Van De Water (1988) The determinants of neuron-sensory cell interactions. Develop Suppl 103.

4. Biology and Function of Neurotrophic Factors

4.1 History

A general definition for a neurotrophic factor is: a substance that will promote the survival of neurons. Included in this definition would be important metabolites such as sodium, potassium, glucose, amino acids, oxygen, etc. While these are essential for neuronal survival, they are not generally considered to be neurotrophic factors. Substances considered to be neurotrophic factors do more than just promote survival. Many neurotrophic factors influence morphological differentiation, for example cell size and neurite elongation. Others direct the physiological maturation of neurons, for example, neurotransmitter synthesis.

To clarify the definition of neurotrophic factors, we briefly review how the concept of neurotrophic factors evolved. The term trophic is derived from the Greek word "trophikos" meaning nourishment. Therefore, "neurotrophic" means "neuron nourishing." While studying peripheral nerve degeneration, Waller (1852) observed that nerves degenerate distal to a lesion, while new fibers "originate from the central portion, which maintained continuity with the 'trophic centers' in the spinal cord." These trophic centers provided the "nourishment" necessary to maintain the integrity of the nerve despite its injury. Distal portions of the nerve no longer in contact with the trophic center lacked trophic support and therefore degenerated (Waller 1852; Ramón y ye Cajal 1928). The neurotrophic hypothesis of nerve growth began with the observation that nerve fibers from the proximal stump of a severed nerve show a preference for entering the distal stump (Forsmann 1898).

The term "tropic" is derived from the Greek word "tropikos" meaning to turn or change direction. A "neurotropic" substance therefore is one that attracts a nerve causing it to turn, or change its path. Most neurotrophic substances are also "neurotropics." Forsmann, experimenting with nerve regeneration, tested the hypothesis that neurotropic substances attracted regenerating nerve fibers. He suggested that the attractive "tropic" substance was a lipoidal product of the degenerating myelin sheath that surrounds the nerve. Santiago Ramón y Cajal disagreed with Forsmann's suggestion. Ramón y Cajal had earlier proposed a "chemotactic hypothesis" to explain a variety of phenomena that he and others in the field had observed in regenerating nerves (Ramón y Cajal 1892), preceding Forsmann's coining of the term "neurotrophic" (Forsmann 1898). Cajal was fascinated by the observation that even nerves separated by significant distances tended to grow together towards their prospective targets (Ramón y Cajal 1892, 1928). Ramón y Cajal's chemotactic hypothesis sought to explain this neural regeneration phenomenon and nerve–target interaction in developing embryos by proposing that "young axons are oriented in their growth by stimuli from attracting . . . substances" and that " . . . the attractive substance is present in the old tubes in the form of a soluble, non-lipoidal substance, and that its elaboration is brought about by the cells of Schwann . . . " (Ramón y Cajal 1928). He summarized his theory in the statement that "The neurotrophic stimulus acts as a ferment or enzyme, provoking protoplasmic assimilation . . . the orienting agent does not operate through attraction, as many have supposed, but by creating a favorable, eminently trophic, and stimulative of the assimilation and growth of the newly formed axons" (Ramón y Cajal 1928).

In the 1930's Hamburger excised the wing bud of a 72 hour chick embryo, and observed that both the anterior spinal cord and the spinal sensory ganglia on the operated side became severely hypoplastic (Hamburger 1934). This finding suggested that neuronal survival promoting neurotrophic activity was also present in the peripheral field. Hamburger (1934) concluded: "The different peripheral structures while growing are in some direct connection with their appropriate centers in the nervous system. Thus they are enabled not only to control the growth of their own centers in general, but even to regulate this growth in quantitative adaptation to their own progressing increase in size."

4.2 Neurotrophin Family of Growth Factors

4.2.1 Nerve Growth Factor (NGF)

Elmer Bueker investigated the nature of trophic signals released by peripheral organs. He removed embryonic chick limbs and transplanted tumor cells of human origin to act as "a uniform histogenetic tissue" substituting for the more complex developing limb. When he implanted mouse sarcoma 180 tissue in the chick embryo at the site of the extirpated limb, he observed a marked hypertrophy of the associated sensory spinal ganglia. The sarcoma tissue was heavily innervated by ingrowing sensory fibers, but there was no ingrowth of motor nerve fibers. Bueker (1948) concluded the following: " . . . sarcoma 180 because of intrinsic physicochemical properties and mechanisms of growth, selectively causes the enlargement of spinal ganglia . . . . The fact that spinal ganglia were enlarged while the lateral motor column was reduced in those segments that innervated sarcoma 180 suggested quite definitely that this tumor selectively favored the development of one and not the other."

Levi-Montalcini and Hamburger (1951) repeated these experiments including sarcoma 37 tissue as well. They not only confirmed the sensory ganglion hypertrophy reported by Bueker, but also observed a similar response from embryonic sympathetic ganglia. Additionally they observed that only a select subpopulation of the sensory ganglion neurons were responding to the tumor tissue.

Stanley Cohen, working with Levi-Montalcini and Hamburger, succeeded in identifying the active factor as either a protein or nucleoprotein (Cohen, Levi-Montalcini, and Hamburger 1954). Further searching among homologous tissues led them to discover potent neurotrophic activity in a tissue extract of the adult male mouse salivary gland (Cohen 1960). The active protein purified from salivary glands was given the name "nerve growth factor" (NGF).

4.2.2 Other Members of the NGF Neurotrophins Family (Brain Derived Neurotrophic Factor, BDNF; Neurotrophins 3 to 6, NT-3, NT-4, NT-5, NT-6)

The realization that NGF does not serve as a neurotrophic factor for all neuronal populations triggered a search for additional factors that might be related to NGF. Barde and colleagues observed neurotrophic activity in glial conditioned medium that could not be blocked by NGF antiserum. In their attempts to purify this factor, it was found to be abundant in the pig (*Sus domesticus*) brain (Barde, Edgar, and Thoenen 1982). The new factor was called brain derived neurotrophic factor (BDNF). It was clear from the start that BDNF was trophic for different populations of neurons than NGF (Sendtner et al. 1992). When BDNF was cloned, the structural similarity between NGF and BDNF became apparent with a greater than 50% homology in amino acid sequence (Leibrock et al. 1989). The amino acid sequence homology was most apparent in the regions responsible for stabilizing the three-dimensional structure of these molecules. These regions contain six cysteine residues that form disulfide bonds between portions of the amino acid chain and are essential for maintaining the biological activity of these molecules.

The high degree of structural similarity between NGF and BDNF initiated a search for other structurally homologous members of the same gene family with six laboratories reporting the identification of a new member of the NGF family of neurotrophins designated as neurotrophin 3 (NT-3) (Ernfors et al. 1990; Hohn et al. 1990; Jones and Reichardt 1990; Kaisho, Yoshimura, and Nakahama 1990; Maisonpierre et al. 1990; Rosenthal et al. 1990). Hallböök and colleagues investigated the evolutionary relationship between the three known neurotrophin molecules (i.e., NGF, BDNF, and NT-3) by constructing phylogenetic trees derived from DNA sequence analysis of the three genes (Hallböök, Ibáñez, and Persson 1991). These investigators discovered another novel member of the neurotrophin family isolated from the african clawed frog (*Xenopus laevis*) and a viper (*Vipera lebetina*). This protein, like NGF, BDNF and NT-3, contains the five highly conserved amino acid regions including the six cysteine residues. Following the established convention, the discoverers named it neurotrophin-4 (NT-4). Berkemeier, Winslow, and Kaplan (1991) reported the discovery of yet another member of the NGF family of neurotrophins isolated from a human placental DNA library that they designated neurotrophin-5 (NT-5).

The NT-5 protein was found to have a greater homology with NT-4 from *Xenopus laevis* than with the other neurotrophins (Berkemeier, Winslow, and Kaplan 1991). This together with the similarity in their biological activity has led many investigators to conclude that NT-5 is in actuality human NT-4. Since NT-4 and NT-5 are likely to be interspecies variations of the same molecule, and have similar activities, they are commonly referred to as NT-4/5.

A new member of the NGF family called neurotrophin 6 (NT-6) has been identified in the platyfish (*Xiphophorus maculatus*) (Gotz et al. 1994). NT-6 is unique in that it is not naturally released into the medium; instead, this neurotrophin requires the addition of heparin. So far, NT-6 appears to have a profile of trophic activity similar to NGF in that it supports both sensory and sympathetic neurons.

4.3 High and Low Affinity Neurotrophin Receptors

4.3.1 NGF Low Affinity Receptor ($p75^{LNGFR}$)

A protein with a molecular weight of 75 kD ($p75^{LNGFR}$) was initially identified as a receptor for NGF (Johnson et al. 1986); however, this receptor bound NGF with a dissociation constant (K_d) of 10^{-9} M, which was a lower binding affinity than appeared necessary to mediate biological activity, i.e., a K_d of 10^{-11} M, high affinity binding (Meakin and Shooter 1992).

Other studies revealed a larger NGF receptor, about 135–140 kD in size, which was structurally unrelated to the 75 kD protein (Massague et al. 1981; Meakin and Shooter 1991). This NGF receptor had tyrosine kinase activity and was shown to bind the NGF ligand with high affinity. It was suggested that the $p75^{LNGFR}$ protein was converted into a 140 kD high affinity receptor ($p140^{trk-A}$) through some type of modification, e.g., by binding to another small protein.

4.3.2 The trk Oncogene (NGF High Affinity Receptor, trk-A)

Oncogenes are genes isolated from tumor cells that have been transformed through mutations and consequently are thought to contribute to malignant transformation. Normal genes from which oncogenes are transformed are called proto-oncogenes. Martin-Zanca et al. (1990) isolated an interesting oncogene from colon carcinoma cells. They named this oncogene trk, with the normal cellular homologue being the trk proto-oncogene. The product of the trk proto-oncogene was a protein receptor about 140 kD in size. At the time of its discovery the ligand that bound the trk-proto-oncogene receptor was unknown.

A study of the pattern of expression of trk proto-oncogene message in mouse embryos revealed a pattern of expression in sensory neurons of neural crest origin (Martin-Zanca, Barbacid, and Parada 1990). These were the same sensory neurons that are responsive to NGF. Additional evidence supporting the idea that the trk proto-oncogene was the high affinity receptor was provided by the observation that addition of exogenous NGF induced phosphorylation of proto-trk on tyrosine in PC-12 cells (Kaplan, Martin-Zanca, and Parada 1991). The final proof was provided by Barbacid's group by demonstrating that the protein coded for by the trk proto-oncogene did in fact bind NGF with high affinity ($K_d = 10^{-11}$ M) (Klein et al. 1991a). These results, combined with the fact that the trk protein product was 140 kD in size, made a compelling argument. Additional studies have demonstrated that trk-A can mediate the biological effects of NGF without the presence of the $p75^{LNGFR}$ (Nebreda et al. 1991; Ip et al. 1993b).

4.3.3 Other Neurotrophin High Affinity Receptors (trk-B, trk-C)

The low affinity NGF receptor ($p75^{LNGFR}$) has been reported to bind BDNF with an affinity roughly equal to that of NGF (Rodriguez-Tebar, Dechant, and Barde 1990). The high affinity receptor, i.e., trk-A, is more discriminating and binds NGF selectively. Subsequently, it has been determined that the $p75^{LNGFR}$ binds all members of the neurotrophin family with low affinity, but its few known activities have only been associated with NGF binding. The new member of the NGF neurotrophin family, NT-6, has yet to be tested for its ability to bind to $p75^{LNGFR}$.

Klein and colleagues reported the identification of trk-B ($p145^{trk-B}$), a second member of the tyrosine protein kinase family, which shared about 57% homology with the trk-A gene (Klein et al. 1989). In situ hybridization studies of mouse embryos revealed that trk-B was expressed throughout the brain, spinal cord, and in some peripheral nervous system ganglia, suggesting that trk-B may code for a cell surface receptor involved in neurogenesis. In subsequent studies, several groups reported that trk-B bound BDNF and NT-3 with comparably high affinities (Klein et al. 1991b; Soppet et al. 1991; Squinto et al. 1991), and later it was found to bind NT-4/5 as well (Klein, Lamballe, and Barbacid 1992). All three neurotrophins can bind to trk-B; however, NT-3 is much less efficient than NGF and BDNF in inducing biological responses through trk-B binding (Glass, et al. 1991; Squinto et al. 1991). Therefore, trk-B probably does not serve as a receptor for NT-3. Similar findings were made with trk-A, which also binds NT-3 but only with low biological activity.

The existence of a third member of the trk family called trk-C ($p145^{trk-C}$) was reported in 1991 (Lamballe, Klein, and Barbacid 1991). This tyrosine kinase receptor bound NT-3 specifically, with high affinity, and mediated its biological activity efficiently, suggesting that trk-C is the high affinity receptor for NT-3. A schematic representing the binding patterns of NGF, BDNF, NT-3 and NT-4/5 with the trk receptors is presented in Figure 3.4.

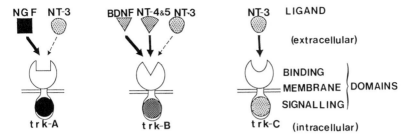

FIGURE 3.4 Schematic representing the binding patterns of nerve growth factor (NGF), brain derived neurotrophic factor (BDNF), neurotrophin-3 (NT-3) and neurotrophin 4/5 (NT-4/5) to the high affinity tyrosine kinase receptors trk-A, trk-B and trk-C. Heavy arrows indicate high affinity binding with a biological response; arrows with broken lines indicate high affinity binding but with no biological response.

It is presently not understood why there is overlap in ligand and receptor binding among the neurotrophins. Similar examples of this type of redundancy exist for other ligands and receptors, e.g., transforming growth factor alpha (TGFα) and epithelial growth factor (EGF) both bind at high affinity to the EGF receptor. Whether there is functional significance to this redundancy of neurotrophin/trk receptor binding is currently unknown.

4.4 Ciliary Neurotrophic Factor (CNTF)

4.4.1 History

Studies of ciliary ganglion–target tissues interactions revealed that the targets express a soluble protein that is trophic for the cholinergic neurons of the ciliary ganglion (Adler et al. 1979). This soluble protein represented a new neurotrophic factor from chick ocular tissues, i.e., ciliary neurotrophic factor (CNTF) (Barbin, Manthorpe, and Varon 1984). CNTF was the third neurotrophic factor to be purified, after NGF and BDNF, but was recognized as being markedly different from these neurotrophins. Additionally, CNTF was the only growth factor discovered that was trophic for parasympathetic neurons, though it had trophic activity for sensory neurons as well (Manthorpe et al. 1982).

4.4.2 CNTF: Structure and Receptor

When CNTF was cloned and its structure fully defined it was discovered that the primary structure of CNTF did not contain an amino terminal signal sequence that is necessary for this protein to be cleaved prior to secretion (Lin et al. 1989). This suggests that CNTF is a cytosolic protein in vivo and that it is not secreted as a target derived growth factor. One interesting feature of CNTF's structure is that it resembles the structure of certain hematopoietic cytokines, suggesting that it is more closely related to the cytokines than to other classical neurotrophic factors, such as the neurotrophin gene family (Bazan 1991).

The CNTF binding protein (CNTF$R\alpha$) confers on the heterodimer gp130/LIFRβ receptor complex specificity for CNTF binding. This binding protein is found almost exclusively in the nervous system (Ip et al. 1993a), suggesting that even though CNTF is structurally related to hematopoietic cytokines it still functions primarily in the nervous system. Interestingly, CNTF is attached to cell membranes by a glycosylphatidyllinositol linkage rather than through the transmembrane domain of its amino acid sequence; therefore, it can be easily released after peripheral nerve injury (Blottner, Bruggemann, and Unsicker 1989). The solubilized form of CNTF$R\alpha$ can bind CNTF and confer responsiveness to cell lines not usually responsive to CNTF, but which express the gp130 and LIFRβ heterodimeric complex.

4.5 Other Growth Factors with Neurotrophic Activity

4.5.1 History

The neurotrophins and CNTF are examples of proteins that were isolated and identified as neurotrophic factors. Their major site of activity is the nervous system despite the fact that many of them, or their receptors, have been isolated in nonneuronal tissue as well. There are other growth factors, originally discovered because of their activity on nonneuronal cells, that have subsequently been shown to possess neurotrophic properties. Some are localized at high concentrations in the brain or other nervous tissues, suggesting a natural role for these factors in the nervous system.

Whether these proteins prove useful in the treatment of nervous system disorders is unclear. Many of these other agents have broad systemic spectrums of action outside the nervous system, thereby rendering their administration more questionable because of the potential for excessive systemic side effects.

4.5.2 Insulin and Its Related Peptides – Insulin-Like Growth Factors I and II

Insulin is a member of a family of structurally and functionally related proteins that also includes insulin-like growth factors I and II (IGF-I and IGF-II). In vitro experiments have shown insulin to be capable of supporting cultured neurons in defined media without a fetal calf serum supplement (Snyder and Kim 1980; Huck 1983; Aizenman and de Vellis 1987). Receptors for insulin are distributed widely in the brain (Baskin 1987), with their highest concentration in the choroid plexus, olfactory bulb, limbic system structures, and hypothalamus. Insulin receptors in the brain are found primarily on neurons, with few receptors localized on glia (Han, Lauder, and D'Ercole 1987). The neurotrophic effects of insulin have been shown to directly affect the neurons without any need for mediation by glial cells (Marks, King, and Baskin 1991). However, insulin has been demonstrated to stimulate DNA and RNA synthesis in glial cells through interactions with the IGF-I or IGF-II receptors. Both receptors are present on glial cells and bind insulin with resultant biological activity (Devaskar 1991).

The insulin-like growth factors have broad somatic activities associated with the regulation of body growth during development. The IGFs bind to a family of at least six binding proteins and three membrane receptors (Neely et al. 1991). Both IGF-I and IGF-II bind to the "IGF-I" receptor with high affinity (Steele-Perkins et al. 1988). Therefore, the IGF-I receptor is commonly referred to as the type-I IGF receptor. The type-I IGF receptor is similar in structure to the insulin receptor (Neely et al. 1991), and insulin can bind to this receptor but with low affinity. The type-II IGF receptor is structurally distinct from both the insulin and the type-I IGF receptor. It binds IGF-II with high affinity but it is not known whether it mediates the biological activities of IGF-II. The insulin-like growth factors and their receptors are found widely expressed throughout the central nervous system (Brown, Graham, and Nissley 1986; Lund et al. 1986), the peripheral nervous system, and skeletal muscle (Lewis et al. 1993). Insulin receptor mRNA in the brain is somewhat more restricted in distribution than the IGF receptor mRNA but tends to be co-localized with it (Baron-Van Evercooren et al. 1991). Many different types of neurons respond to the IGFs. Both IGF-I and IGF-II promote neuronal survival, neurite elongation and gene activation in different types of cultured brain cells (Lenoir and Honneger 1983; Aizenman and de Vellis 1987; Knusel et al. 1990; Cheng and Mattson 1992). Increased expression of the type-I IGF receptor during embryogenesis has suggested that the IGFs may be most active during development (Brown, Graham, and Nissley 1986; Lund et al. 1986). However, IGFs probably do not have a traditional neurotrophic role with a specific population of responsive neurons. The spectrum of neuronal populations responsive to the IGFs is too broad. Instead, the IGFs probably play a nonselective role in a general promotion of the growth and maturation of components of the central nervous system.

In the peripheral nervous system, the activity of the IGFs appears to be more selective. Here, too, the IGFs and their receptors are widely expressed (Hansson et al. 1988; Edwall et al. 1989), but in this system they may specifically be active in promoting neuronal regeneration. The IGFs promote neuritogenesis in cultured sensory, motor, and sympathetic neurons (Recio-Pinto, Reichler, and Ishii, 1986). In vivo, local infusion of IGF-II promotes the regeneration of both motor (Near et al. 1992) and sensory

(Glazner et al. 1993) nerve fibers following a crush injury to the sciatic nerve. Local infusion of IGF-I can also promote the regeneration of sensory nerves (Kanje et al. 1989) and sprouting by motor neurons in denervated skeletal muscle (Caroni and Grandes 1990). Additionally, antiserum to the IGFs reduces both motor (Near et al. 1992) and sensory (Glazner et al. 1993) nerve regeneration providing strong support that a major physiological role of the IGFs in the peripheral nervous system is the enhancement of neural regeneration.

Whether the IGFs also serve as survival-promoting factors for neurons of the peripheral nervous system is not clear. Both insulin and IGF-II can promote the survival of sensory and sympathetic neurons in culture (Recio-Pinto, Reichler, and Ishii 1986). IGF-I promotes the survival of parasympathetic ciliary ganglion neurons in vitro, but only when a protein kinase C activator is added (Crouch and Hendry 1991). IGF-I has not been shown to promote the survival of sensory neurons in culture. IGF-I also promotes the survival of cultured motor neurons but only in the presence of astrocytes, suggesting an indirect mechanism of trophic support (Ang et al. 1992). The in vivo administration of IGF-I has been shown to attenuate motor neuron death following axotomy (Lewis et al. 1993), and improve motor function in a model of vincristine motor neuropathy (Apfel et al. 1993).

4.5.3 Heparin Binding Growth Factor Family of Fibroblast Growth Factors (FGFs 1-7)

Fibroblast growth factors (FGFs), initially isolated from crude brain homogenates, promoted mitogenic activity in cultured fibroblasts. A partial purification of what was believed to be a single FGF protein (Gospodarowicz 1974) was later shown to be two closely related factors that could be further separated by isoelectric focusing (Lemmon et al. 1982; Thomas, Rios-Candelore, and Fitzpatrick 1984). One FGF possessed an acidic isoelectric point and the other FGF had a basic isoelectric point therefore, they were designated as acidic (aFGF) and basic (bFGF) fibroblast growth factors. The amino acid sequences for these two FGF proteins share about a 55% homology (Gospodarowicz et al. 1984; Esch et al. 1985; Gimenez-Gallego et al. 1986). Despite their close homology, aFGF (FGF-1) and bFGF (FGF-2) are the products of different genes. Both aFGF and bFGF bind with high affinity to heparin complexes that are widely distributed over cell surfaces and in basement membranes. Binding to heparin substantially amplifies the biological activities of aFGF and to a lesser extent bFGF (Gimenez-Gallego et al. 1986; Gospodarowicz et al. 1986) and serves to protect these FGF proteins from heat inactivation (Gospodarowicz and Cheng 1986). Other members of the FGF family all share in common the feature of binding to heparin, resulting in the designation "heparin binding growth factor" (HBGF) family (Burgess and Maciag 1989). Apart from aFGF (FGF-1) and bFGF (FGF-2), other members of this family include a product of the int-2 oncogene (FGF-3), a product of the hst/k-fgf oncogene (FGF-4), FGF-5, FGF-6 and keratinocyte growth factor (FGF-7) (Sensenbrenner 1993).

Both acidic and basic FGF proteins have similar systemic activities. They are both potent mitogens for cells of mesodermal and neuroectodermal derivation. The cell types responsive to aFGF and bFGF include fibroblasts, vascular endothelial cells, chondrocytes, osteoblasts, myoblasts, neuroblasts, and astrocytes (Gospodarowicz et al. 1986; Perraud et al. 1988). Without heparin, bFGF has a greater degree of activity as a mitogen than aFGF, but in the presence of heparin, these two FGFs are of approximately equal activity (Bohlen et al. 1985). Basic FGF has been shown to increase the survival rate and neurite extension activities of both cultured fetal rat hippocampal neurons (Walicke et al. 1986) and cultured rat cortical neurons (Morrison et al. 1986). Basic FGF also stimulates neurite outgrowth in PC12 cells similarly to NGF (Togari et al. 1985). Both aFGF and bFGF promote proliferation and neurite outgrowth by adrenal chromaffin cells, but do not effect their survival in vitro (Claude et al. 1988). Other neural cell lines have also been shown to respond to the FGFs during periods of early differentiation and development. Additionally, bFGF promotes the proliferation, survival, and differentiation of early stage neuroepithelial cells from E10 mice (Murphy, Drago, and Bartlett 1990). Putative stem cells from mouse striatum, which proliferate in response to epithelial growth factor (EGF), also express receptors for bFGF and some of these stem cells have demonstrated a proliferative response to exogenous bFGF (Vescovi et al. 1993). Basic FGF has been shown to promote the proliferation of neuroblasts from embryonic rat spinal cord (Ray and Gage 1994). Both acidic and basic FGF have been shown to have neurotrophic activity when administered in vivo, e.g., promoting the survival of retinal ganglion neurons following optic nerve section (Sievers et al. 1987). Similarly, bFGF prevents photoreceptor degeneration in a model of an inherited retinal dystrophy (Faktorovich et al. 1990) and protects retinal ganglion cells from ischemic injury (Unoki and LaVail 1994). Administration of bFGF has also prevented septal cholinergic neuronal cell death following fimbria-fornix transection in adult rats (Anderson et al. 1988).

In the peripheral nervous system bFGF enhances the survival of dorsal root ganglion neurons (Unsicker et al. 1987), parasympathetic ciliary ganglion neurons (Dreyer et al. 1989), sympathetic neurons (Eckenstein et al. 1990), and motor neurons (Arakawa, Sendtner, and Thoenen 1990). The ability of bFGF to promote motor neuron survival can be enhanced by CNTF (Arakawa, Sendtner, and Thoenen 1990). Both bFGF and aFGF are able to promote motor and sensory nerve regeneration across a nerve gap, but further evidence for their role as promoters of neural regeneration is needed (Aebischer, Salessiotis, and Winn 1989; Cordeiro, Seckel, and Lipton 1989).

The physiological role of the FGFs in the nervous system remains uncertain. Because bFGF mRNA is in higher abundance in the adult brain than in the periphery, it probably plays a more important physiological role in the central nervous system (Burgess and Maciag 1989; Gospodarowicz, Neufeld, and Schweigerer 1987). Basic FGF levels increase after injury to central axons such as the optic nerve and fimbria–fornix, but remain low after injury to peripheral nerve (Gomez-Pinilla, Lee, and Cotman 1992; Eckenstein, Shipley, and Nishi 1991). Acidic FGF is found at higher levels of expression in the sciatic nerve, but decreases rapidly following nerve transection (Eckenstein, Shipley, and Nishi 1991). Therefore, what role, if any, either aFGF or bFGF play in response to injury remains unclear.

4.5.4 Transforming Growth Factor Beta Subfamily (TGFβ1-5)

TGFβ exists in at least five distinct isoforms that have significant structural homology but differ in their function and in their pattern of expression in the nervous system (Roberts and Sporn 1990). The occurrence of TGFβ_4 and β_5 has not been observed in mammals. TGFβ_2 and β_3 mRNA are distributed throughout the mammalian brain and spinal cord (Unsicker et al. 1991). Only TGFβ_3 mRNA is found in the sciatic nerve. Both TGFβ_2 and β_3 have been reported to have identical patterns of immunostaining in both neurons and glia throughout the nervous system (Unsicker et al. 1991). In the peripheral nervous system, dorsal root ganglia were heavily immunostained for the presence of TGFβ_1 and β_2 (Unsicker et al. 1991). TGFβ_1 has been demonstrated to promote the survival of motor neurons in culture (Martinou et al. 1990), though this increase in survival may be the result of indirect effects (Flanders, Ludecke, and Engels 1991). TGFβ_1 and β_2 have also been reported to promote neurite outgrowth from cultured hippocampal neurons (Ishihara, Saito and Abe 1994) and to exert mitogenic effects on cultured Schwann cells (Ridley et al. 1989). TGFβ_2 and β_3 appear to have an inhibitory effect on the survival of sympathetic and ciliary ganglion neurons (Flanders, Ludecke, and Engels 1991).

5. The Role of Neurotrophic Growth Factors in Auditory Neuron Survival and Repair

5.1 Neurotrophins

The response of murine auditory neurons to NGF as well as binding of ^{125}I labelled NGF to embryonic days 11 to 14 (E11-14) statoacoustic ganglion was clearly demonstrated in cultures of E11-14 ganglia (Lefebvre et al. 1991c). These observation were extended to include BDNF and NT-3, which were shown to enhance survival and neuritogenesis of embryonic chick (Avila et al. 1993) and murine (Vasquez et al. 1994) auditory neurons in vitro. The biological responsiveness of developing murine auditory neurons to NT-3 is illustrated in Figure 3.5 where an untreated E13 spiral ganglion explant that shows no observable neuritic outgrowth response (Fig. 3.5A) is contrasted to the robust neuritogenesis response of a comparable E13 spiral ganglion explant exposed to exogenous NT-3 (Fig. 3.5B). The presence of transcripts for both BDNF and NT-3 and their high affinity tyrosine kinase receptors (trk-B and trk-C) have been described for the developing neuroepithelium of the otocyst using in situ hybridization (Ernfors, Merlio, and Persson 1992; Pirvola et al. 1992, 1994; Schecterson and Bothwell 1994). The actions of neurotrophins have also been documented clearly during the early postnatal period, at which time ongoing development and fine tuning of the auditory system is taking place. In situ hybridization studies carried out by Ylikoski et al. (1993) on cochlear tissues of 7PP rats have shown that there were high levels of expression of NT-3 mRNAs in the inner hair cells with lower levels of expression in the outer hair cells. BDNF mRNA expression was localized to the hair cells of the early postnatal organ of Corti, but overall was expressed at higher levels in the sensory epithelium of the vestibular system.

This study failed to detect any NGF mRNA expression in either the inner or outer hair cells. As a correlate, hybridization for neurotrophin receptors showed that both trk-B and trk-C, the high-affinity receptors for BDNF and NT-3, localized to the neuronal populations of both the postnatal spiral and vestibular ganglia (Ernfors, Merlio, and Persson 1992; Ylikoski et al. 1993). These studies failed to detect the presence of NGF or trk-A in the postnatal rat organ of Corti. Hybridization patterns for neurotrophin mRNAs were found to be slightly different in the adult rat organ of Corti. Expression of BDNF message was not detected in the adult rat organ of Corti and expression of NT-3 transcripts were limited to the inner hair cells. There was also a difference in NT-3 expression levels in the organ of Corti tissue in a base to apex gradient pattern (Ylikoski et al. 1993). The changes in BDNF expression roughly corresponded to the timing of programmed neuronal cell death and the differentiation of the neuronal population of spiral ganglion into type I and type II auditory neurons. Type I neurons compose 90% of the neuronal population of the spiral ganglion and innervate the inner hair cells, thus suggesting that NT-3 may exert a tropic influence on the afferent type I neurons. A change in BDNF expression also was observed between the postnatal and adult stages. This again suggests that changes in expression of this neurotrophin may affect the process of cochlear maturation. BDNF has been shown to be a survival factor for the embryonic neurons of the cochleovestibular ganglion (Avila et al. 1993; Vasquez et al. 1994). Thus, a decrease in target-derived neurotrophin expression may be related to the onset of neuronal cell death in the maturing cochlea.

Although placing an isolated auditory neuron in vitro in itself is a model of neuronal injury, very few studies specifically address the subject of the

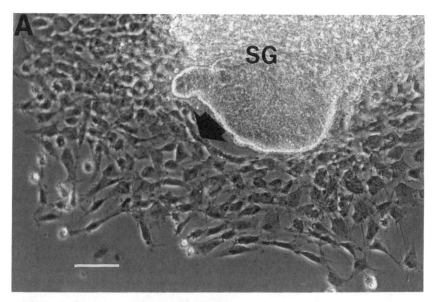
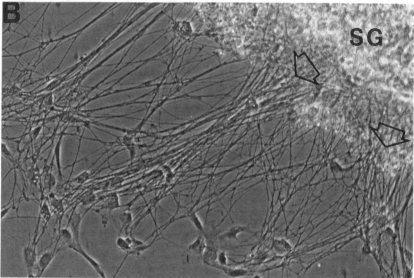

FIGURE 3.5 Spiral ganglion explants (SG) from E13 mouse embryos after 24 hr in vitro. (A) control spiral ganglion explants in defined medium only, note the outgrowth of many fibroblast-type cells (solid arrow). (B) spiral ganglion explants in defined medium supplemented with exogenous NT-3 (10 ng/ml), note the dense neuritic outgrowth (open arrows) in response to the presence of this neurotrophin. Bar = 25 µm.

response of injured auditory neurons to the addition of exogenous neurotrophins. Recently the in vitro responses of adult auditory neurons to neurotrophins have been tested. NT-3 and BDNF were found to actively promote neuronal survival, whereas NGF, while not acting as a survival factor for these neurons, was found to promote neuritic outgrowth when used in combination with a survival promoting neurotrophic factor (Lefebvre et al. 1991a; 1992b; 1994). Recent experiments carried out in chinchillas have shown that vestibular neurectomy results in upregulation of the p75^{LNGFR} while the high affinity receptors for BDNF or NT-3 (trk-B and trk-C) were not affected (Fina, Popper, and Honrubia 1994). The low affinity receptor is thought to play a role in the sequestration and presentation of neurotrophins (Lee et. al. 1992).

This study suggests that neurotrophins play a role in central target derived trophic support of auditory neurons in vivo. At present no studies have assessed the molecular biology of neurotrophin receptor expression in response to peripheral injury.

In support of a central role for BDNF and NT-3 in the developing and mature auditory and vestibular systems are recent reports that describe the morphological consequences to the sensory ganglia of the inner ear of null mutations (gene knockout experiments) where either BDNF or NT-3 genes have been deleted and, therefore, the mice develop in the absence of the targeted neurotrophin. The most profound effect on developing auditory neurons is observed in NT-3 ($-/-$) homozygous mice where NT-3 protein is lacking. There is a greater than 80% reduction in the neuronal population of the spiral ganglion (Farinas et al. 1994; Ernfors et al. 1995a, 1995b), and the afferent but not the efferent innervation of the inner hair cells of the NT-3 ($-/-$) homozygote is affected (Ernfors et al. 1995a, 1995b). These findings are in agreement with in situ hybridization observations that show a decrease in trk-B expression and a predominance of trk-C transcripts in the spiral ganglion as it matures (Ernfors, Merlio, and Persson 1992; Ylikoski et al. 1993). The effects of a lack of NT-3 on the developing and mature vestibular system is less pronounced with only approximately a 30% loss of vestibular neurons. This results in no noticeable effect on the pattern of afferent innervation of vestibular sensory receptors of the NT-3 ($-/-$) homozygote (Ernfors et al. 1995a, 1995b). When BDNF is eliminated through a null mutation quite a different pattern of effects is observed in the developing and mature inner ears of the homozygotes (BDNF $-/-$). There is only a minor (i.e., 7%) reduction in the neuronal population of the spiral ganglion, and there was no detectable change in the afferent innervation pattern to the inner hair cells. In striking contrast to the lack of any change in the inner hair cell afferents, no afferent nerve fibers were detectable in the vicinity of the three rows of outer hair cells of the BDNF ($-/-$) homozygotes. Furthermore, neuronal counts revealed that the small loss of neuronal cells observed in the spiral ganglion were predominantly type II auditory neurons (Ernfors et al. 1995a, 1995b). The analysis of

BDNF null mutants, however, clearly demonstrated that the predominant effect of a lack of BDNF protein was on the survival of vestibular ganglion neurons, with approximately 80% of these neurons dying in the inner ears of the homozygotes (BDNF $-/-$). Additionally, there was a greater than 30% reduction in the neuronal population of the vestibular ganglia in heterozygotes (BDNF, $+/-$) (Ernfors, Lee, and Jaenisch 1994; Jones et al. 1994; Ernfors et al. 1995a, 1995b). These observations are supported by the expression patterns of BDNF mRNA in the developing and mature inner ears of mice and rats (Ernfors, Merlio, and Persson 1992; Pirvola et al. 1992, 1994; Ylikoski et al. 1993; Schecterson and Bothwell 1994; Wheeler et al. 1994). Taken together with the in vitro findings (Lefebvre et al. 1994) these gene knockout results (Ernfors, Lee, and Jaenisch 1994; Ernfors et al. 1995a, 1995b; Farinas et al. 1994; Jones et al. 1994) strongly suggest that both NT-3 and BDNF play integral roles in both the development and maintenance of auditory neurons, and that BDNF is an essential factor for the development and maintenance of the vestibular system.

5.2 Ciliary Neurotrophic Factor (CNTF)

There is at present only preliminary evidence that CNTF has any effect on auditory neurons. An in vitro study has shown that CNTF can stimulate neuritogenesis from chick cochleovestibular ganglion explants (Bianchi and Cohan 1993). Because CNTF is considered a central trauma factor (Lo 1993), and trauma to the central processes of the auditory nerve can result in the initiation of cell death in the auditory neurons of mammals (Spoendlin 1971), this neurotrophic factor was tested for its ability to promote neuronal survival in traumatized auditory neurons in vitro. In dissociated cell cultures of early postnatal rat spiral ganglia, CNTF alone proved to have little effect on neuronal survival; however, in combination with BDNF the survival-promoting ability of CNTF was greatly enhanced (Van De Water et al. 1994). Only in explants of adult rat spiral ganglia was CNTF proven to be an effective trophic factor for the support of neuronal survival (Staecker et al. 1994). It appears that CNTF will play a role in the survival of traumatized auditory neurons, but the exact nature of that role needs additional clarification. Investigations need to be undertaken on both in vitro and in vivo levels to define the distribution of this growth factor and its receptor in situ, and its regulation during the process of injury and repair in the auditory system.

5.3 Insulin and the Insulin-Like Growth Factors (IGFs)

Nothing is known about the IGFs during injury and repair in the auditory system. It is well known that insulin is a required growth factor supplement for the culture of many different types of neurons (Bottenstein and Sato 1979) and is required for the successful serum-free culture of both developing (Lefebvre et al. 1991c) and adult (Lefebvre et al. 1991a) auditory neurons in a defined medium. Although the IGFs have been implicated as repair enhancing factors during neural regeneration in the peripheral nervous system, there are at present no studies to determine if these growth factors are involved in repair and regeneration of the VIIIn.

5.4 Fibroblast Growth Factor (FGF) Family

Three members of the FGF family can be expected to play a role in the development and maintenance of the auditory neurons and their axonal projections. These FGFs are aFGF (FGF-1), bFGF (FGF-2) and int-2 (FGF-3). Transcripts for aFGF have been shown to be abundant in both the peripheral (i.e., cochlea; Luo et al. 1993) and central (i.e. cochlear nucleus; Luo et al. 1995) targets of the spiral ganglion as well as in the auditory neurons themselves (Luo et al. 1995). At present there have been no reported in vitro experiments to test if in fact aFGF will promote the survival of auditory neurons. However, there is preliminary evidence that aFGF can support the survival of these neurons in spiral ganglion explants from adult rats (Staecker and Van De Water, unpublished results). Exogenous bFGF has been shown to be ineffective in promoting neuronal survival in dissociated cell cultures of adult auditory neurons unless very high levels of ligand were used (i.e., 10 μg/ml). However, when the bFGF was bound to the substrate, levels as low as 10 ng/ml were very effective in supporting neuronal survival in these cultures (Lefebvre et al. 1991a). The expression of int-2 mRNA in areas of differentiating sensory epithelium (Wilkinson, Bhatt, and McMahon 1989) and the fact that bFGF and int-2 proteins generally target the same cell populations, with the int-2 protein generally being active at much lower concentrations, strongly suggests that int-2 will be effective in supporting the survival of auditory neurons. At present, this conjecture remains unproven. There have been no experiments testing the remaining members of the FGF family (i.e., FGFs 4–7) and no indirect evidence that would suggest any of these as trophic agents for the auditory system.

5.5 Transforming Growth Factor β Subfamily

Unlike the direct neurotrophic actions of either NT-3 or bFGF on auditory neurons, transforming growth factor β appears to exert its survival-promoting effect by enhancing a neuron's ability to respond to other survival-promoting factors. Therefore, cultures of adult auditory neurons treated with only $TGF\beta_1$ showed an increase in their neuronal survival response only if bFGF was added in combination with $TGF\beta_1$ (Lefebvre et al. 1991a). The mechanism of TGFβ's ability to enhance bFGF mediated neuronal survival was further defined in an in vitro study where it was demonstrated that treatment of dissociated cell cultures of auditory neurons with TGFβ_1 resulted in an increase in transcripts encoding for bFGF receptor, and correspondingly, an increase in bFGF receptors on the neurons, thereby rendering these TGFν_1 treated neurons more sensitive to the survival-promoting effects of bFGF (Lefebvre et al. 1991b). These findings gained added importance when it was discovered in a series of in vivo experiments that trauma to the auditory nerve (i.e., sectioning of the central axons) resulted in an upregulation of the expression of TGFβ_1 mRNA in the injured auditory neurons (Lefebvre et al. 1992a). These findings strongly suggest that TGFβ_1 is an injury response factor in auditory neurons. This neurotrophic factor acts by modulating the ability of the injured neurons to respond to other neurotrophic molecules, thereby increasing their ability to survive following a traumatic injury. At present only one neurotrophic factor receptor has been studied, and there are many other factors and receptors that can affect the survival of traumatized neurons that need to be assayed in an injured auditory system that is responding to increases in endogenous TGFβ_1. Additionally, the effects of exogenous TGFβ_2 and β_3 have not been tested in the in vitro model of injured auditory neurons (Lefebvre et al. 1991a) nor have either of these trophic molecules been studied in response to trauma in vivo (Lefebvre et al. 1992a). The observations described above suggest that the TGFβs are injury response molecules for injured auditory neurons. At present there is insufficient information to define the exact nature of the roles that these neurotrophic factors play in repair and regeneration in the auditory system other than to state that one such role appears to be that of a modulator of the traumatized neurons' ability to survive.

6. Delivery Systems

6.1 Approach

One of the major difficulties in applying neurotrophic factor to the inner ear is accessibility. Virtually all of the known neurotrophic factors are large proteins that will not easily cross the blood-perilymph barrier. This fact limits the practicality of systemic administration of neurotrophic factors to treat the auditory neurons of the spiral ganglion. Therefore, strategies for administering neurotrophic factors to the auditory neurons have focused primarily on bypassing the blood-perilymph barrier. A look at cochlear anatomy shows that there is access to the extracellular spaces that surround the auditory neurons of the spiral ganglion via communications between the peripheral axonal processes and the perilymphatic fluids of the scala tympani. Most current strategies for the delivery of pharmacologically effective amounts of neurotrophic factors to the auditory neurons involve the use of the communications between the perilymph and these neurons.

6.2 Direct Infusion of Neurotrophic Factors

Currently, infusing drugs directly into the fluid spaces of the scala tympani is the most practical method of bypassing the blood–perilymph barrier. The technique necessitates surgically placing a catheter through either the round window membrane or through the lateral wall of the basal turn of the cochlea into the fluid space of the scala tympani (Brown et al. 1993). The catheter is linked up to an infusion device placed subcutaneously where it can be refilled as necessary.

This technique has been used to deliver a mixture of two neurotrophins (i.e., BDNF and NT-3) into the scala tympani of a guinea pig following the destruction of the auditory hair cells of the organ of Corti by a single dose of ototoxin and loop diuretic (Staecker et al. 1995). This study has provided preliminary evidence that neurotrophins delivered to the cochlea via this route of infusion supports increased survival of the targeted auditory neurons.

Alternatively, if only a single administration of growth factor(s) is required then a single delivery of a growth factor(s) can be administered via the catheter or through the round window membrane. The general problem with the delivery of a single dose of growth factor is that the local concentration of this factor is quickly reduced to low levels probably long before the growth factor has had an opportunity to penetrate the cochlear duct tissues over a depth of more than a couple of millimeters. Long term chronic infusion via a miniosmotic pump is preferable since this method of delivery is designed to compensate for the washout phenomena by maintaining continuous high levels of factor for prolonged periods of time.

6.3 Carrier Drugs

Small molecules that bind to specific receptors on endothelial cells that are transported across these cells have been suggested as vectors for carrying growth factors across this barrier. Growth factor(s) could be covalently bonded to a carrier molecule and, following transport across the endothelial cell, the bond could be cleaved, releasing the growth factor(s) to bind to appropriate high affinity receptors on the targeted cell population within the inner ear. NGF has been conjugated to antitransferrin antibodies, and it has been demonstrated that this conjugation did not reduce the trophic activity of NGF. Subsequent intravenous injection of a radiolabeled NGF-antitransferrin receptor complex demonstrated a time dependent migration of NGF from brain capillaries into the brain tissue while there was no such transfer of radiolabeled unconjugated NGF (Pardridge 1986; Fishman et al. 1987).

In addition to receptor mediated transcytosis, growth factors may be transported by a mechanism termed absorptive mediated trancytosis. Since the capillary endothelial cells in the inner ear contain numerous negative charges on their surface membranes, proteins that are positively charged can trigger this phenomenon (Kumagai, Eisenberg and Pardridge 1987).

6.4 Implantable Polymers

Embedding a growth factor within a biocompatible polymer is a potentially useful approach since the polymer complex can both protect the factor from degradation by its environment and achieve a long term sustained release at the site of implantation. The growth factor is combined with a polymeric resin that forms a matrix whose porosity and degradation rates can be controlled (Leong, Brott, and Langer 1985). This allows the specific design of polymers with a predetermined rate of growth factor release (Leong et al. 1986; Tomargo et al. 1989; Powell, Sobarzo, and Salzman 1990). Biorelease polymers can be shaped to fit the application site and, when produced as microspheres containing growth factor, these polymers can be administered by injection. A disadvantage of polymer implantation is that if growth factor release is to be discontinued for any reason, additional surgery is required to remove the polymer.

6.5 Cell Transplantation

One method of avoiding the problems inherent in both infusion of drugs and the implantation of biorelease polymers would be to implant growth factor producing factories that could obtain the raw materials needed for the production of growth factors from the local environment (Selden et al. 1987). Mammalian cells can be thought of as small manufacturing plants that can be genetically engineered to produce almost any growth factor required either singly or in combinations. The raw materials, such as amino acids, are freely available in perilymphatic fluid of the scala tympani. There are already two preliminary studies (Ryan and Luo 1995; Staecker et al. 1995) showing that genetically engineered cells can be used to deliver growth factors to the tissues of the cochlear duct. The grafting of genetically engineered fibroblasts secreting neurotrophins has been used for a number of years in the brain with quite good success (Gage et al. 1987; Olson et al. 1988; Rosenberg et al. 1988). A recent report has combined the features of polymer release characteristics with genetically engineered cells by encapsulating NGF secreting cells with a polymer cage and then introducing these into the lesioned brain (Winn et al. 1994). This study resulted in the protection of septal cholinergic neurons from degeneration due to a loss of trophic support from the lesioned target tissue. One advantage of polymer-embedded cells is ease of localization and later removal, if required. Unencapsulated growth factor secreting cells can spread to many sites and present a serious problem of their removal, if required for any reason.

6.6 Viral Vector Gene Therapy

Viruses that have been rendered replication deficient can be used to carry and express inserted genes that encode for neurotrophic factors. The expression of the inserted genes can be driven either by the promoters that usually coordinate viral replication within the infected cell or any inserted promoters that may be used to obtain cell specific gene expression (e.g., a neurofilament gene promoter to obtain gene expression primarily in neurons). There are several types of viruses that are currently used: retroviruses that can target actively proliferating cells, herpes viruses that have adequate space to insert several gene constructs and infect postmitotic differentiated cells, especially neurons; adenoviruses that infect postmitotic cells and will maintain gene expression for extended periods of time within in vivo infected cells; and adenoassociated viruses that will infect postmitotic cells, maintain gene expression for extended periods of time in vivo, and have negligible long term toxicity effects. The advantage of using viral vectors to deliver growth factors to auditory neurons is that initiation of therapy with a viral vector requires only one disturbance of inner ear structures when the virus carrying and expressing the desired growth factor gene(s) is delivered to the scala tympani where the replication deficient infectious agent has access to the cells of the spiral ganglion. A modified *Herpes simplex* virus expressing the high affinity receptor for NGF (pHSVtrkA) has been used to infect neurons that would not normally respond to NGF and change their responsiveness to this neurotrophin (Hong et al. 1994). The high affinity receptor for NGF (trk-A) is not present in maturing and adult auditory neurons, therefore, the same pHSVtrkA vector mentioned above was used in cultures of these neurons to change their ability to respond to the addition of exogenous NGF by transducing the expression of trk-A (Van De Water et al. 1992). More recently a *Herpes simplex* viral vector containing a BDNF gene, (pHSVbdnf) was used to treat postnatal rat spiral ganglion explants and to evoke a robust neuritogenesis response comparable to responses obtained from the addition of exogenous BDNF to spiral ganglion explants (Hartnick et al. 1995). These studies show that a herpes viral vector can be used to infect spiral ganglion cells and produce biological effects on these auditory neurons in culture. It now remains to be demonstrated that these or other viral vectors can be used to express growth factor genes effectively in the inner ear in vivo and to achieve a biological response from the targeted populations of cells. In support of in vivo applications is the report of a *Herpes simplex* viral vector used to express a nerve growth factor gene (pHSVngf) in the peripheral nervous system and to protect neurons of a peripheral ganglion (i.e., dorsal root ganglion) from the deleterious consequences of a traumatic injury (Federoff et al. 1992).

An example of a *Herpes simplex* virus construct containing a nerve growth factor mini-gene whose expression is driven by the viral promoter genes is shown in Figure 3.6. The benefits of a viral vector gene therapy

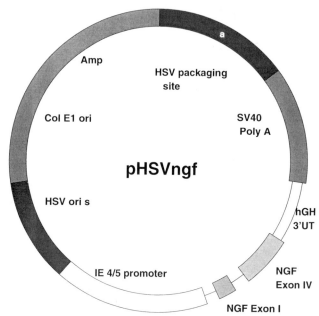

FIGURE 3.6 The structure of a *Herpes simplex* viral amplicon vector containing an inserted nerve growth factor mini-gene (pHSVngf). pHSVngf contains two genetic elements from HSV1, the "ori_s" and "a" sequences (HSV packaging site), that are sufficient for packaging into viral particles. It also contains a transcription unit composed of the HSV-1 IE 4/5 promoter (unfilled segment) and the NGF mini-gene (stippled segments). For replication and selection in E. coli, pHSVngf contains the origin of replication (Col E1 ori) and the β-lactamase genes (Amp). Other components of the pHSVngf construct include a three prime untranslated region of human growth hormone (hGH3'UT) and a poly A tail of simian virus type forty (SV40Poly A).

approach is that growth factor therapy can be delivered to a restricted area and that it requires a minimum of intervention (Breakfield and Geller 1987; Geschwind et al. 1994). As this type of approach to somatic complementation gene therapy progresses, the problems of greater control over the viral vector and expression of the gene construct that it carries will have to be solved.

7. Clinical Applications

Over the last 50 years significant discoveries have been made that have drastically changed how we view the neurobiology of neuronal damage and recovery. Possible clinical applications for neurotrophins after injury of auditory hair cells are numerous. Neurotrophin therapy may be useful in ameliorating overall damage to the auditory system after treatment with aminoglycosides or other ototoxic agents such as cis-platinum. Furthermore, neurotrophin therapy may also improve existing cochlear implant technology by improving neuronal survival or possibly even repair and regrowth of damaged neuronal processes after cochlear injury. If mammalian cochlear hair cell regeneration becomes a reality, as has been shown in the avian system, neurotrophin treatment will be required initially to support neuronal survival and maintain neuronal health until a new target population of regenerated–repaired hair cells are ready for reinnervation by the surviving auditory neurons.

8. Summary

The auditory nerve in lower vertebrates (e.g., amphibians) when transected or damaged can regenerate and restore both the correct pattern of innervation and normal physiologic function (Section 2.1). In contrast, the auditory neurons of higher vertebrates (e.g., mammals) degenerate in response to loss of either their peripheral (i.e., organ of Corti) or central (i.e., cochlear nucleus) targets, and although the damaged neurons attempt to repair themselves, neither reinnervation nor restoration of function of the auditory system is accomplished (Section 2.2). In addressing the issue of a lack of an adequate regeneration–repair response by injured auditory neurons, the field of neurotrophic factors is examined and candidates likely to affect the survival and regeneration–repair processes of auditory neurons are presented in Section 4 on the biology and function of neurotrophic factors. This list of candidate neurotrophic factors for the auditory system (Section 5) include: the neurotrophins NGF, BDNF, NT-3, NT-4/5 and NT-6; CNTF; the heparin binding family of growth factors FGF-1, 2 and 3; Insulin and the insulin like growth factors IGF-I and II; and the transforming growth factor beta subfamily TGFβs 1, 2, and 3. Of this diverse group of growth factors the following have been shown to affect either the survival or regeneration–repair processes or both processes by auditory neurons in vitro: NGF, BDNF, NT-3; CNTF; Insulin; FGF-1, FGF-2; and TGFβ_1. Null mutations for BDNF and NT-3 have verified the hypothesis that BDNF and NT-3 are the natural neurotrophic factors that the survival of the auditory neurons, and in situ hybridization studies have confirmed that transcripts for these two neurotrophins are found in both the developing and the postnatal maturing organ of Corti. It has also been demonstrated that the mature auditory neurons that have lost their peripheral target derived trophic support (i.e., the auditory hair cells) can be supported by the local infusion of a mixture of exogenous BDNF and NT-3. Novel methods for the delivery of trophic support to injured auditory neurons were discussed in Section 6 and are as follows: localized application by direct infusion, growth factor incorporation into implantable biorelease polymers, implantation of cells genetically engineered to produce growth factors, and infection with replication deficient viral vector that contains a growth factor mini-gene. All of the above information is presented with a view toward the eventual application of growth factors as therapeutic agents in the clinic (Section 7). The challenge that faces investigators in this exciting field is further definition of basic mechanisms of action in the auditory system, the interactions between growth factors from different classes of factors (e.g., NT-3, a neurotrophin, and CNTF, a cytokine) on auditory neurons and the much needed translation research that will act to bridge the gap between basic laboratory studies and clinical applications.

Acknowledgements. The authors thank Rose Imperati for the many weeks spent in the word processing of this chapter, and Dr. Howard Federoff for kindly providing Figure 3.6.

References

Adler R, Landa KB, Manthorpe M, Varon S (1979) Cholinergic neuronotrophic factors: Intraocular distribution of trophic activity for ciliary neurons. Science 204:1434–1436.

Aebischer P, Salessiotis AN, Winn SR (1989) Basic fibroblast growth factor released from synthetic guidance channels facilitates peripheral nerve regeneration across long nerve gaps. J Neurosci Res 23:282–289.

Aizenman Y, de Vellis J (1987) Brain neurons develop in a serum and glial free environment: effects of transferrin, insulin, insulin like growth factor-I and thyroid hormone on neuronal survival, growth and differentiation. Brain Res 406:32–42.

Anderson KJ, Dam D, Lee S, Cotman CW (1988) Basic fibroblast growth factor prevents death of lesioned cholinergic neurons in-vivo. Nature 332:360–361.

Ang LC, Bhaumick B, Munoz DG, Sass J, Juurlink BHJ (1992) Effects of astrocytes, insulin, and insulin like growth factor-I on the survival of motor neurons in-vitro. J Neurol Sci 109:168–172.

Anniko M (1988) Maturation of inner ear ganglion cells: Morphometric analyses of cross-sectional areas. J Oto Rhino Laryngol 50:103–109.

Apfel SC, Arezzo JC, Lewis ME, Kessler JA (1993) The use of insulin like growth factor-I in the prevention of vincristine neuropathy in mice. Ann N Y Acad Sci 692:243–245.

Arakawa Y, Sendtner M, Thoenen H (1990) Survival effect of ciliary neurotrophic factor (CNTF) on chick embryonic motoneurons in culture: Comparison with other neurotrophic factors and cytokines. J Neurosci 10:3507–3515.

Ard M, Morest D, Haugher S (1985) Trophic interactions between the cochleovestibular ganglion of the chick embryo and its synaptic targets in culture. Neurosci 16:151–170.

Avila M, Varela-Nieto I, Romero G, Mato J, Van De Water TR, Represa J (1993) Brain derived neurotrophic factor and neurotrophin-3 support the survival and neuritogenesis response of developing cochleovestibular ganglion neurons. Devel Biol 159:266–275.

Barbin G, Manthorpe M, Varon S (1984) Purification of the chick eye ciliary neurotrophic factor. J Neurochem 43:1468–1478.
Barde YA, Edgar D, Thoenen H (1982) Purification of a new neurotrophic factor from mammalian brain. EMBO J 1:549–553.
Baron-Van Evercooren A, Olichon-Berthe C, Kowalski A, Visciano G, Van Obberghen E (1991) Expression of IGF-I and insulin receptor gene in the rat central nervous system: a developmental regional and cellular analysis. J Neurosci Res 28:244–253.
Baskin DG (1987) Insulin in the brain. Ann Rev Physiol 49:335–347.
Bazan JF (1991) Neuropoietic cytokines in the hematopoietic fold. Neuron 7:197–208.
Berkemeier LR, Winslow JW, Kaplan DR (1991) Neurotrophin-5: A novel neurotrophic factor that activates trk and trk-B. Neuron 7:857–866.
Bianchi LM, Cohan CS (1993) Effects of the neurotrophins and CNTF on developing statoacoustic neurons: Comparison with an otocyst derived factor. Devel Biol 159:353–365.
Bichler E, Spoendlin H, Rauchegge H (1983) Degeneration of cochlear neurons after amikacin intoxication in the rat. Arch Otolaryngol 237:201–207.
Blottner D, Bruggemann W, Unsicker K (1989) Ciliary neurotrophic factor supports target deprived preganglionic sympathetic spinal cord neurons. Neurosci Lett 105:316–320.
Bohlen P, Esch F, Baird A, Gospodarowicz D (1985) Acidic fibroblast growth factor (aFGF) from bovine brain: Amino terminal sequence and comparison with basic FGF. EMBO J 4:1951–1956.
Bohne B, Harding G (1992) Neuronal regeneration in the noise damaged chinchilla cochlea. Laryngoscope 102:693–703.
Bottenstein JE, Sato GH (1979) Growth of a rat neuroblastoma cell line in serum-free supplemented medium. Proc Natl Acad Sci USA 76:514–517.
Breakefield XO, Geller AI (1987) Gene transfer into the nervous system. Mol Neurobiol Rev 1:339–371.
Brown AL, Graham DE, Nissley SP (1986) Developmental regulation of insulin like growth factor-I mRNA in different rat tissues. J Biol Chem 261:13144–13150.
Brown JN, Miller JM, Atlschuler RA, Nutall AL (1993) Osmotic pump implant for chronic infusion of drugs into the inner ear. Hear Res 70:167–172.
Bueker ED (1948) Implantation of tumors in the hind limb field of the embryonic chick and the developmental response of the lumbosacral nervous system. Anat Rec 102:369–390.
Burgess WH, Maciag T (1989) The heparin binding (FGF) growth factor family of proteins. Ann Rev Biochem 58:575–606.
Caroni P, Grandes P (1990) Nerve sprouting in innervated adult skeletal muscle induced by exposure to elevated levels of insulin like growth factors. J Cell Biol 110:1307–1317.
Cheng B, Mattson MP (1992) IGF-I and IGF-II protect cultured hippocampal and septal neurons against calcium mediated hypoglycemic damage. J Neurosci 12:1558–1566.
Claude P, Parada IM, Gordon KA, D'Amore PA, Wagner JA, (1988) Acidic fibroblast growth factor stimulates adrenal chromaffin cells to proliferate and to extend neurites, but is not a long term survival factor. Neuron 1:783–790.
Cohen S (1960) Purification of a nerve growth promoting protein from the mouse

salivary gland and its neurocytotoxic antiserum. Proc Natl Acad Sci USA 46:302–311.
Cohen S, Levi-Montalcini R, Hamburger V (1954) A nerve growth stimulating factor isolated from sarcomas 37 and 180. Proc Natl Acad Sci USA 40:1014–1018.
Cordeiro PG, Seckel BR, Lipton SA (1989) Acidic fibroblast growth factor enhances peripheral nerve regeneration in-vivo. Plast Reconstr Surg 83:1013–1019.
Crouch MF, Hendry IA (1991) Co-activation of Insulin like growth factor-I receptors and protein kinase C results in parasympathetic neuronal survival. J Neurosci Res 28:115–120.
Devaskar SU (1991) A review of insulin/insulin like peptide in the central nervous system. In: Raizada MK, LeRoith D. (eds) Molecular Biology and Physiology of Insulin and Insulin Like Growth Factors. New York: Plenum Press.
Dreyer D, Lagrange A, Grothe C, Unsicker K (1989) Basic fibroblast growth factor prevents ontogenetic neuron death in-vivo. Neurosci Lett 99:35–38.
Eckenstein FP, Esch F, Holbert T, Blacher RW, Nishi R (1990) Purification and characterization of a trophic factor for embryonic peripheral neurons: Comparison with fibroblast growth factors. Neuron 4:623–631.
Eckenstein FP, Shipley GD, Nishi R (1991) Acidic and basic fibroblast growth factors in the nervous system: Distribution and differential alteration of levels after injury of central versus peripheral nerve. J Neurosci 11:412–419.
Edwall D, Schalling M, Jennische E, Norstedt G (1989) Induction of insulin like growth factor-I messenger ribonucleic acid during regeneration of rat skeletal muscle. Endocrinol 124:820–825.
Emmerling MR, Sobkowicz HM (1988) Differentiation and distribution of acetylcholinesterase molecular forms in the mouse cochlea. Hear Res 468:233–242.
Emmerling MR, Sobkowicz HM (1990) Intermittent expression of GAP-43 during the innervation of the cochlea in the mouse. Abstracts of the Thirteenth Midwinter Meeting of the Association for Research in Otolaryngology 61, Abstr. 72.
Emmerling MR, Sobkowicz HM, Levenick CV, Scott GL, Slapnick SM, Rose JE (1990) Biochemical and morphological differentiation of acetylcholinesterase-positive efferent fibers in the mouse cochlea. J Electron Micros Techn 15:123–143.
Ernfors P, Ibanez CF, Ebendal T, Olson L, Persson H (1990) Molecular cloning and neurotrophic activities of a protein with structural similarities to nerve growth factor: Developmental and topographical expression in the brain. Proc Natl Acad Sci USA 87:5454–5458.
Ernfors P, Merlio J-P, Persson H (1992) Cells expressing mRNA for neurotrophins and their receptors during embryonic rat development. Eur J Neurosci 4:1140–1158.
Ernfors P, Lee K-F, Jaenisch R (1994) Mice lacking brain-derived neurotrophic factor develop with sensory deficits. Nature 368:147–150.
Ernfors P, Loring J, Jaenisch R, Van De Water TR (1995a) Function of neurotrophins in the auditory and vestibular systems: Analysis of BDNF and NT-3 gene knockout mice. Abstracts of the Eighteenth Midwinter Meeting of the Association for Research in Otolaryngology, p. 190, Abstr. 759.
Ernfors P, Van De Water T, Loring J, Jaenisch R (1995b) Complementary roles of BDNF and NT-3 in vestibular and auditory development. Neuron 14:1153–1164.

Esch F, Baird A, Ling N, Ueno N, Hill F, Denoroy L, Kleppe R, Gospodarowicz D, Bohlen P, Guillemin R (1985) Primary structure of bovine pituitary basic fibroblast growth factor (FGF) and comparison with the amino terminal sequence of bovine brain acidic FGF. Proc Natl Acad Sci USA 82:6507–6511.

Faktorovich EG, Steinberg RH, Yasumura D, Matthes MT, LaVail MM (1990) Photoreceptor degeneration in inherited retinal dystrophy delayed by basic fibroblast growth factor. Nature 347:83–86.

Farinas I, Jones KR, Backus C, Wang X-Y, Reichardt LF (1994) Severe sensory and sympathetic deficits in mice lacking neurotrophin-3. Nature 369:658–661.

Federoff HJ, Geschwind MD, Geller AI, Kessler JA (1992) Expression of NGF in vivo, from a defective HSV-1 vector, prevents effects of axotomy on sympathetic ganglia. Proc Natl Acad Sci USA 89:1636–1640.

Fina M, Popper A, Honrubia V (1994) Survival and cell biologic changes in the vestibular ganglion cells following vestibular neuronectomy in the chinchilla. Abstracts of the Seventeenth Midwinter Meeting of the Association for Research in Otolaryngology, p. 35, Abstr. 139.

Fishman JB, Rubin JB, Handrahan JV, Connor JR, Fine RE (1987) Receptor-mediated transcytosis of transferrin across the blood-brain barrier. J Neurosci Res 18:299–304.

Flanders KC, Ludecke G, Engels S (1991) Localization and actions of transforming growth factor-s in the embryonic nervous system. Develop 113:183–191.

Forsmann J (1898) Uber die ursachen welche die wachsthumsrichtung der peripheren nervenfasern bei der regeneration bestimmen. Beitr Path Anat Allg Path 24:56–100.

Gage FH, Wolff JA, Rosenberg MB, Xu L, Yee JF, Shults C, Friedmann T (1987) Grafting genetically modified cells to the brain: Possibilities for the future. Neurosci 23:795–807.

Galinovic-Schwartz V, Peng D, Chiu FC, Van De Water TR (1991) Temporal pattern of innervation in the developing mouse embryo: An immunohistochemical study of a 66 kD subunit of mammalian neurofilament. J Neurosci Res 30:124–130.

Geschwind MD, Kessler JA, Geller AI, Federoff HJ (1994) Transfer of the nerve growth factor gene into cell lines and cultured neurons using a defective Herpes Simplex virus vector. Transfer of the NGF gene into cells by a HSV-1 vector. Mol Brain Res 24:327–335.

Gimenez-Gallego G, Conn G, Hatcher VB, Thomas KA (1986) Human brain derived acidic and basic fibroblast growth factors: Amino terminal sequences and specific mitogenic activities. Biochem Biophys Res Comm 135:541–548.

Glass DJ, Nye SH, Hantzopoulos P, Macchi MJ, Squinto SP, Goldfarb M, Yancopoulos GD (1991) Trk B mediates BDNF/NT-3 dependent survival and proliferation in fibroblasts lacking the low affinity NGF receptor. Cell 66:405–413.

Glazner GW, Lupien S, Miller JA, Ishii DN (1993) Insulin like growth factor-II increases the rate of sciatic nerve regeneration in rats. Neurosci 54:791–797.

Gleisner L, Wersäll J (1975) Experimental studies on the nerve-sensory cell relationship during degeneration and regeneration in ampullar nerve of the frog labyrinth. Acta Otolaryngol (Stockh) Suppl 333:1–28.

Gomez-Pinilla F, Lee JWK, Cotman CW (1992) Basic FGF in adult rat brain: Cellular distribution and response to entorhinal lesion and fimbria fornix transection. J Neurosci 12:345–355.

Gospodarowicz D (1974) Localization of a fibroblast growth factor and its effect alone and with hydrocortisone on 3T3 cell growth. Nature 249:123–127.

Gospodarowicz D, Cheng J (1986) Heparin protects basic and acidic FGF from inactivation. J Cell Physiol 128:475–484.

Gospodarowicz D, Cheng J, Lui GM, Baird A, Bohlen P (1984) Isolation of brain fibroblast growth factor by heparin-Sepharose affinity chromatography: identity with pituitary fibroblast growth factor. Proc Natl Acad Sci USA 81:6963–6967.

Gospodarowicz D, Massoglia SL, Cheng J, Fujii DK (1986) Effect of fibroblast growth factor and lipoproteins on the proliferation of endothelial cells derived from bovine adrenal and brain cortex, bovine corpus luteum, and brain capillaries. J Cell Physiol 127:121–136.

Gospodarowicz DK, Neufeld G, Schweigerer L (1987) Fibroblast growth factor: structural and biological properties. J Cell Physiol (Suppl) 5:15–26.

Gotz R, Koster R, Winkler C, Raulf F, Lottspeich F, Sehartl M, Thoenen H (1994) Neurotrophin-6 is a new member of the nerve growth factor family. Nature 372:266–269.

Hafidi A, Romand R (1989) First appearance of type II neurons during ontogenesis in the spiral ganglion of the rat: An immunocytochemical study. Devel Brain Res 48:143–9.

Hallböök F, Ibáñez CF, Persson H (1991) Evolutionary studies of the nerve growth factor family reveal a novel member abundantly expressed in *Xenopus* ovary. Neuron 6:845–858.

Hamburger, V (1934) The effects of wing bud extirpation on the development of the central nervous system in chick embryos. J Exp Zool 68:449–494.

Han VKM, Lauder JM, D'Ercole J (1987) Characterization of somatomedin/insulin like growth factor receptors and correlation with biologic action in cultured neonatal rat astroglial cells. J Neurosci 7:505–511.

Hansson HA, Nilsson A, Isgaard J, Billig H, Isaksson O, Skottner A, Andersson IK, Rozell B (1988) Immunohistochemical localization of insulin like growth factor I in the adult rat. Histochem 89:403–410.

Hartnick CJ, Geschwind MD, Federoff HJ, Van De Water TR (1995) Use of a defective herpes simplex virus to deliver BDNF to spiral ganglion explants. Abstracts of the Association for Research in Otolaryngology, p. 108, Abstr. 432.

Hohn A, Leibrock J, Bailey K, Barde YA (1990) Identification and characterization of a novel member of the nerve growth factor/brain derived growth factor family. Nature 344:339–341.

Hong X, Federoff H, Maragos J, Parada LF, Kessler JA (1994) Viral transduction of trkA into cultured nodose and spinal motor neurons conveys NGF responsiveness. Devel Biol 163:152–161.

Huck S (1983) Serum free medium for cultures of the postnatal mouse cerebellum: only insulin is essential. Brain Res Bull 10:667–674.

Ip NY, McClain J, Barrezueta NX, Aldrich TH, Pan L, Li Y, Wiegand J, Friedman B, Davis S, Yancopoulos GD (1993a) The α component of the CNTF receptor is required for signaling and defines potential CNTF targets in the adult and during development. Neuron 10:89–102.

Ip NY, Stitt TN, Tapley P, Klein R, Glass DJ, Fandl J, Greene LA, Barbacid M, Yancopoulos GD (1993b) Similarities and differences in the way neurotrophins interact with the trk receptors in neuronal and non-neuronal cells. Neuron 10:137–149.

Ishihara A, Saito H, Abe K (1994) Transforming growth factor-beta 1 and -beta 2

promote neurite sprouting and elongation of cultured rat hippocampal neurons. Brain Res 639:21-25.
Johnson D, Lanahan A, Buck CR, Sehgal A, Morgan C, Mercer E, Bothwell M, Chao M (1986) Expression and structure of the human NGF receptor. Cell 47:545-554.
Jones KR, Reichardt LF (1990) Molecular cloning of a human gene that is a member of the nerve growth factor family. Proc Natl Acad Sci USA 87:8060-8064.
Jones KR, Farinas I, Backus C, Reichardt LF (1994) Targeted disruption of the BDNF gene perturbs brain and sensory neuron development but not motor neuron development. Cell 76:989-999.
Kaisho Y, Yoshimura K, Nakahama K (1990) Cloning and expression of a cDNA encoding a novel human neurotrophic factor. FEBS Lett 266:187-191.
Kanje M, Skottner A, Sjoberg J, Lundborg G (1989) Insulin like growth factor-I (IGF-I) stimulates regeneration of the rat sciatic nerve. Brain Res 486:396-398.
Kaplan DR, Martin-Zanca D, Parada LF (1991) Tyrosine phosphorylation and tyrosine kinase activity of the trk proto-oncogene product induced by NGF. Nature 350:158-160.
Klein R, Parada LF, Coulier F, Barbacid M (1989) Trk B, a novel tyrosine protein kinase receptor expressed during mouse neural development. EMBO J 8:3701-3709.
Klein R, Jing SQ, Nanduri E, O'Rourke E, Barbacid M (1991a) The trk proto-oncogene encodes a receptor for nerve growth factor. Cell 65:189-197.
Klein R, Nanduri V, Jing SA, Lamballe F, Tapley P, Bryant S, Cordon-Cardo C, Jones KR, Reichardt LF, Barbacid M (1991b) The trkB tyrosine protein kinase is a receptor for brain derived neurotrophic factor and neurotrophin-3. Cell 66:395-403.
Klein R, Lamballe F, Barbacid M (1992) The trkB tyrosine protein kinase is a receptor for neurotrophin-4. Neuron 8:947-956.
Knusel B, Michel PP, Schwaber JS, Hefti F (1990) Selective and non-selective stimulation of central cholinergic and dopaminergic development in-vitro by nerve growth factor, basic fibroblast growth factor, epidermal growth factor, insulin and insulin like growth factors I and II. J Neurosci 10:558-570.
Kumagai AK, Eisenberg J, Pardridge WM (1987) Absorptive-mediated endocytosis of cationized albumin and a β-endorphin cationized albumin chimeric peptide by isolated brain capillaries: Model system of blood-brain barrier transport. J Biol Chem 262:15214-15219.
Lamballe F, Klein R, Barbacid M (1991) TrkC, a new member of the trk family of tyrosine protein kinases, is a receptor for neurotrophin-3. Cell 66:967-979.
Lefebvre PP, Van De Water TR, Weber T, Rogister B, Moonen G (1991a) Growth factor interactions in cultures of dissociated adult acoustic ganglia: neuronotrophic effects. Brain Res 567:306-312.
Lefebvre PP, Staecker H, Weber, T, Van De Water TR, Rogister B, Moonen G (1991b) TGFβ^1 modulates bFGF receptor message expression in cultured adult auditory neurons. NeuroReport 2:305-308.
Lefebvre PP, Van De Water TR, Represa J, Liu W, Bernd P, Modlin S, Moonen G, Mayer MB (1991c) Temporal pattern of nerve growth factor (NGF) binding in-vivo and the in-vitro effects of NGF on cultures of developing auditory and vestibular neurons. Acta Otolaryngol (Stockh) 111:304-311.
Lefebvre PP, Martin D, Staecker H, Weber T, Moonen G, Van De Water TR (1992a) TGFβ_1 expression is initiated in adult auditory neurons by sectioning of

the auditory nerve. NeuroReport 3:295–298.

Lefebvre PP, Van De Water TR, Staecker H, Weber T, Galinovic-Schwartz V, Moonen G, Ruben RJ (1992b) Nerve growth factor stimulates neurite regeneration but not survival of adult auditory neurons in-vitro. Acta Otolaryngol (Stockh) 112:288–293.

Lefebvre PP, Malgrange B, Staecker H, Moghadassi M, Van De Water TR, Moonen G (1994) Neurotrophins affect survival and neuritogenesis by adult injured auditory neurons in-vitro. NeuroReport 5:865–868.

Leibrock J, Lottspeich F, Hohn A, Hofer M, Hengerer B, Masiakowski P, Thoenen H, Barde Y-A (1989) Molecular cloning and expression of brain derived neurotrophic factor. Nature 341:149–152.

Lemmon SK, Riley MC, Thomas KA, Hoover GA, Maciag T, Bradshaw RA (1982) Bovine fibroblast growth factor: Comparison of brain and pituitary preparations. J Cell Biol 95:162–169.

Lenoir D, Honneger P (1983) Insulin like growth factor-I (IGF-I) stimulates DNA synthesis in fetal rat brain cell cultures. Dev Brain Res 7:205–213.

Lenoir M, Shnerson A, Pujol R (1980) Cochlear receptor development in the rat with emphasis on synaptogenesis. Anat Embryol 160:253–262.

Leong KV, Brott BD, Langer R, (1985) Bioerodible polyanhydrides as drug-carrier matrices I: Characterization, degradation and release characteristics. J Biomed Mater Res 19:941–955.

Leong KV, D'Amore P, Marletta M, Langer R. (1986) Bioerodible polyanhydrides as drug-carrier matrices II: Biocompatibility and chemical reactivity. J Biomed Mater Res 20:51–64.

Levi-Montalcini R, Hamburger V (1951) Selective growth stimulating effects of mouse sarcoma on the sensory and sympathetic nervous system of the chick embryo. J Exp Zool 116:321–362.

Lewis ME, Neff NT, Contreras PC, Stong DB, Oppenheim RW, Grebow PE, Vaught JL (1993) Insulin like growth factor-I: Potential for treatment of motor neuronal disorders. Exp Neurol 124:73–88.

Lin L-FH, Mismer D, Lile JD, Armes LG, Butler ET, Vanmice JL, Collins F (1989) Purification, cloning, and expression of ciliary neurotrophic factor (CNTF). Science 246:1023–1025.

Lo DC (1993) A central role for ciliary neurotrophic factor? Proc Natl Acad Sci USA 90:2557–2558.

Luo L, Coutnouyan H, Baird A, Ryan A (1993) Acidic and basic FGF mRNA expression in the adult and developing rat cochlea. Hear Res 69:182–193.

Luo L, Moore J, Baird A, Ryan A (1995) Expression of acidic FGF mRNA in rat auditory brainstem during postnatal maturation. Devel Brain Res (IN PRESS).

Lund PK, Moats-Staats BM, Hynes MA, Simmons JG, Jansen M, D'Ercole AJ, Van Wyk JJ (1986) Somatomedin-C/insulin like growth factor-I and insulin like growth factor II mRNAs in rat foetal and adult tissues. J Biol Chem 261:14539–14544.

Maisonpierre PC, Belluscio L, Squinto S, Ip NY, Furth ME, Lindsay RM, Yancopoulos GD (1990) Neurotrophin-3: A neurotrophic factor related to NGF and BDNF. Science 247:1446–1451.

Manthorpe M, Skaper SD, Barbin G, Varon S. (1982) Cholinergic neuronotrophic factors (CNTF's): VII. Concurrent activities on certain nerve growth factor responsive neurons. J Neurochem 38:415–421.

Marks JL, King MG, Baskin DG (1991) Localization of insulin and type I IGF receptors in rat brain by in-vitro autoradiography and in-situ hybridization. In: Raizada MK and LeRoith D (eds) Molecular Biology and Physiology of Insulin and Insulin Like Growth Factors. New York: Plenum Press.

Martinou JC, Le Van Thai A, Valette A, Weber MJ (1990) Transforming growth factor beta$_1$ is a potent survival factor for rat embryo motoneurons in culture. Dev Brain Res 52:175–181.

Martin-Zanca D, Barbacid M, Parada LF (1990) Expression of the trk proto-oncogene is restricted to the sensory cranial and spinal ganglia of neural crest origin in mouse development. Genes Devel 4:683–694.

Massagué J, Guillette BJ, Czech MP, Morgan CJ, Bradshaw RA (1981) Identification of a nerve growth factor receptor protein in sympathetic ganglia membranes by affinity labelling. J Biol Chem 256:9419–9424.

Meakin SO, Shooter EM (1991) Molecular investigations of the high affinity nerve growth factor receptor. Neuron 6:153–163.

Meakin SO, Shooter EM (1992) The nerve growth factor family of receptors. Trends Neurosci 15:323–331.

Morrison RS, Sharma A, De Vellis J, Bradshaw RA (1986) Basic fibroblast growth factor supports the survival of cerebral cortical neurons in primary culture. Proc Natl Acad Sci USA 83:7537–7541.

Murphy M, Drago J, Bartlett PF (1990) Fibroblast growth factor stimulates the proliferation and differentiation of neural precursors cells in-vitro. J Neurosci Res 25:463–475.

Near SL, Whalen LR, Miller JA, Ishii DN (1992) Insulin like growth factor-II stimulates motor nerve regeneration. Proc Natl Acad Sci USA 89:11716–11720.

Nebreda AR, Martin-Zanca D, Kaplan DR, Parada LF, Santos E (1991) Induction by NGF of meiotic maturation of *Xenopus* oocytes expressing the trk proto-oncogene product. Science 252:558–563.

Neely EK, Beukers MW, Oh Y, Cohen P, Rosenfeld RG (1991) Insulin like growth factor receptors. Acta Ped Scand Suppl 372:116–123.

Newman A, Kuruvilla A, Pereda A, Honrubia V (1986) Regeneration of the VIIIth cranial nerve I: Anatomic verification in the bullfrog. Laryngoscope 96:484–493.

Newman A, Honrubia V, Bell T (1987) Regeneration of the eight cranial nerve II: Physiologic verification in the bullfrog. Laryngoscope 97:1219–1232.

Olson L, Ernford P, Ebendal T, Mouton P, Stronebey I, Persson H (1988) The establishment and use of stable cell lines that overexpress a transfected beta-nerve growth factor (NGF) gene: Studies in-vitro, in oculo, and intracranially. Abstract of the Annual Meeting of the Society for Neuroscience, 276.6.

Pardridge WM (1986) Receptor-mediated peptide transport through the blood brain barrier. Endocrinol Rev 7:314–330.

Perraud F, Labourdette G, Miehe M, Loret C, Sensenbrenner M (1988) Comparisons of the morphological effects of acidic and basic fibroblast growth factors on rat astroblasts in culture. J Neurosci Res 20:1–11.

Pirvola U, Ylikoski J, Palgi J, Lehtonen E, Arumae U, Saarma M (1992) Brain-derived neurotrophic factor and neurotrophin-3 mRNAs in the peripheral target fields of developing inner ear ganglia. Proc Natl Acad Sci USA 89:9915–9919.

Pirvola U, Arumae U, Moshnyakov, M, Palgi U, Saarma M, Ylikoski J (1994) Coordinated expression and function of neurotrophins and their receptors in the rat inner ear during target innervation. Hear Res 75:131–144.

Powell EM, Sobarzo MR, Saltzman WM (1990) Controlled release of nerve growth factor from a polymeric implant. Brain Res 515:309–311.

Ramón y Cajal S (1892) La rétine des vertébrés. La Cellule 9:121–246.

Ramón y Cajal S (1928) Degeneration and Regeneration of the Nervous System. May RM, translator. London: Oxford Press.

Ray J, Gage FH (1994) Spinal cord neuroblasts proliferate in response to basic fibroblast growth factor. J Neurosci 14:3548–3564.

Recio-Pinto E, Reichler MM, Ishii DN (1986) Effects of insulin, insulin like growth factor-II, and nerve growth factor on neurite formation and survival in cultured sympathetic and sensory neurons. J Neurosci 6:1211–1219.

Ridley AJ, David JB, Stroobant P, Land H (1989) Transforming growth factors-1 and 2 are mitogens for rat Schwann cells. J Cell Biol 109:3419–3424.

Roberts AB, Sporn MB (1990) The transforming growth factors. In: Sporn MB, Roberts AB (eds) Handbook of Experimental Pharmacology "Peptide Growth Factors and Their Receptors" Heidelberg: Springer.

Rodriguez-Tebar A, Dechant G, Barde YA (1990) Binding of brain derived neurotrophic factor to the nerve growth factor receptor. Neuron 4:487–492.

Romand MR, Romand R (1982) The ultrastructure of spiral ganglion cells in the mouse. Acta Otolaryngol (Stockh) 104:29–39.

Romand MR, Romand R (1990) Development of spiral ganglion cells in the mammalian cochlea. J Electron Micros Techn 15:144–154.

Rosenberg MB, Friedmann T, Robertson RC, Tuszynski M, Wolff JA, Breakefield XO, Gage FH (1988) Grafting genetically modified cells to the damaged brain; restorative effects of NGF expression. Science 242:1575–1578.

Rosenthal A, Goeddel DV, Lewis NM, Lewis M, Shih A, Laramee GR, Nicholics K, Winslow JW (1990) Primary structure and biological activity of a novel human neurotrophic factor. Neuron 4:676–773.

Ruben R (1967) Development of the inner ear of the mouse: A radioautographic study of terminal mitoses. Acta Otolaryngol (Stockh) Suppl 220:1–44.

Rueda J, De la Sen C, Juiz J, Merchan JA (1987) Neuronal loss in the spiral ganglion of young rats. Arch Otolaryngol 104:417–421.

Rugh, R (1968) The Mouse: Its Reproduction and Development. Minneapolis: Burgess Publishing Co.

Ryan AF, Luo L (1995) Delivery of a recombinant growth factor into the mouse inner ear by implantation of a transfected cell line. Abstracts of the Eighteenth Midwinter Meeting of the Association for Research in Otolaryngology, p 117, Abstr. 185.

Schecterson LC, Bothwell M (1994) Neurotrophin and neurotrophin receptor mRNA expression in developing inner ear. Hear Res 73:92–100.

Selden RF, Skoskievicz MJ, Howie KB, Russell PS, Goodman HM (1987) Implantation of genetically engineered fibroblasts into mice: implications for gene therapy. Science 236:714–718.

Sendtner M, Holtmann B, Kolbeck R, Thoenen H, Barde Y-A (1992) Brain-derived neurotrophic factor prevents the death in motoneurons in newborn rats after nerve section. Nature 360:24–31.

Sensenbrenner M (1993) The neurotrophic activity of fibroblast growth factors. Prog Neurobiol 41:683–704.

Sher AE (1971) Embryonic and postnatal development of the inner ear of the mouse. Acta Otolaryngol (Stockh) Supp 285:1–77.

Shnerson A, Devigne C, Pujol R (1982) Age-related changes in the C57BL/6J mouse cochlea. II. Ultrastructural findings. Devel Brain Res 2:77–88.

Sievers J, Hausmann B, Unsicker K, Berry M (1987) Fibroblast growth factors promote the survival of adult rat retinal ganglion cells after transection of the optic nerve. Neurosci Lett 76:157-162.

Snyder EY, Kim SU (1980) Insulin: Is it a nerve survival factor? Brain Res 196:565-571.

Sobkowicz H (1992) The development of innervation in the organ of Corti. In: Romand R (ed) Development of Auditory Vestibular Systems. Amsterdam: Elsevier, pp 59-100.

Sobkowicz HM, Emmerling MR (1989) Development of acetylcholinesterase-positive neuronal pathways in the cochlea of the mouse. J Neurocytol 18:209-224.

Sobkowicz HM, Rose J, Scott G, Kuwanda S, Hind J, Oertel D, Slapnick S (1980) Neuronal growth in the organ of Corti in culture. In: Giacobini E (ed) Tissue Culture in Neurobiology. New York: Raven Press, pp. 253-275.

Sobkowicz HM, Rose JE, Scott GL, Slapnick SM (1982) Ribbon synapses in the developing intact and cultured organ of Corti in the mouse. J Neurosci 2:942-957.

Sobkowicz HM, Rose JE, Scott GL, Levenick CV (1986) Distribution of synaptic ribbons in the developing organ of Corti. J Neurocytol 15:693-714.

Sobkowicz HM, Emmerling MR, Whitlon DS (1988) Development of cochlear efferents in the postnatal mouse. Abstract of the Eleventh Midwinter Meeting of the Association for Research in Otolaryngology, p 131, Abstr. 16P.

Soppet D, Escandon E, Maragos J, Middlemas DS, Reid SW, Blair J, Burton LE, Stanton BR, Kaplan DR, Hunter T, Nikolics K, Parada LF (1991) The neurotrophic factors, brain derived neurotrophic factor and neurotrophin-3 are ligands for the trkB tyrosine kinase receptor. Cell 65:895-903.

Speidel CC (1948) Correlated studies of sense organs and nerves of the lateral line in living frog tadpoles. II. The trophic influence of specific nerve supply as revealed by prolonged observations of denervated and reinnervated organs. Am J Anat 82:277-320.

Sperry R (1945) Centripetal regeneration of the VIII cranial nerve root with systematic restoration of vestibular reflexes. Amer J Physiol 144:735-741.

Spoendlin H (1971) Degeneration behavior of the cochlear nerve. Arch Klin Exp Ohr Nas Heilk Heilk 200:275-291.

Spoendlin H (1988) Biology of the vestibulocochlear nerve. In: Albert PW, Ruben RJ (eds) Otologic Medicine and Surgery Volume 1. New York: Churchill Livingstone, pp. 117-150.

Spoendlin H, Suter R (1976) Regeneration in the VIII Nerve. Acta Otolaryngol (Stockh) 81:228-236.

Squinto SP, Stitt TN, Aldrich TH, Davis S, Bianco SM, Radziejewski C, Glass DJ, Masiakowski P, Furth ME, Valenzuela DM, DiStefano PS, Yancopoulos GD (1991) Trk B encodes a functional receptor for brain derived neurotrophic factor and neurotrophin-3 but not nerve growth factor. Cell 65:885-893.

Staecker H, Liu W, Hartnick C, Lefebvre PP, Moonen G, Van De Water TR (1994) Ciliary neurotrophic factor is a central target tissue derived neurotrophic factor for auditory neurons. Abstracts of the Seventeenth Midwinter Meeting of the Association for Research in Otolaryngology, p. 138, Abstr. 549.

Staecker H, Kopke R, Lefebvre PP, Malgrange B, Liu W, Moonen G, Van De Water TR (1995) The effects of neurotrophins on adult auditory neurons in-vitro and in vivo. Abstracts of the Eighteenth Midwinter Meeting of the Association for Research in Otolaryngology, p. 177, Abstr. 705.

Steele-Perkins G, Turner J, Edman JC, Hari J, Pierce SB, Stover C, Rutter WJ, Roth RA (1988) Expression and characterization of a functional human insulin like growth factor-I receptor. J Biol Chem 263:11486-11492.

Sun JC, Bohne BA, Harding GW (1995) Age at the time of acoustical injury affects the magnitude of nerve-fiber regeneration. Abstracts of the Eighteenth Midwinter Meeting of the Association for Research in Otolaryngology, p. 74. Abstr. 294.

Terrayama Y, Kaneko Y, Kawanto K, Akai N (1977) Ultrastructural changes of the neuronal elements following destruction of the organ of Corti. Acta Otolaryngol (Stockh) 83:291-302.

Thomas KA, Rios-Candelore M, Fitzpatrick S (1984) Purification and characterization of acidic fibroblast growth factor from bovine brain. Proc Natl Acad Sci USA 81:357-361.

Togari A, Dickens G, Kuzuya H, Guroff G (1985) The effect of fibroblast growth factor on PC12 cells. J Neurosci 5:307-316.

Tohyama Y, Kiyama H, Kitajiri M, Yamashita T, Kumazawa T, Tohyama M (1989) Ontogeny of calcitonin gene-related peptide in the organ of Corti of the rat. Devel Brain Res 45:309-312.

Tomargo RJ, Epstein JI, Reinhard CS, Chasin M, Brem H (1989) Brain biocompatibility of a biodegradable controlled release polymer in rats. J Biomed Mat Res 23:253-266.

Unoki K, LaVail MM (1994) Protection of the rat retina from ischemic injury by brain derived neurotrophic factor, ciliary neurotrophic factor, and basic fibroblast growth factor. Invest Opthomol Vis Sci 35:907-915.

Unsicker K, Reichert-Preibsch H, Schmidt R, Pettmann B, Labourdette G, Sensenbrenner M (1987) Astroglial and fibroblast growth factors have neurotrophic functions for cultured peripheral and central nervous system neurons. Proc Natl Acad Sci USA 84:5459-5463.

Unsicker K, Flanders KC, Cissel DS, Lafyatis R, Sporn MB (1991) Transforming growth factor beta isoforms in the adult rat central and peripheral nervous system. Neurosci 613-625.

Van De Water TR (1986) Determinants of neuron-sensory receptor cell interaction during the development of the inner ear. Hear Res 22:265-277.

Van De Water TR (1988) Tissue interactions and cell differentiation: Neuron-sensory cell interaction during otic development. Devel 103:185-193.

Van De Water TR, Ruben RJ (1984) Neurotrophic interactions during in-vitro development of the inner ear. Ann Otol Rhinol Laryngol 93:558-564.

Van De Water TR, Galinovic-Schwartz V, Ruben RJ (1989) Determinants of ganglion-receptor cell interaction during development of the inner ear: A heterochronic ganglia study. Acta Otolaryngol (Stockh) 108:227-237.

Van De Water TR, Frenz DA, Giraldez F, Represa J, Lefebvre PP, Rogister B, Moonen G (1991) Growth factors and development of the stato-acoustic system. In: Romand R (ed) Development of Auditory and Vestibular Systems II: Amsterdam, Elsevier Science BV, pp. 1-32.

Van De Water TR, Lefebvre PP, Liu W, Xu H, Moonen G, Kessler JA, Federoff H (1992) Infection of dissociated cochlear ganglion cell cultures with pHSVtrkA changes the survival response of auditory neurons to exogenous NGF. Abstracts of the Annual Meeting of the Society for Neuroscience, Abstr. 401.21.

Van De Water TR, Staecker H, Lefebvre PP, Moonen G (1994) Homeostasis in the spiral ganglion: The role of growth factors. Abstracts of the Seventeenth

Midwinter Meeting of the Association for Research in Otolaryngology, p. 3, Abstr. 12.

Vasquez E, Van De Water TR, Del Valle M, Staecker H, Vega JA, Giraldez F, Represa J (1994) Pattern of trkB protein-like immunoreactivity in-vivo and the in-vitro effects of brain-derived neurotrophic factor (BDNF) on developing cochlear and vestibular neurons. Anat Embryol 189:157-167.

Vescovi AL, Reynolds BA, Fraser DD, Weiss S (1993) bFGF regulates the proliferative fate of unipotent (neuronal) and bipotent (neuronal/astroglial) EGF generated CNS progenitor cells. Neuron 11:951-966.

Walicke P, Cowan WM, Ueno N, Baird A, Guillemin R (1986) Fibroblast growth factor promotes survival of dissociated hippocampal neurons and enhances neurite extension. Proc Natl Acad Sci USA 83:3012-3016.

Waller AV (1852) Sur la reproduction des nerfs et sur la structure et les fonctions des ganglions spinaux. Arch Anat Physiol Wiss Med 392-401.

Wheeler EF, Bothwell M, Schecterson LC, von Bartheld CS (1994) Expression of BDNF and NT-3 mRNA in hair cells of the organ of Corti: Quantitative analysis in developing rats. Hear Res 73:46-56.

Whitlon DS, Sobkowicz HM (1988) Neuron-specific enolase during the development of the organ of Corti. Internat J Devel Neurosci 6:77-87.

Whitlon DS, Sobkowicz HM (1989) GABA-like immunoreactivity in the cochlea of the developing mouse. J Neurocytol 18:505-518.

Wilkinson DG, Bhatt S, McMahon AP (1989) Expression pattern of the FGF-related proto-oncogene int-2 suggests multiple roles in fetal development. Devel 105:131-136.

Winn SR, Hammang JP, Emerich DF, Lee A, Palmiter RD, Baetge ED (1994) Polymer-encapsulated cells genetically modified to secrete human nerve growth factor promote the survival of axotomized septal cholinergic neurons. Proc Natl Acad Sci USA 91:2324-2328.

Ylikoski J, Pirvola U, Moshnyakov M, Palgi J, Arumae U, Saarma M (1993) Expression patterns of neurotrophin and their receptor mRNAs in the rat inner ear. Hear Res 56:69-78.

Zakon H (1986) Regeneration and recovery of function in the amphibian auditory system. In: Ruben RJ, Van De Water TR, Rubel EW (eds) The Biology of Change in Otolaryngology. Amsterdam: Elsevier, pp. 305-317.

Zakon H, Capranica R (1981a) Reformation of organized connections in the auditory system after regeneration of the VIIIth nerve. Science 213:242-244.

Zakon H, Capranica R (1981b) An anatomical and physiological study of regeneration of the VIIIth nerve in the leopard frog. Brain Res 209:325-338.

Zhou X, and Van De Water TR (1987) The effect of target tissues on survival and differentiation of mammalian statoacoustic ganglion neurons in organ culture. Acta Otolaryngol (Stockh) 104:90-98.

4

Auditory Deprivation and Its Consequences: From Animal Models to Humans

JUDITH S. GRAVEL AND ROBERT J. RUBEN

Introduction

Mounting evidence demonstrates the adverse consequences of various forms of auditory deprivation in both animals and humans. The term "auditory deprivation" is a specific form or perversion of auditory input that deviates from what is expected and/or needed for the optimization of auditory function in the organism—be it mouse, or human. Auditory deprivation appears particularly deleterious when the input restrictions occur within the developmental period, although recent research findings suggest that even the mature auditory system will respond to afferent restrictions. The immediate adaptation of the nervous system to the restriction of auditory input, as well as the plasticity of the system resulting in its reorganization in the longer term, are topics important to scientists and clinicians alike (Moore 1993). Questions on the effects of onset time and duration of the deprivation and whether partial or complete reversal of deleterious consequences can be achieved through the restoration of normal or near-normal auditory experiences have both scientific and social ramifications.

It is well known that animals utilize auditory input for life-sustaining purposes: their survival often depends upon the use of audition for movement, food seeking, and avoidance of predators. In humans, partial hearing impairments or even total deafness are not considered life threatening. Auditory deprivation in humans, however, does dramatically impact the most significant and unique function of our species; namely, the development and maintenance of auditory/oral communication (Ruben and Rapin 1980).

The neural response to deprivation appears exacerbated by auditory restriction imposed unilaterally. Dramatic reorganization of the auditory system occurs when deprivation results in asymmetric auditory input, since the binaural system is organized to serve a "comparator function" (Trahiotis 1992). Increasingly apparent through empiric study is the marked influence of monaural deprivation particularly at the level of the brain stem.

Monaural auditory restriction results in asymmetric input to higher-order (central) auditory nuclei that are normally responsive to binaural afferent activity (Trahiotis 1992). Reorganization of the auditory system ensues to accommodate the lack of binaural input.

Delineating the quantity (degree and spectrum), duration, quality (type), and timing (the point in development) of the auditory deprivation that is of consequence to the organism is important to our understanding of both normal and deviant auditory development. Currently, long-held notions that auditory deprivation, adaptation, plasticity, and reorganization are issues pertinent only for the immature developing nervous system have been challenged by new evidence suggesting that the mature auditory system is both susceptible and adaptable to auditory deprivation. These findings impact upon our approaches and timetables for specific screening, assessment, and intervention practices, as well as more global health care policy planning.

We approach this chapter with these basic, clinical and societal issues in mind. This chapter reviews our current knowledge of the consequences of monaural and binaural auditory deprivation in animals and humans; restrictions that are the result of congenital, acquired or environmentally imposed deprivation of auditory input. We review some of the research describing the anatomical, physiological, and behavioral consequences of binaural and monaural deprivation in immature and mature auditory systems. We examine issues that are relevant to the development and application of recent research to the identification and treatment of human communication disorders. At the conclusion of the chapter, we pose questions for consideration and future study. Abbreviations used in this chapter are presented in Table 4.1.

2. Auditory Deprivation

Adequate sensory experience is critical to the developing nervous system; important for both the expression of certain sensory functions as well as to their maintenance, even when such functions are innately determined (Knudsen, 1985). In both animals and humans, total sensory deprivation has profound consequences for the organism, reflected in an inability to function normally in the native environment. Moreover, severe environmental language deprivation has been shown to irreversibly effect the development of communication in humans with normal hearing (Ruben and Rapin 1980).

Auditory deprivation may result from either temporary or permanent peripheral hearing loss or a restricted, impoverished acoustic environment. Deprivation can occur before or after the onset of hearing, and before, during, or after the complete maturation of the organism. Various types of auditory deficits may be imposed on an organism. Conductive hearing loss

TABLE 4.1 Abbreviations

Anatomical structures:
AVCN	anteroventral cochlear nucleus
CN	cochlear nucleus
CNS	central nervous system
DCN	dorsal cochlear nucleus
IC	inferior colliculus
LSO	lateral superior olive
NM	nucleus magnocellularis
MNTB	medial nucleus of the trapezoid body
MSO	medial superior olivary nucleus
SOC	superior olivary complex
VCN	ventral cochlear nucleus

Auditory terminology:
ABR	auditory brainstem response
CAP	compound action potential
dB	decibel (intensity)
Hz	Hertz (frequency)
MLD	masking level difference
OM	otitis media
OME	otitis media with effusion
SoNo	signal in phase; masker in phase
SπNo	signal out of phase; masker in phase

affects only peripheral mechanisms (outer and middle ear) that conduct sound waves to an intact organ of hearing. Conductive losses may be temporary or permanent, the result of pathology, obstruction, or structural abnormalities. A complete conductive hearing loss does not completely eliminate the possibility of auditory stimulation, however, as higher intensity auditory signals and sounds produced internally are audible (via bone conduction). Cochlear hearing loss results from damage to, or abnormal development of, the inner ear (sensory endorgan). In humans, a cochlear deficit is frequently referred to as "sensorineural" hearing loss. Conductive and cochlear deficits may exist simultaneously and are termed mixed hearing loss. Environmental deprivation results from aberrant auditory experiences in the presence of an otherwise normal peripheral hearing mechanism.

Various methods (e.g., surgical ablation of the cochlea, ossilectomy, ear plugging, canal suturing, ototoxic drug treatment, rearing in sound-treated environments) have been utilized to impose auditory deprivation in experimental animals. Certain limitations to achieving and maintaining and/or completely reversing a specific type of auditory deprivation, however, can compromise experimental results. Readers are cautioned that the methods used to impose auditory deficits should be regarded carefully in considering the outcome of studies of auditory deprivation.

The timing of the onset of hearing varies among species. Moreover, the time period over which hearing function matures is species-specific (Moore

1985). An unanswered question critical to the issue of auditory deprivation is whether the various levels (peripheral and central) of the auditory system develop in series or in parallel (Moore 1985). While some species such as humans and guinea pigs (Moore 1985) are precocial (have hearing function prenatally), others such as rats (Clopton 1980) and ferrets (Moore 1985) are altricial with the onset of hearing occurring postnatally days, weeks, or even months after birth (Moore 1985). Experimental outcomes may be influenced by when the deprivation was imposed relative to the onset of hearing and may make generalizations across species problematic.

Clearly, the quantity, quality, duration, and timing of the auditory deprivation may or may not be amenable to systematic study in a particular species, particularly humans. Moreover, certain types of auditory deprivations may not be completely reversible—a desirable condition for the study of the response of the nervous system to the normalization of auditory experiences after a period of restricted input. The alleviation of some amount of auditory restriction may be accomplished using prosthetic devices, surgery, or enhanced acoustic environments, thereby providing researchers the opportunity to study the influence of partial restoration of "normal" auditory experiences.

A general problem throughout both animal and human studies in auditory deprivation has been the frequent lack of delineation of the hearing loss imposed. That is, the audiometric profile, specifically the degree, spectral configuration, and symmetry of the auditory deficit over time, has not always been characterized. Consequently, important questions regarding the specifics of the auditory deprivation leading to sequelae thus far have not been fully delineated. Auditory sensitivity may be determined using auditory evoked potentials (i.e., electrocochleography or EcoG and the auditory brain stem response or ABR; Kraus and McGee 1992) and/or behavioral methods. Quantification of any auditory restriction can only be made when for each subject (individual) direct, frequency-specific, and periodic measurements of auditory sensitivity are completed; assumed or average values may over- or under-estimate the actual restriction. Moreover, occult impairments may exist (conductive or cochlear) during or after the deprivation period.

Thus, assessment of auditory sensitivity is critical for understanding the effects of particular degrees and configurations of auditory impairments, to document the stability of the hearing deficit, and to ensure the completeness of any reversal of the deprivation. Only when such data are available can the influence of any residual peripheral impairment (conductive or cochlear) on outcome be recognized. Not all studies reviewed in this chapter accounted for the previously mentioned factors. However, the research presented in the following sections, in our opinion, forms a reasonable basis for the reader's appreciation of the issues surrounding auditory deprivation in animals and humans.

2.1 Concept of a Critical Period

As will be documented in later sections of this chapter, a period in development appears to exist for most animals during which normal auditory input (experience) is crucial for the later development of optimal auditory function. This time interval, occurring before the complete maturation of the organism, has been termed a "critical" or more recently, a "sensitive" period for normal anatomical, physiological, and behavioral auditory development. When deprivation has been imposed before or during this time period, the resulting consequences were thought irreversible by later experience with sound (see for example Ruben and Rapin 1980). Moreover, it has long been held that if auditory deprivation were imposed after this time period no effects on the neural system would be evident. However, as suggested previously, current reports reveal that the adult auditory system is responsive to auditory deprivation and that neural plasticity (reorganization) is present in the mature auditory system (Moore 1993).

Auditory input during some period of early development appears critical. If the organism has experienced even brief exposure to normal auditory input prior to the onset of deprivation, a more normal expression of auditory function is achieved than if no experience with normal input had ever been available. Clinical experiences, as well as some empiric evidence, suggest that a critical, or more likely, a sensitive period for auditory development may exist in humans. It is frequently posited that the first three years of life are important for language development (e.g., Menyuk 1986). Systematic studies that delineate the parameters of any sensitive period for language development in humans, however, are lacking. The existence of one or more critical periods for language development is highly relevant for the timing of medical/surgical intervention, hearing screening programs, amplification/cochlear implant provision, and other therapeutic initiatives. We begin by examining the behavioral, physiological and anatomical consequences of auditory deprivation in animals. Such work serves as the foundation for our appreciation of the possible underlying mechanisms subserving similar consequences in humans.

3. Behavioral Consequences of Auditory Deprivation — Animals

An accurate and complete representation of auditory space (derived from the timing and intensity differences that exist among acoustic cues arriving at each ear) is often critical for the survival of the species. Knudsen (1985) suggests that the association made by animals between a particular pattern of auditory cues and their location in space is species-specific. Acoustic input transduced by the peripheral mechanism (including the head, external ear position and morphology, and middle ear) provide spectral shaping of the arriving input. Processing of binaural auditory cues, however, is a function of the central auditory system beginning at the level of the brainstem.

Moreover, the establishment and maintenance of communication is also dependent upon normal auditory function as well as the availability of a normal auditory environment that provides the requisite experiences with species-specific sound patterns. Studies that have examined the behavioral consequences of binaural, environmental (experiential), and monaural auditory deprivation are reviewed below.

3.1 Binaural Deprivation

In early studies of the behavioral consequences of auditory deprivation, rats were deprived of auditory input in early life (i.e., through ear plugging) and then had their hearing restored. Subsequently, these rodents were found to be less accomplished food competitors when the signal cueing food availability was auditory rather than visual (Wolf 1943; Gauron and Becker 1959).

Later, Tees (1967a) examined the ability of rats raised with bilateral ear plugging and normally reared controls to learn two-tone pattern discrimination (e.g., high-low-high versus low-high-low). Plugs were removed for the experimental task. The normally reared control animals took significantly fewer trials to reach criterion than did the deprived. Tess (1967b) also examined the abilities of early deprived rats on a task that required both groups to detect a change in the duration of a series of pulsed tones. Again, experimental animals required more trials to learn the durational discrimination. Thus, complex auditory tasks were adversely influenced by early auditory deprivation that resulted in a lack of early experience with temporally patterned events.

Interestingly, Tees also examined simple frequency (2000 Hz versus 4000 Hz tones; 1967a) and intensity (a 15 dB increase in the intensity of a pulsing tone; 1967b) discrimination in early deprived and control rats. Tees found that there was no difference in learning rates between the groups. Thus, while complex auditory pattern learning was influenced negatively by early bilateral conductive deprivation, basic frequency and intensity discrimination were not.

Defining and maintaining societal space is also subserved by auditory cues (Strasser and Dixon 1986). Normally hearing mice who had formed territories within an enclosure were deprived of normal sound input via ear plugging. This resulted in a reduced capacity to track (locate) and chase an intruder mouse entering the rodent's established territory. Mere detection of the intruder (that is, alerting to its presence and location) was significantly curtailed by the imposed deprivation (Strasser and Dixon 1986).

3.2 Environmental Deprivation

Deprivation resulting from a restricted or deviant auditory environment has also been demonstrated to result in certain behavioral consequences in various species. As with imposed binaural impairments, complex auditory pattern learning rather than simple auditory abilities (e.g., detection and discrimination) appear more influenced by environmental deprivation.

Rats reared in white noise beginning at birth were deprived of the features typical of their normal acoustic environment (Patchett 1977). The rats deprived of varied auditory experiences, however, were able to perform a simple frequency discrimination task similar to control animals reared in a normal auditory environment.

Mallard ducklings deprived of normal input by both devocalization in the embryonic period and rearing in isolation were examined for their call preference (Gottlieb 1975a). When compared to controls, deprived ducks were more likely to prefer (select) a maternal mallard call that was low-pass filtered than a full spectrum call. The effects of this type of deprivation was, for this species, frequency specific in that the early deprivation (lack of exposure to normally high-frequency vocalizations) apparently resulted in a later insensitivity for those higher frequency spectral components within the maternal call and an actual preference for a predominantly low frequency signal (to which they may have had some early exposure). Deprived ducks were, however, as adept at discriminating a full-spectrum mallard call from a low-pass filtered call, as were normally reared and vocal ducklings.

Gottlieb (1975b, 1975c) suggested that discrimination of the call of one's species is mediated both exogenously by experience or exposure as well as endogenously by normal neural maturation. Early exposure to the normal duck call is critical to normal selective performance in the longer term. In Peking ducklings devocalized embryologically, Gottlieb (1978) demonstrated that preference for the normal temporal pattern of the species-specific call was disrupted by early auditory deprivation; deprived ducklings, however, did respond to lower-than-normal rates of the call pattern. He also demonstrated that devocalized ducklings showed no preference for a duck call over that of a chicken differing in repetition rate.

Kerr, Ostapoff, and Rubel (1979) examined two groups of chicks who were repeatedly exposed (habituated) to a tone: one group at one day posthatch, the second group at 3 to 4 days posthatch. The chicks were then exposed to the habituating stimulus and four other tones of different frequencies. One-day-old chicks demonstrated flatter frequency generalization gradients than found in the older group. In a related experiment, these researchers examined chicks deprived of acoustic input from the embryonic period through 3 to 4 days post hatch. Deprived birds and two other groups (normal and sham-operated controls) were examined. Early auditory deprivation caused a frequency generalization deficit in only the deprived chicks, a pattern similar to that seen in their first experiment.

3.3 Monaural Deprivation

Knudsen, Knudsen, and Esterly (1984) examined the abilities of barn owls (*Tyto alba*) to recover from early monaural deprivation. These authors examined sound localization accuracy and precision in birds who had been deprived of sound at various onset times in early life. Their preliminary study (Knudsen, Esterly, and Knudsen 1984) demonstrated that young owls receiving monaural ear plugging learned to localize sound with the plug in place. Next, they examined the recovery of normal localization abilities in the birds after the ear plug was removed. Results indicated that the recovery of accurate localization abilities was dependent on the time period in which the deprivation was imposed and subsequently reversed, suggesting a critical period for the development of auditory localization in young barn owls.

Importantly, if the birds experienced even a brief amount of normal binaural cues prior to the onset of deprivation, normal localization abilities were recovered quickly following earplug removal. This suggests that some amount of normal experience is important to the establishment of the processes that underlie normal auditory localization abilities. The precision and accuracy of sound localization recovery was dependent upon the age at which the plug was removed; the younger the age, the better the outcome. Beyond the critical period, recovery of normal abilities was prolonged or never achieved. Finally, these authors found that if the monaural plug was placed after the critical period (i.e., in mature birds), adult owls never learned to accurately localize sound (Knudsen, Esterly, and Knudsen 1984).

4. Physiological Consequences of Auditory Deprivation — Animals

Aberrant neural physiology likely subserves the behavioral consequences of auditory deprivation discussed above; however, a one-to-one relationship between physiology and behavioral function must be studied directly in order to support the association. Auditory evoked response measures allow researchers to examine physiological function within single units or for neural pathways. Recording of activity of individual neurons provides an indication of function of specific neural units. The auditory brain stem response (ABR; traditionally elicited using broadband clicks and surface recordings) latency measures (in msec) of specific components (wave peaks) and inter-peak intervals, provides information regarding the integrity of the transmission of neural activity through the brain stem pathway (Kraus and McGee 1992). Normally, response latency decreases as a function of increasing frequency and intensity. Response latency is reflective of both travel time along the basilar membrane of the cochlea and the transmission of neural activity through the brainstem.

4.1 Binaural Deprivation

Silverman and Clopton (1977), Clopton and Silverman (1978), and Clopton (1980) found that in rats receiving early binaural conductive deprivation (and later reversed), the physiological activity recorded from neuronal units of the inferior colliculus was no different than that recorded from the same units in control animals. These authors reported higher (elevated) thresholds and broader tuning curves (indicative of poor frequency resolution) for units in the cochlear nucleus (CN) compared to normally reared animals. The broader tuning curves and elevated thresholds were similar to those recorded in the CN of immature rats, perhaps indicative of arrested or slowed maturation of the CN units.

Doyle and Webster (1991) found that bilateral conductive hearing loss imposed in rhesus monkeys (*Macaca mulatta*) early in development did not affect the absolute latencies or inter-peak intervals of animals' ABR when age and degree of hearing loss were considered. They reported that bilateral conductive hearing loss did not adversely influence the maturation (development) of the auditory brainstem pathway in monkeys, at least as revealed through serial ABR recordings.

Sohmer and Friedman (1992) recorded the ABR in rats deprived of hearing in early life; recordings were accomplished both with and without the ears plugged. The Wave I-IV peak latency of the ABR was significantly shorter in early auditory deprived rats after plug removal than in control animals. These authors found no difference in inter-wave latencies of adult animals deprived after maturation when compared to controls. They speculated that the shorter ABR latencies were the result of the smaller brain stem neurons found in anatomical investigations of auditory deprivation (see Section 5.2).

4.2 Environmental Deprivation

In keeping with the Sohmer and Friedman (1992) report, Evans, Webster, and Cullen (1983) had earlier examined both the physiological and anatomical (see below) consequences of environmental auditory deprivation in mice. These researchers also found that mean click-ABR latencies (both peak and inter-peak) were shorter in deprived animals than controls. Specifically, Wave I as well as the I-IV, I-V, and III-V inter-wave intervals were significantly shorter in deprived than control animals. Click threshold sensitivity, however, did not differ between groups.

4.3 Monaural Deprivation

Silverman and Clopton (1977), Clopton and Silverman (1978), and Clopton (1980) examined the effects of early and late monaural conductive deprivation on the neurophysiology of central processes underlying binaural interaction. Single neuron responses were recorded from the inferior colliculus (IC) of rats.

Normally, clicks presented to the ear opposite (contralateral) to the IC recording site facilitate activity, while clicks presented in the ear ipsilateral to the recording site inhibit activity elicited by the contralateral stimulation. This is the case even if the ipsilateral clicks are only several decibels more intense than those presented to the contralateral ear. This phenomenon is termed ipsilateral suppression—a mechanism critical to the accurate localization of sound in auditory space. Suppression of single unit activity (when compared to the monaural-alone condition) is indicative of binaural interaction in brainstem structures (Clopton 1980).

In animals who received monaural conductive deprivation, the outcome was dramatic. Greater-than-normal ipsilateral suppression was revealed when unit activity was recorded from neurons driven by the contralateral deprived ear. In marked contrast, when activity was recorded from the IC ipsilateral to the ear that received deprivation, no ipsilateral suppression was apparent even when the signal was increased by 40 dB over the stimuli presented to the contralateral nondeprived ear.

Clopton and Silverman (1978) then examined the latency and duration of single neuron responses within the IC contralateral to the stimulated ear in normally reared control rats and in a group with early monaural auditory deprivation. In rats experiencing early auditory deprivation, response latency measured in the IC was the same as in control animals for frequencies below 10 kHz. However, delayed latencies in units with characteristic frequencies above 10 kHz were recorded in deprived versus control animals. They suggested that aberrant binaural analysis may be the result of the prolonged response latencies found for high-frequency (above 10 kHz) IC units that adversely influenced the coding of time and intensity cues.

Knudsen (1985) obtained physiological recordings from bimodal (auditory-visual/spatial) neuronal units in the optic tectum (above the level of the inferior colliculus) of the barn owl. A bimodal fraction was derived to differentiate units that were driven by both auditory and visual input versus those most responsive to only one form of sensory input.

Knudsen (1985) suggested that bimodal units are "tuned": alignment of specific auditory cues with appropriate visual-spatial locations occurs with experience. When owls were raised with a monaural earplug, alignment of bimodal units was evident only when the plug was in place. Removal of the earplug caused a misalignment of the units. Interestingly, young owls whose auditory-visual space had been represented in one way early in development and then "normalized" before maturity, demonstrated normal alignment of optic tectum units with time. Regardless of the subsequent length of normal experience, however, owls deprived throughout the critical period and then unplugged as mature birds never demonstrated normal bimodal unit alignment. When an adult owl was deprived (i.e., initially at maturity), there was no change in the auditory-spatial tuning of the units, even after one year of monaural plugging. With the plug in place, the bimodal tectum units demonstrated misalignment; when the plug was removed the units were aligned normally.

5. Anatomical Consequences of Auditory Deprivation — Animals

Anatomical consequences of deprivation in animals have been reported at various levels of the auditory system and appear to result from both cochlear and conductive hearing loss. The issues regarding the symmetry of the deprivation are as significant when examining studies at this level of analysis. Again, the relationship between morphological changes (in one or more cell groups) to electrophysiological or behavioral indices must be directly examined.

Some anatomical studies have demonstrated that the volume of nuclei and size of neurons in the brainstem are adversely affected by cochlear, conductive, and environmental deficits imposed in the neonatal period in animals. However, the binaural symmetry of neuronal projections is preserved in brain stem and midbrain nuclei when early-onset deprivation occurs equally to both ears. Similar to the behavioral and electrophysiologic studies previously discussed, it appears that monaural deprivation has anatomical consequences for brain stem and higher order auditory structures.

5.1 Binaural Cochlear Deprivation

The anatomical effects of bilateral cochlear deafness on the peripheral mechanism have been demonstrated by Moore (1990a, 1990b) who found nearly complete atrophy of the cochlear nerves after destruction of the cochlea in young ferrets (*Mustela putorius*). Any remaining cochlear nerves were dependent upon the number of spiral ganglion neurons that remained after cochlear lesioning.

Complete bilateral cochlear hearing loss results in changes in brain stem nuclei anatomy in animals deprived of auditory input in infancy. Anniko, Sjostrom, and Webster (1989) measured the volume of the ventral cochlear nucleus (VCN) and dorsal cochlear nucleus (DCN) in congenitally deaf mice and controls. DCN volume was similar between groups, while the volume of VCN was smaller in congenitally deaf versus normal animals. Similarly, Fleckeisen, Harrison, and Mount (1991) examined the cochlear nucleus of chinchillas deafened in the neonatal period as well as in animals adventitiously deafened after maturity. The volume of the cochlear nucleus (specifically, VCN) was significantly reduced when deafness was induced in infancy but not in animals experiencing late-onset bilateral deafness.

Moore (1990b) examined the cochlear nucleus of ferrets who received bilateral cochlear destruction in infancy. The overall volume of the CN was reduced as was the volume of each division of the CN (DCN, posterior and anterior VCN); moreover, the shape of the cochlear nucleus was distorted in comparison to control animals. These findings were consistent for both right and left brain stem nuclei. Although CN volume was reduced, the symmetry of neuronal projections from each of the CN to the ipsilateral IC were similar to those of normal hearing animals. A complete cochlear hearing loss in the neonatal period did not alter the symmetry of brainstem neuronal projections from either CN to the midbrain IC of ferrets.

5.2 Binaural Conductive Deprivation

The brain stem nuclei of mice experiencing early-onset conductive deficits (i.e., before the onset of hearing) were examined anatomically by Webster and Webster (1977, 1979) and by Evans, Webster, and Cullen (1983). In general, these authors found smaller neurons in the VCN and the lateral superior olive (LSO) and the medial nucleus of the trapezoid body (MNTB) of the superior olivary complex (SOC; i.e., the first nucleus receiving binaural input in the mammalian brain), as well as fewer neurons (as evidenced by a lower packing density) in the DCN. While overall CN neuron sizes in binaurally deprived rats were smaller than cells in the anterior VCN (AVCN) of control animals and from ipsilateral nondeprived ears, there was no difference in neuron size between right and left cochlear nuclei of binaurally deprived rats (Coleman and O'Connor 1979).

Anatomical analyses of the volumes of the DCN and VCN were completed on the conductively deprived rhesus monkeys studied with the ABR by Doyle and Webster (1991). Neuron size was also determined for specific cell types within the cochlear nucleus (VCN), SOC (lateral and medial), MNTB, and the IC. There were no significant anatomical differences between the brain stem nuclei studied from normal versus conductively impaired ears.

Feng and Rogowski (1980) examined bipolar neurons of the medial superior olivary nucleus (MSO). They found no differences in dendritic extensions on either side of MSO neurons and symmetrical (balanced) dendritic growth in binaurally deprived young rats compared to control animals.

5.3 Environmental Deprivation

Examinations of brain stem nuclei in mice experiencing environmental sound deprivation imposed in the neonatal period have demonstrated smaller neurons (soma lengths, cross-sectional areas) in the VCN, and the

LSO and MNTB of the SOC, and fewer neurons (i.e., lower packing density) in the DCN (Webster and Webster, 1977; 1979; Evans, Webster, and Cullen, 1983). These anatomical findings from environmental deprivation were not reversible by later normal auditory experiences (Webster and Webster 1979).

5.4 Monaural Cochlear Deprivation

Monaural cochlear hearing loss imposed in ferrets resulted in a reduction in the volume of the whole CN in ferrets; greater for the AVCN ipsilateral to the lesioned cochlea (Moore and Kowalchuk 1988). Discrete cochlear damage imposed without mechanical damage to other auditory structures (i.e., cochlear loss only without a confounding conductive overlay) also results in a change in the chick nucleus magnocellularis (NM; i.e., the avian equivalent of the mammalian AVCN) manifested as a reduction in volume of the nucleus rather than a reduction in cell size or number of neurons (Tucci and Rubel 1985).

Unilateral cochlear lesion in the neonatal period results in a significant increase in the number of CN neurons projecting to the ipsilateral IC—the unlesioned side (Moore and Kowalchuk 1988). Generally, about 50 times as many CN neurons project to the contralateral than ipsilateral IC of normal hearing ferrets. Observed changes in brainstem connectivity in the case of monaural deprivation might reflect an alteration in either the level or balance of neuronal activity between the two ears. A competition for synaptic space on IC neurons may be created by ipsilateral deprivation and a "rewiring" of the binaural auditory pathway (Moore and Kowalchuk 1988).

McMullen and coworkers (1988) examined the distribution of cell types in the contralateral auditory cortex of rabbits deafened neonatally in one ear. They found the distribution to be similar to that of control animals. There was no significant increase in the number of dendrites or in the total number of dendritic branches, although cross-sectional cell areas were reduced. These workers observed increased dendritic branch lengths in experimental versus control animals, however, describing these dendritic branches as "wandering". They speculated that this type of auditory pathway reorganization was the result of reduced afferent input to auditory cortex contralateral to the deafened ear. While binaural input results in competitive interaction, monaural deprivation causes dendrites to grow unchecked, or axons to seek to fill areas lacking input due to the peripheral deprivation (McMullen et al. 1988).

5.5 Monaural Conductive Deprivation

While Webster (1983) reported small neuron sizes in the spiral ganglion of mouse ears that were conductively impaired in the neonatal period, Moore and his colleagues (Moore et al. 1989) reported normal cochleas and spiral ganglion neurons in ferrets similarly deprived during the same period. The difference in these findings may be the result of methods used to impose the conductive impairment rather than species-specific differences (see below).

Beyond the cochlea, monaural conductive deprivation imposed in the neonatal period has resulted in reports of smaller neurons in the AVCN ipsilateral to the impaired ear with no effect seen in the same structure on the opposite nonimpaired side. No difference in neuron size was seen in the DCN of rats. Effects were reported to be greater in rats impaired prior to the onset of hearing versus those deprived after the onset of hearing (Coleman and O'Connor 1979).

Similar findings were demonstrated in chicks where neuron size in NM was significantly smaller on the deprived side. Chicks were studied at several ages post hearing onset. NM neurons appeared to grow normally initially and then "leveled off" after a particular time period. No compensatory hypertrophy was found in neurons in NM on the nondeprived side (Conlee and Parks 1981). In rats, an effect of age of onset was demonstrated when spherical cells within the AVCN were examined. The effect of time of deprivation was evident as were regional differences: cells within the dorsal AVCN were more affected than those in medial and ventral areas. Each area of AVCN is a tonotopically organized frequency region (Blatchley, William, and Coleman 1983).

In mice, morphological changes in all cells of VCN and in one cell type of DCN on the conductively deprived side were reported. The volume of the VCN ipsilateral to the side of the deprivation was found to be significantly smaller than VCN on the nondeprived side (Webster 1983). In SOC, size asymmetries were noted between the LSO and MNTB that received afferent input from the deprived VCN versus those same structures receiving input primarily from the nondeprived ear. The MSO that received input from both VCN (from the deprived and nondeprived side) was not significantly different. Webster (1983) also demonstrated that IC neurons receiving input from the deprived contralateral VCN were smaller than IC neurons that received contralateral input from the nondeprived VCN.

Contradictory findings (in another species) have been reported by Moore and colleagues (1989). These researchers examined the effect of auditory deprivation in ferrets who had received monaural deprivation (conductive, frequency specific early hearing loss). Electrophysiological recordings were completed with the earplug removed after three to fifteen months of deprivation. Assessment of pure-tone hearing sensitivity was examined via neuronal recordings from the contralateral IC. Conductive hearing loss was documented for the majority of the deprivation period. After the animals were sacrificed, their cochleas were reportedly normal and neurons within the spiral ganglion were normal in terms of morphology and number. Moore and colleagues found the volume of CN in ferrets and size of neuron units in the AVCN were similar when the conductively deprived and nondeprived sides were compared (Moore et al. 1989). They suggested that previous findings of anatomical changes in CN in conductively deprived ears may have been influenced by the presence of occult cochlear pathology.

Trune and Morgan (1988a) reported fewer auditory nerve terminals in the AVCN of deprived mice. Feng and Rogowski (1980) found increased neuron branching of neurons in the MSO of rats that received afferent input from the nondeprived ear versus the MSO from the deprived side. Similarly, in chicks, Smith, Gray and Rubel (1983) found dendrites of cells in nucleus laminaris (i.e., the avian equivalent of the mammalian MSO) that received afferent input from the deprived ear were longer that dendrites of cells from the nondeprived side. These findings were for low-frequency regions of the nucleus. Conversely, deprived dendrites were shorter in nuclear regions responsive to high-frequency input.

Trune and Morgan (1988b) examined neuronal metabolism (by examination of the cytoplasm of the cell) in mice that received early monaural conductive deprivation. They found that cells were larger in low- versus high-frequency regions of the CN. This size disparity was reportedly due to reduction in the cytoplasm of cells receiving input from the deprived ear. The nuclei of neurons in the CN of deprived and nondeprived ears were no different. These authors suggest that normal metabolic function requires adequate stimulation; deprivation results in reduced metabolic events and thus, a reduced amount of cytoplasm. Trune and Morgan (1988a) also examined intracellular mitochondria in AVCN and found them smaller and darker in deprived neurons; therefore, appearing less active metabolically.

Fukushima, White, and Harrison (1990) examined hair cell damage in one portion of the basilar membrane of chinchillas after bilateral acoustic trauma using scanning electron microscopy. One ear of the animal was conductively impaired by ossiculectomy after the presence of bilateral noise trauma was confirmed using compound action potential (CAP) audiograms. Results demonstrated that the ears that experienced conductive deprivation during varied recovery periods demonstrated greater hair cell damage than in the nondeprived cochlea. These authors suggest that auditory stimulation may serve to inhibit any degenerative process after acoustic trauma or to support the "repair" of damaged hair cells.

6. Auditory Deprivation in Humans

The study of auditory deprivation in humans has focused on the consequences of hearing loss in children and adults; specifically, unilateral and bilateral congenital and acquired conductive and cochlear (conventionally termed sensorineural) hearing impairments, as well as, to a lesser degree, the influence of auditory experience (i.e., the quality and quantity of the aural language environment). Notable parallels often exist between the studies of human and animal responses to auditory deprivation demonstrated in both behavioral and, more recently, electrophysiological investigations. It is tempting to directly apply the results of the animal work previously reviewed to humans. Numerous reasons suggest that while the results of animal studies serve as valuable models to further our understanding of auditory deprivation in humans, the species-specific consequences of hearing loss on function must be examined directly.

6.1 Binaural Cochlear Hearing Loss — Early Onset

The communication consequences of bilateral congenital deafness of severe to profound degree have been well documented, literally, for centuries (Ruben and Rapin 1980). Severe and profound congenital cochlear hearing loss deprives the individual of essentially all exposure to the acoustic cues of spoken language input and results in the lack of development of aural/oral communication unless remediation through amplification and specific auditory/speech-language training is provided. Moreover, there is a relationship between congenital severe-to-profound hearing loss and below average academic and vocational outcomes (Boothroyd 1982).

Normal hearing during development is a prerequisite for the emergence of oral communication (Skinner 1978; Boothroyd 1982; Oller and Eilers 1988). The intelligibility of spoken language is considerably enhanced in children who experience a period of normal hearing before the onset of deafness (Boothroyd 1993). Speech quality is usually markedly affected by the lack of both a normal input model and an auditory feedback loop (Boothroyd 1982).

The critical nature of early auditory experiences on aural/oral language abilities has been further supported by recent findings in children who have received cochlear implants. Cochlear implantation has provided an intervention alternative for some children with profound hearing impairments who do not benefit from conventional acoustic amplification (Boothroyd 1993). Numerous examples exist of the measured benefit of cochlear implant use in children with acquired (e.g., due to meningitis) profound hearing loss who have had some experience with normal auditory cues (have established an "auditory memory") versus those with congenital deafness (Boothroyd 1993; Tyler 1993). Moreover, it appears that children receiving their cochlear implant soon after the onset of their deafness (the current minimal allowable age that a child may be implanted is twenty-four months) rather than after a prolonged period without auditory cues may be better users of the device. However, long-term follow-up of larger groups of children is necessary to clarify this issue (Tyler 1993).

Congenital mild and moderate hearing loss does not totally preclude speech and language development since some speech cues are available to the child even without amplification. However, depending upon the degree and configuration of the individual deficit, varying effects on the aural/oral language are usually evident beginning early in life (Oller and Eilers 1988).

Older children with congenital bilateral cochlear hearing loss of mild to moderate degree also experience speech–language deficits, difficulties in listening in competitive background noise, and, on average, lower academic achievement and poorer social–emotional development than peers with normal hearing (Davis et al. 1986; Crandell 1993). Similar degrees of mild impairment would not necessarily cause such significant impact in an older child or adult with hearing loss acquired after the full emergence of language and speech (Bess 1985).

Regardless of the degree of hearing loss, it appears that the age at which early intervention (e.g., provision of hearing aids and therapeutic programs) is initiated is related to speech and language outcome (Levitt, McGarr, and Geffner 1987). Children who received special education before the age of three years had better communication outcomes than those who began receiving remediation at older ages.

Therefore, advocates of infant and preschool hearing screening programs support the early identification of cochlear hearing loss and the initiation of remediation strategies as soon as the loss is detected, preferably within the first year of life (Joint Committee on Infant Hearing 1990 Position Statement; U.S. Department of Health and Human Services, 1990; National Institutes of Health Consensus Statement 1993). Such proponents maintain that a strong relationship exists between the age of detection of the hearing loss, the initiation of intervention, and later communication competency. Currently, while clinical experience supports such a notion, little research evidence exists for the delineation of a critical or sensitive period for audition in early childhood. Direct empiric studies of this issue, however, may be precluded by ethical constraints.

6.2 Binaural Cochlear Hearing Loss — Late Onset

Psychoacoustic studies have demonstrated that adult listeners with acquired cochlear hearing loss have poor frequency resolution, reduced temporal processing and a restricted dynamic range (Humes 1982). Numerous studies have demonstrated that speech perception is adversely affected by late-onset cochlear hearing loss (e.g., Bilger and Wang 1976). Listeners with cochlear hearing loss experience particular problems in understanding speech in background noise (Dirks, Morgan, and Dubno 1982). This difficulty is of particular consequence to the optimum utilization of hearing aids (Fabry 1991).

In particular, the acquired hearing loss associated with senescence (presbycusis) may be accompanied by central auditory deficits that may exacerbate the communication deficits usually associated with a particular degree of peripheral hearing loss (Jerger, et al. 1989).

Recent interest in the effects of auditory deprivation in the mature auditory system has arisen as a result of the seminal work of Silman, Gelfand, Silverman and their colleagues (Silman, Gelfand, and Silverman

1984; Gelfand, Silman, and Ross 1987; Silverman and Silman 1990; Silman et al. 1992; Gelfand and Silman 1993). In a series of reports, these researchers demonstrated the existence of what they consider "late-onset auditory deprivation" in adults and older children fit with a single hearing aid in the presence of bilaterally symmetrical cochlear hearing loss. After a period of hearing aid use, speech recognition scores (percent correct monosyllabic word identification) for aided and unaided ears were compared to performance scores for the same ears obtained prior to hearing aid fitting. The time elapsed between tests was frequently years in duration. The coworkers reported speech understanding deteriorated below pre-fitting performance in the nonamplified ear, but was maintained, or in some cases improved, in the aided ear over pre-fit scores. Moreover, when a hearing aid was provided to the previously unfit ear, the speech recognition score improved to equal that of the aided ear. Hood (1984, 1990) has also supported the existence of late-onset auditory deprivation in individuals with cochlear hearing loss.

Gatehouse (1989, 1992), however, has questioned whether these observations represent late-onset auditory deprivation or merely an adaptation effect. That is, in the case of monaural amplification users, the long-term aided ear adapts to the supra-threshold speech spectrum provided by the hearing aid, while the unaided ear does not. Gatehouse has suggested that when the unaided ear is fit with a hearing aid, after a period of "acclimatization" or experience with amplified speech, performance scores in the previously unaided ear would be expected to equal that of the long-term aided ear. The length of such an acclimatization period remains to be delineated.

Case reports of the sometimes deleterious consequence of providing amplification to both ears have emerged recently (Jerger et al. 1993). Reportedly, in some elderly individuals, the provision of binaural amplification resulted in a negative outcome. Instead of the binaural enhancement usually derived from two hearing aids, the input from both ears actually interfered with these listeners' speech perception abilities. Thus, in some older adults, the provision of binaural cues following long-standing monaural hearing aid use may result in a significant reduction in speech understanding, mediated by central binaural processing deficits (Jerger et al. 1993).

6.3 Monaural Cochlear Hearing Loss

Research suggests that early onset unilateral hearing loss places children at risk for auditory, communication, and academic sequelae. An extensive case control study has examined the effects of long-standing unilateral sensorineural hearing loss in school-aged children (Bess 1986a). Prior to this investigation, children with unilateral hearing loss were not considered disabled, as long as hearing remained normal in the unaffected ear

(Northern and Downs 1974). However, Bess (1986a) found that a high proportion (one-third) of the sample of children with unilateral hearing loss studied had repeated a grade in school and/or required resource help. Interestingly, children with unilateral loss affecting the right ear had poorer test outcomes than children who had left ear impairments, and those with lesser degrees of hearing loss did better than those with severe and profound impairments.

A subsequent study by Oyler, Oyler, and Matkin (1987) confirmed the findings of Bess (1986a). These workers also demonstrated that unilateral cochlear hearing loss should be considered a risk factor for academic, language, auditory, and behavioral problems in some children.

6.4 Conductive Hearing Loss—Early Onset

Conductive hearing loss in children may result from congenital craniofacial malformations, but most commonly occurs secondary to otitis media with effusion (OME). While the consequences of the moderate to moderate-severe hearing loss resulting from bilateral aural atresia (complete absence or closure of the external ear canal) are undisputed, the question of whether developmental sequelae result from OME is still under debate.

At issue is whether OME should be considered a form of auditory deprivation, since the hearing loss associated with the condition is often temporary, fluctuant, and is generally considered "mild" in degree. Most often cases of OME occur bilaterally, but the disease may fluctuate, causing monaural and asymmetric binaural impairments. Several characteristics of OME make it both difficult to study, diagnose, treat, and manage. To date, such fluctuating hearing loss has not been examined in an animal model.

6.4.1 Otitis Media with Effusion

Otitis media (OM: inflammation of the middle ear) is a common and fairly ubiquitous condition of early childhood with the prevalence of otitis media greatest in the first three years of life (Teele, Klein, and Rosner 1980). Numerous endogenous and exogenous factors are thought to place children at risk for repeated OM episodes (Todd 1986). The hearing levels associated with otitis media range from no threshold elevation to moderate degrees of impairment (Fria, Cantekin, and Eichler 1985; Bess 1986b; Gravel 1989). While the configuration of the hearing loss is typically characterized as equal (flat) across the speech–frequency range, varying audiometric patterns may occur. Moreover, the degree, configuration, and symmetry of the hearing loss may vary within as well as between episodes.

Models of the influence of OME on development are predicated on the hypothesis that restricted auditory input during early life results in disordered emerging communication (receptive and expressive language), higher order (central) auditory processing deficits, attentional deficits, behavioral problems, and, ultimately, below average academic achievement (see for example, Needleman 1977; Feagans 1986; Roberts et al. 1986, 1989). Proponents of these models suggest that the continual instability (fluctuation) of auditory input during development is detrimental to language development (Skinner 1978; Menyuk 1986) and that mild hearing loss in infants is particularly detrimental to speech discrimination (Nozza 1988). Others argue, however, that early studies (see for example, Gottlieb, Zinkus, and Thompson 1979; Zinkus and Gottlieb 1980; Brandes and Ehinger 1981; Sak and Ruben 1981) that have related OM to developmental deficits have suffered from numerous methodological shortcomings, in particular the use of retrospective research designs (Ventry 1980; Paradise 1981).

Some recent prospective investigations, designed to control for previous design weaknesses, have demonstrated a relationship between early persistent otitis media and communication development in infants and preschool children. Expressive and/or receptive language, speech production, behavior, and attentional deficits have been demonstrated in some children with OM histories before school age (Teele et al. 1984; Pearce et al. 1988; Wallace et al. 1988a, 1988b; Friel-Patti and Finitzo 1990). Moreover, higher-order auditory abilities, specifically, understanding speech in a background of competition, have been found to be adversely influenced by an early history of OM (Jerger et al. 1983; Gravel and Wallace 1992). Other prospective studies, however, have failed to find differences in the communication skills of preschoolers with and without early histories of otitis media (Roberts et al. 1986, 1988; Wright et al. 1988).

The question remains unanswered as to whether any early deficiencies persist into later life. Several studies that have followed children in the longer term suggest that as a group, early OM-positive children do not experience language deficits later in childhood (Roberts et al. 1986, 1988, 1989). Others report language, speech, and academic problems at school age in children with early, persistant OM histories (Klein et al. 1988) and auditory-based behavioral/attentional deficits in children with early otitis media histories compared to their OM-free peers (Feagans et al. 1987; Roberts et al. 1989). Haggard and colleagues found that listening difficulties are reported by college-aged students who related having extensive OM histories (Haggard et al. 1993).

Of significance is the recent report that the effects of OM on child language may be mediated by the language environment of the home. Empiric evidence suggests that the quality and quantity of the language to which normally hearing children are exposed mediates their language competency (Peterson and Sherrod 1982; Huttenlocker et al. 1991). In a prospective investigation, Wallace and her colleagues examined the consequences of maternal language on the language of two groups of two-year-old children with and without first-year otitis media histories (Wallace et al. 1993). Within the early OM-positive subgroup, children whose caregiver used abundant language to seek and provide information had better language outcomes than children whose caregiver provided a less-than-enriched language input.

6.4.1.1 Psychoacoustic Studies of OME.

A particularly useful psychoacoustic technique is the measurement of the masking level difference (MLD) — an indicator of binaural auditory brainstem processing (Hall and Grose 1993a). In brief, thresholds for a signal presented binaurally are obtained in the presence of a noise masker. In the first condition, the signal (S) and masker (N) are presented in phase to both ears (referred to as the SoNo condition). A second masked threshold is then obtained most commonly with the signal to one ear delivered 180 degrees out of phase relative to the signal presented to the opposite ear; the masker remains in phase to both (the SπNo condition). In the SπNo condition, the listener usually experiences a release from masking, that is, the individual's threshold for the masked signal is lower (better) than the threshold obtained for the SoNo condition. The disparate interaural phase relationship of the signal provides the auditory system with a facilitating binaural cue. The MLD is determined by subtracting the SπNo threshold from the SoNo threshold. Normally, the MLD increases with age (Nozza, Wagner, and Crandell 1988).

Pillsbury, Grose, and Hall (1991) demonstrated reduced MLDs in children just prior to surgical intervention for prolonged histories of OME in comparison to MLDs obtained from children of the same age. While reduced MLDs could have been the result of presurgical conductive hearing loss, many children continued to demonstrated lower-than-normal MLDs after surgical restoration of their hearing. In the longer term, children returned to expected levels, presumably demonstrating their incorporation of normal binaural auditory cues with experience. Similarly, Moore and colleagues demonstrated smaller MLDs for children with reported histories of persistent early otitis media versus children without such backgrounds (Moore, Hutchings, and Meyer 1991).

Morrongiello (1989) studied the immediate effect of unilateral acute otitis media on localization abilities of infants. The infants' horizontal plane localization abilities were adversely affected by the presence of unilateral OM; following resolution, localization returned to normal.

6.4.1.2 Electrophysiological Studies of OME.

Several reports have found brainstem abnormalities as indicated through ABR recordings in some children with OME histories. Table 4.2 specifies the ABR findings of these studies while Table 4.3 suggests the brainstem structures currently considered to underlie the components of the electrophysiological response.

Only the report of Gunnarson and Finitzo (1991) examined groups of children whose OM histories had been documented prospectively from

TABLE 4.2 Studies examining the ABR in children with histories of otitis media.

Authors	Design	Result
Dobie and Berlin 1979	Retrospective Single case study	Abnormal BI
Folsom, Weber, and Thompson 1983	Retrospective 6 to 11-year olds	Inc. absolute lat.: III & V Inc. interpeak lat.: I-III III-V
Lenhardt, Shaia, and Abedi 1985	Retrospective two case reports	Inc. absolute lat: III & V Inc. interpeak lat.: I-III
Anteby, Hafner, Pratt, and Uri 1986	Retrospective + current cases of OME 4 to 12-year-olds 205 ears	Inc. interpeak lat.: I-III III-V I-V
Gunnarson and Finitzo 1991	Prospective 5 to 7-year-olds 27 children: 3 groups	Inc. absolute lat: III & V Inc. interpeak lat: I-III I-V Abnormal BI
Hall and Grose 1993a	5 to 9-year-olds followed 4-6 months prior to surgery (myringotomy and tubes) 13 controls; 14 with OME histories	Inc. absolute lat. III & V Inc. interpeak lat: I-III I-V

Inc. = Increased; lat = latency; BI = binaural interaction. l

birth. These authors examined absolute, inter-wave, and inter-aural wave latencies, and binaural interaction in a group of children who had been followed from birth. These authors found increased absolute and inter-wave intervals and reduced binaural interaction as measured by the ABR, in children with histories of OM and concomitant mild hearing loss (threshold sensitivity estimated by the click ABR on several occasions in early life). Dobie and Berlin (1979) were the first to report a case of absent binaural interaction in an individual with an extensive early OM history.

Recently, Hall, and Grose (1993a) demonstrated a small but significant relationship between the MLD and interaural ABR wave asymmetries in children with histories of otitis media. Their work demonstrated abnormal brainstem processing, utilizing both psychoacoustic and electrophysiological measures in the same children with early otitis media histories. The functional consequences of these findings remains to be determined.

6.5 Conductive Hearing Loss—Late Onset

Florentine (1976) demonstrated the effect of unilateral conductive hearing loss on the binaural system using adult volunteers who utilized a monaural

TABLE 4.3 Delineation of the human ABR waves (I-V) and their currently acceepted anatomical basis

Human ABR Wave	Presumed Anatomic Bases
I	Distal VIIIth nerve fibers leaving cochlea and entering IAC
II	Proximal VIIIth nerve entering the BS (first-order neurons)
III	Cochlear nucleus (CN) (second-order neurons)
IV	Primarily superior olivary complex (SOC) with contributions from CN and NLL (pontine third-order neurons)
V	Inferior colliculus (IC)

(Based on Hall 1992).
Note: Except for Waves I and II, remaining peaks (III, IV, and V) likely have multiple generators (Hall 1992).

ear plug for approximately one week. Late-onset monaurally impaired listeners adapted to the new acoustic cues and were able to lateralize and or localize sound accurately with the ear plug in place. Interestingly, after plug removal, the period of time required to re-learn normal binaural cues was more prolonged than was the time required to learn the previously novel monaural cues.

Hall and Derlacki (1986) reported reduced MLDs in adults with surgically corrected cases of conductive hearing loss. The reduced MLDs persisted two to three months after the restoration of normal hearing. Hall and Grose (1993b) examined adults with surgically restored hearing who had experienced monaural conductive hearing loss for extended periods secondary to otosclerosis. These patients presumably had all experienced normal binaural cues during development prior to the onset of stapes fixation. These authors found that MLDs were abnormally low prior to surgery and when tested one month after surgery in all but one of the subjects studied. At follow-up one year after surgery, MLDs had essentially returned to expected values for all but two subjects. Hall and Grose (1993b) suggest the parsimonious explanation is that adaptation occurred over the one year period with listeners "re-learning" the normal binaural auditory cues.

7. Research Questions

Numerous issues remain to be addressed in auditory deprivation. The issues surrounding adaptation and plasticity of the auditory system are important for our understanding of the duration of tolerable deprivation in the developing, mature, and aging auditory system. The outcome of such research will support and guide hearing screening initiatives, amplification fitting strategies, and the initiation of medical and surgical intervention.

If a critical or sensitive period exists in humans, it must be delineated as to time course (developmental period of onset) and the degree and configuration of loss that will have detrimental consequences. Moreover, the determination and characterization of any synergistic interaction between age of onset, duration, and degree of deficit is critical.

The effects of monaural versus binaural deprivation and the consequences that such losses have on the functional capabilities of the organism require further study.

The levels of the auditory system that respond to particular types of deprivation must be further explored. For example, neurons at the level of the brainstem may evidence little plasticity once deprivation has been imposed (i.e., their morphology or physiology may be irreversibly altered). However, plastic cortical auditory areas that reorganize in response to deprivation, may with learning, be capable of functional compensation (Weinberger and Diamond 1987). Such information will be critical in determining the timing and duration of rehabilitation efforts.

The question of the existence of late-onset deprivation is important for the justification of rehabilitation initiatives in adults, as well as children with hearing loss. Moreover, if young children are initially fit with one rather than two hearing aids, is auditory processing irreversibly affected even with later binaural experience?

Importantly, we must also understand the role of the environment on outcome. That is, can enriched language experiences serve to overcome some of the effects of auditory deprivation? It is apparent in humans that environment influences outcome; children with identical audiograms frequently function differently as a consequence of their auditory/language experiences.

Answers to these questions will have a profound impact on the conduct of research and clinical practice, as well as public and private health care policy formulation in the decades to come.

Acknowledgements. Work was supported by NIDCD Center Grant 2P50 DC00223. We are indebted to Maria Murino for her assistance in preparation of this manuscript.

References

Anniko M, Sjostrom B, Webster D (1989) The effects of auditory deprivation on morphological maturation of the ventral cochlear nucleus. Arch Otorhinolaryngol 246:43-47.

Anteby I, Hafner H, Pratt H, Uri N (1986) Auditory brainstem evoked potentials in evaluating the central effects of middle ear effusion. Int J Pediatr Otorhinolaryngol 12:1-11.

Bess, FH (1985) The minimally hearing-impaired child. Ear Hear 6:43-47.

Bess FH (1986a) The unilaterally hearing-impaired child: A final comment. Ear Hear 7:52-54.

Bess FH (1986b) Audiometric approaches used in the identification of middle ear disease in children. In: J Kavanagh (ed) Otitis Media and Child Development. Parkton MD: York Press, pp. 70-82.

Bilger RC, Wang M (1976) Consonant confusions in patients with sensorineural hearing loss. J Speech Hear Res 19:718-748.

Blatchley BJ, Williams JE, Coleman JR (1983) Age dependent effects of acoustic deprivation on spherical cells of the rat anteroventral. Cochlear nucleus. Exp Neurol 80:81-93.

Boothroyd A (1982) Hearing Impairments in Young Children. Englewood Cliffs, NJ: Prentice-Hall.

Boothroyd A (1993) Profound deafness. In: R Tyler (ed) Cochlear Implants. San Diego CA: Singular, pp 1-34.

Brandes PJ, Ehinger DM (1981) The effects of early middle ear pathology on auditory perception and academic achievement. J Speech Hear Dis 46:301-307.

Clopton BM (1980) Neurophysiology of auditory deprivation. Birth defects. 16:271-288.

Clopton BM, Silverman MS (1978) Changes in latency and duration of neural responding following developmental auditory deprivation. Exp Brain Res 32:39-47.

Coleman JR, O'Connor P (1979) Effects of monaural and binaural sound deprivation on cell development in the anteroventral cochlear nucleus of rats. Exp Neurol 64:553-566.

Conlee JW, Parks TN (1981) Age-and position-dependent effects of monaural acoustic deprivation in nucleus magnocellularis of the chicken. J Comp Neurol 202:373-384.

Crandell CC (1993) Speech recognition in noise by children with minimal degrees of sensorineural hearing loss. Ear Hear 14:210-216.

Davis JM, Elfenbein J, Schum R, Bentler R (1986) Effects of mild and moderate hearing impairments on language, educational and psychosocial behavior. J Speech Hear Dis 51:53-62.

Dirks DD, Morgan DE, Dubno JR (1982) A procedure for quantifying the effects of noise on speech recognition. J Speech Hear Dis 47:114-123.

Dobie RA, Berlin CI (1979) Influence of otitis media on hearing and development. Ann Otol Rhinol Laryngol (Suppl) 88:48-53.

Doyle WJ, Webster DB (1991) Neonatal conductive hearing loss does not compromise brainstem auditory function and structure in rhesus monkeys. Hear Res 54:145-151.

Evans WJ, Webster DB, Cullen JK (1983) Auditory brainstem responses in neonatally sound deprived CBA/J mice. Hear Res 10:269-277.

Fabry DA (1991) Programmable and automatic noise reduction in existing hearing aids. In: Studebaker G, Bess, F, Beck L (eds) The Vanderbilt Hearing Aid Report II. Parkton MD: York Press, pp. 65-78.

Feagans L (1986) Otitis media: A model for long-term effects with implications for intervention. In: Kavanaugh J (ed) Otitis Media and Child Development. Parkton MD: York Press, pp. 139-159.

Feagans L, Sanyal M, Henderson F, Collier A, Appelbaum M (1987) Relationship of middle ear disease in early childhood to later narrative and attention skills. J Pediatr Psych 12:581-594.

Feng AS, Rogowski BA (1980) Effects of monaural and binaural occlusions on the morphology of neurons in the medial superior olivary nucleus of the rat. Brain Res 189:530-534.

Fleckeisen CE, Harrison RV, Mount RJ (1991) Effects of total cochlear hair cell loss on integrity of cochlear nucleus. A quantitative study. Acta Otolaryngol (Stockh) Suppl 489:23-31.

Florentine M (1976) Relation between lateralization and loudness in asymmetrical hearing losses. J Am Audiol Soc 1:243-251.

Folsom RC, Weber BA, Thompson G 1983) Auditory brainstem responses in children with early recurrent middle ear disease. Ann Otol Rhinol Laryngol 92:249-253.

Fria TJ, Cantekin EI, Eichler J (1985) Hearing acuity of children with effusion. Arch Otolaryngol 111:10-16.

Friel-Patti S, Finitzo T (1990) Language learning: A prospective study of otitis media with effusion in the first two years of life. J Speech Hear Res 33:188-194.

Fukushima N, White P, Harrison RV (1990) Influence of acoustic deprivation on recovery of hair cells after acoustic trauma. Hear Res 50:107-118.

Gatehouse S (1989) Apparent auditory deprivation effects of late onset: The role of presentation level. J Acoust Soc Am 86:2103-2106.

Gatehouse S (1992) The time course and magnitude of perceptual acclimatization to frequency responses: Evidence from monaural fitting of hearing aids. J Acoust Soc Am 92:1258-1268.

Gauron EF, Becker WC (1959) The effects of early sensory deprivation on adult rat behaviour under competitive stress. J Comp Physio Pschol 52:322-328.

Gelfand SA, Silman S (1993) Apparent auditory deprivation in children: Implications of monaural versus binaural amplification. J Am Acad Audiol 4:313-318.

Gelfand SA, Silman S, Ross L (1987) Long-term effects of monaural, binaural and no amplification in subjects with bilateral hearing loss. Scan Audiol 16:201-207.

Gottlieb G (1975a) Development of species identification in ducklings: I. Nature of perceptual deficit caused by embryonic auditory deprivation. J Comp Physiol Psychol 89:387-399.

Gottlieb G (1975b) Development of species identification in ducklings: II. Experimental prevention of perceptual deficit caused by embryonic auditory deprivation. J Comp Physiol Psychol 89:675-684.

Gottlieb G (1975c) Development of species identification in ducklings: III. Maturational rectification of perceptual deficit caused by auditory deprivation. J Comp Physiol Psychol 89:899-912.

Gottlieb G (1978) Development of species-specific perception caused by auditory deprivation. J Comp Physiol Psychol 92:375-387.

Gottlieb MI, Zinkus PW, Thompson A (1979) Chronic middle ear disease and auditory perceptual deficits. Clin Pediatr 18:725-732.

Gravel JS (1989) Behavioral assessment of auditory function. Sem Hear 10:216-228.

Gravel JS, Wallace IF (1992) Listening and language at 4 years of age: Effects of early otitis media. J Speech Hear Res 35:588-595.

Gunnarson AD, Finitzo T (1991) Conductive hearing loss during infancy: Effects on later auditory brainstem electrophysiology. J Speech Hear Res 34:1207-1215.

Haggard MP, Lim MJ, Smith D, Fantini DA (1993) Long-term OME sequelae in binaural hearing. In: Lim D, Bluestone C, Klein J, Nelson J, Ogra P (eds) Recent Advances in Otitis Media. Toronto: Decker, pp. 546-549.

Hall JW (1992) Handbook of Auditory Evoked Responses. Boston MA: Allyn and Bacon.

Hall JW, Derlacki EL (1986) Effect of conductive hearing loss and middle ear surgery on binaural hearing. Ann Otol Rhinol Laryngol 95:525-530.

Hall JW, Grose JH (1993a) The effect of otitis media with effusion on the masking level difference and the auditory brainstem response. J Speech Hear Res 36:210-217.

Hall JW, Grose JH (1993b) Short-term and long-term effects on the masking level difference following middle ear surgery. J Am Acad Audiol 4:307-312.

Hood JJ (1984) Speech discrimination in bilateral and unilateral hearing loss due to Meniere's disease. Br J Audiol 18:173-177.

Hood JJ (1990) Problems in central binaural integration in hearing loss cases. Hear Instr 41:6, 8, 11, 56.

Humes LE (1982) Spectral and temporal resolution by the hearing impaired. In: Studebaker G, Bess F (eds) The Vanderbilt Hearing Aid Report. Upper Darby PA: Monographs in Contemporary Audiology pp. 16-31.

Huttenlocker J, Haight W, Bryh A, Seltzer M, Lyont T (1991) Early vocabulary growth: Relation to language input and gender. Devel Psych 27:236-248.

Jerger J, Silman S, Lew HL, Chmiel R (1993) Case studies in binaural interference: converging evidence from behavioral and electrophysiologic measures J Am Acad Audiol 4:122-131.

Jerger J, Jerger S, Oliver T, Pirozzolo F (1989) Speech understanding in the elderly. Ear Hear 12:103-109.

Jerger S, Jerger J, Alford BR, Abrams S (1983) Development of speech intelligibility in children with recurrent otitis media. Ear Hear 4:138-145.

Joint Committee on Infant Hearing (1991) 1990 Position Statement. ASHA 5:3-6.

Kerr LM, Ostapoff EM, Rubel EW (1979) The influence of acoustic experience on the ontogeny of frequency generalization gradients in the chicken. J Exp Psych 5:97-115.

Klein J, Chase C, Teele D, Menyuk P, Rosner B et al. (1988) Otitis media and the development of speech, language, and cognitive abilities at seven years of age. In Lim D et al. (eds) Recent Advances in Otitis Media. Toronto: BC Decker, pp. 396-397.

Knudsen EI (1985) Experience alters the spatial tuning of auditory units in the optic tectum during a sensitive period in the barn owl. J Neurosci 5:3094-3109.

Knudsen EI, Esterly SD, Knudsen PF (1984) Monaural occlusion alters sound localization during a sensitive period in the barn owl. J Neurosci 4:1001-1011.

Knudsen EI, Knudsen PF, Esterly SD (1984) A critical period for the recovery of sound localization accuracy following monaural occlusion in the barn owl. J Neurosci 4:1012-1020.

Kraus N, McGee T (1992) Electrophysiology of the human auditory system. In: Popper AN, Fay RR (eds) The Mammalian Auditory Pathway: Neurophysiology. NY: Springer-Verlag, pp. 335-403.

Lenhardt M, Shaia FT, Abedi E (1985) Brainstem evoked response waveform variation associated with recurrent otitis media. Arch Otolaryngol 111:315-316.

Levitt H, McGarr NS, Geffner D (1987) Development of language and communication skills in hearing impaired children. ASHA Monograph 26. Washington DC: American Speech-Language-Hearing Assoc.

Marchant D, Shurin P, Turczyk V, Waiskowski D, Tutihasi M, Kinney S (1984) Course and outcome of otitis media in early infancy: A prospective study. J Pediat 104:826-831.

McMullen NT, Goldberger B, Suter CM, Glaster EM (1988) Neonatal deafening alters nonpyramidal dendrite orientation in auditory cortex: a computer microscope study in the rabbit. J Comp Neurol 267:92-106.

Menyuk P (1986) Predicting speech and language problems with persistent otitis media. In: Kavanagh J F (ed) Otitis Media and Child Development. Parkton, MD: York Press, pp. 192-208.

Moore DR (1985) Postnatal development of the mammalian central auditory system and the neural consequences of auditory deprivation. Acta Otolaryngol (Stockh.) Suppl 421:19-30.

Moore DR (1990a) Auditory brainstem of the ferret: Early cessation of developmental sensitivity of neurons in the cochlear nucleus to removal of the cochlea. J Comp Neurol 302:810-823.

Moore DR (1990b) Auditory brainstem of the ferret: Bilateral cochlear lesions in infancy do not affect the number of neurons projecting from the cochlear nucleus to the inferior colliculus. Devel Brain Res 54:125-130.

Moore DR (1993) Plasticity of binaural hearing and some possible mechanisms following late-onset deprivation. J Am Acad Audiol 4:277-283.

Moore DR, Kowalchuk NE (1988) Auditory brainstem of the ferret: Effects of unilateral cochlear lesions on cochlear nucleus volume and projections to the inferior colliculus. J Comp Neurol 272:503-515.

Moore DR, Hutchings ME, King AJ, Kowalchuk NE (1989) Auditory brainstem of the ferret: Some effects of rearing with a unilateral earplug on the cochlea, cochlear nucleus, and projections to the inferior colliculus. J Neurosci 9:1213-1222.

Moore DR, Hutchings ME, Meyer SE (1991) Binaural masking level differences in children with a history of otitis media. Audiology 30:91-101.

Morrongiello BA (1989) Infants' monaural localization of sounds: Effects of unilateral ear infection. J Acoust Soc Am 86:597-602.

National Institutes of Health (1993) Early identification of hearing impairment in infants and young children: Consensus Development Conference Statement. Bethesda, MD.

Needleman H (1977) Effects of hearing loss from early recurrent otitis media on speech and language development. In: Jaffe B (ed) Hearing Loss in Children. Baltimore: University Park Press, pp. 640-649.

Northern JL, Downs MP (1974) Hearing In Children. Baltimore: Williams & Wilkins.

Nozza R (1988) Auditory deficit in infants with otitis media with effusion: More than a "mild" hearing loss. In Lim D, Bluestone C, Klein J, Nelson J: (eds) Recent Advances in Otitis Media. Toronto: BC Decker, pp. 376-379.

Nozza RJ, Wagner EF, Crandall MA (1988) Binaural release from masking for speech sound in infants, preschoolers, and adults. J Speech Hear Res 31:212-218.

Oller DK, Eilers R (1988) The role of audition in infant babbling. Child Devel 59:441-449.

Oyler RF, Oyler AL, Matkin ND (1987) Warning: A unilateral hearing loss may be detrimental to a child's academic career. Hear J 40(9):18, 22.

Paradise J (1981) Otitis media during early life: How hazardous to development? A critical review of the evidence. Pediatr 68:869-873.

Patchett RF (1977) Auditory frequency discrimination in patterned sound-naive albino rats. Percept Mot Skills 44:127-136.

Pearce PS, Saunders MA, Creighton DE, Sauve RS (1988) Hearing and verbal-cognitive abilities in high-risk pre-term infants prone to otitis media with effusion. Devel Behav Pediatr 9:346-351.

Peterson GA, Sherrod KB (1982) Relationship of maternal language to language development and language delay in children. Am J Mental Def 86:391–398.

Pillsbury HC, Grose JH, Hall JW (1991) Otitis media with effusion in children: Binaural hearing before and after corrective surgery. Arch Otolaryngol 117:718–723.

Roberts JE, Sanyal MA, Burchinal MA, Collier AM, Ramey CT, Henderson FW (1986) Otitis media in early childhood and its relationship to later verbal and academic performance. Pediatr 78:423–430.

Roberts J, Burchinal M, Koch M, Footo M, Henderson F (1988) Otitis media in early childhood and its relationship to later phonological development. J Speech Hear Dis 53:416–424.

Roberts JE, Burchinal MR, Collier AM, Ramey CT, Koch MA, Henderson FW (1989) Otitis media in early childhood and cognitive, academic, and classroom performance of the school-aged child. Pediatr 83:477–485.

Ruben RJ, Rapin I (1980) Plasticity of the developing auditory system. Ann Otol Rhinol Laryngol 89:303–311.

Sak RJ, Ruben RJ (1981) Recurrent middle ear effusion in childhood: Implications of temporary auditory deprivation for language and learning. Ann Otol Rhinol Laryngol 90:546–551.

Silman S, Gelfand SA, Silverman CA (1984) Late-onset auditory deprivation: Effects of monaural versus binaural hearing aids. J Acoust Soc Am 76:1357–1362.

Silman S, Silverman CA, Emmer MB, Gelfand SA (1992) Adult onset auditory deprivation. J Am Acad Audiol 3:390–396.

Silverman MS, Clopton BM (1977) Plasticity of binaural interaction. I. Effect of early auditory deprivation. J Neurophysiol 40:1266–1274.

Silverman CA, Silman S (1990) Apparent auditory deprivation from monaural amplification and recovery with binaural amplification: Two case studies. J Am Acad Audiol 1:175–180.

Skinner M (1978) The hearing of speech during language acquisition. Otolaryngol Clin N Am 11:631–650.

Smith Z, Gray L, Rubel E (1983) Afferent influences on brainstem auditory nuclei of the chicken: N. laminaris dendritic length following monaural conductive hearing loss. J Comp Neurol 220:199–205.

Sohmer H, Friedman I (1992) Prolonged conductive hearing loss in rat pups causes shorter brainstem transmission time. Hear Res 61:189–196.

Strasser S, Dixon AK (1986). Effects of visual and acoustic deprivation on agonistic behaviour of the albino mouse. (M. musculus L.) Physiol Behav 36:773–778.

Teele DW, Klein JO, Rosner BA (1980) Epidemiology of otitis media in children. Ann Otol Rhinol Laryngol 89:5–6.

Teele DW, Klein JO, Rosner BA and the Greater Boston Collaborative (1984) Otitis media with effusion during the first three years of life and development of speech and language. Pediatr 74:282–287.

Tees RC (1967a) Effects of early auditory restriction in the rat on adult pattern discrimination. J Comp Physiol Psychol 63:389–393.

Tees RC (1967b) The effects of early auditory restriction in the rat on adult duration discrimination. J Aud Res 7:195–207.

Todd NW (1986) High risk populations for otitis media. In: Kavanagh J (ed) Otitis Media and Child Development. Parkton MD: York Press, pp 52–59.

Trahiotis C (1992) Developmental considerations in binaural hearing experiments. In: Werner LA, Rubel EW (eds) Developmental Psychoacoustics. Washington

DC: American Psychological Association, pp. 281-292.

Trune DR, Morgan CR (1988a) Influence in developmental auditory deprivation on neuronal ultrastructure in the mouse anteroventral cochlear nucleus. Brain Res 470:304-308.

Trune DR, Morgan CR (1988b) Stimulation-dependent development of neuronal cytoplasm in mouse cochlear nucleus. Hear Res 33:141-149.

Tucci DL, Rubel EW (1985) Afferent influences on brainstem auditory nuclei of the chicken: Effects of conductive and sensorineural hearing loss on n. magnocellularis. J Comp Neurol 238:371-381.

Tyler RS (1993) Speech perception by children. In: Tyler R (ed) Cochlear Implants. San Diego CA: Singular, pp. 191-256.

U.S. Department of Health and Human Services (1990) Healthy people 2000: National health promotion and disease prevention objective. Washington DC: Public Health Service.

Ventry I (1980) Effects of conductive hearing loss: Fact or fiction? J Speech Hear Dis 45:143-156.

Wallace IF, Gravel JS, Ganon EC, Ruben RJ (1993) Two-year language outcomes as a function of otitis media and parental linguistic styles. In: Lim D, Bluestone C, Klein J, Nelson J, Ogra P (eds) Recent Advances in Otitis Media. Toronto: Decker, pp. 527-530.

Wallace IF, Gravel JS, McCarton CM, Ruben RJ (1988a) Otitis media and language development at one-year of age. J Speech Hear Dis 53:245-251.

Wallace IF, Gravel JS, McCarton CM, Stapells DR, Bernstein RS, Ruben, RJ (1988b) Otitis media auditory sensitivity and language outcomes at 1-year. Laryngoscope 98:64-70.

Webster DB (1983) Auditory neuronal sizes after a unilateral conductive hearing loss. Exp Neurol 79:130-140.

Webster DB, Webster M (1977) Neonatal sound deprivation affects brain stem auditory nuclei. Arch Otolaryngol 103:392-396.

Webster DB, Webster M (1979) Effects of neonatal conductive hearing loss on brain stem auditory nuclei. Ann Otol 88:684-688.

Weinberger NM, Diamond DM (1987) Physiological plasticity in auditory cortex: Rapid induction by learning. Prog Neurobiol 29:1-55.

Wolf A (1943) The dynamics of selective inhibition of specific function in neurosis. Psychosom Med 5:27-38.

Wright PE, Sell SH, McConnell KB, Sitton AB, Thompson J, Vaughn WK, Bess FH (1988) Impact of recurrent otitis media on middle ear function, hearing and language. J Pediatr 113:581-587.

Zinkus PW, Gottlieb MI (1980) Patterns of perceptual and academic deficits related to early chronic otitis media. Pediatr 66:246-253.

5
Ototoxicity: Of Mice and Men

SUSAN L. GARETZ AND JOCHEN SCHACHT

1. Introduction

1.1 Scope of the Review

Clinically used drugs are well-known causes of ototoxicity, defined as the property of exerting a harmful effect on the end organs of hearing and balance. Additional noxious influences are becoming increasingly important in the modern world, among them industrial chemicals and solvents (Table 5.1) and the exposure to traumatic levels of noise. For example, presbycusis, or hearing loss with advancing age, may be in part the sum of lifelong exposure of our sensory systems to toxic insults (Hawkins 1973a).

This review will focus on drugs in clinical use whose ototoxic effects have been well established in both humans and animals and whose mechanisms of action have been investigated. Descriptions of other miscellaneous ototoxic agents can be found in previous reviews (Hawkins 1976; Kisiel and Bobbin 1981; Spandow, Anniko, and Møller 1988; Huang and Schacht 1989; Hoeffding and Fechter 1991).

TABLE 5.1 Selected ototoxic agents.

Therapeutic or Chemical Class	Representative	Effect on Auditory Perception
Antipyretic, analgesic	Salicylates (Aspirin)	TTS, tinnitus
Antimalarial	Quinine, chloroquine	TTS, PTS, tinnitus
Loop diuretics	Ethacrynic acid, furosemide	TTS
Antineoplastic	Nitrogen mustards, vinblastine, cisplatin	PTS
Antibacterial	Aminoglycosides, peptide-antibiotics	PTS
Heavy metals	Arsenicals, mercurials	PTS
Chemical agents	Organotins, butyl nitrite, toluene, potassium bromate	TTS, PTS
Noise		TTS, tinnitus, PTS

PTS = permanent threshold shift; TTS = temporary threshold shift.

1.2 Ototoxic Drugs

Adverse effects on the auditory or vestibular system are associated with several widely used classes of drugs. Some ototoxic agents are merely of historic interest (e.g., the anthelmintic Oil of Chenopodium, ascaridole) or are in limited use as research tools (e.g., mercurials and arsenicals). Many, however, continue to play a critical role in the treatment of serious or life-threatening disease (aminoglycosides, cisplatin) or provide therapeutic gains that by far outweigh the risk of therapy (loop diuretics, salicylates). In addition, some of these highly effective and low-cost drugs are freely prescribed without adequate monitoring in developing countries, exacerbating the problems associated with their side effects.

Ototoxicity is a property of chemically diverse compounds (Table 5.1), and there is no correlation between the molecular structure and the ototoxic potential of different classes of drugs and, thus, no predictability. While members of a given group of drugs, for example, the chemically closely related aminoglycoside antibiotics, may share a similar spectrum of toxicity, qualitative and quantitative differences still exist in their toxic actions. Therefore, it is not surprising that individual patterns of pathology and mechanisms of action are associated with each drug. The targets of damage may be auditory or vestibular hair cells or secretory tissues such as the stria vascularis. For reasons yet to be established, ototoxic drugs frequently are also nephrotoxic. Examples include polypeptide and aminoglycoside antibiotics, antineoplastics, and heavy metals. Renal toxicity may complicate drug therapy since reduced renal excretion of a drug elevates its serum levels and may enhance its side effects.

Symptoms of ototoxicity vary widely in both scope and magnitude. Auditory impairments can include tinnitus or hearing loss, ranging in severity from barely perceptible temporary threshold shifts to profound bilateral deafness. Likewise, vestibular disturbances can range from mild ataxia to incapacitating vertigo. These symptoms and their underlying mechanisms will be described when we consider individual drugs. First, however, we will briefly introduce animal models and measures of auditory and vestibular deficits upon which most of our knowledge of ototoxicity is based. This is not an exhaustive discussion of these aspects of toxicity research; rather, we would like to alert the reader to the possibilities and limitations of experimental designs.

2. Models for the Study of Ototoxicity

2.1 General Considerations

Data concerning the prevalence and magnitude of drug-induced ototoxicity in humans comes largely from retrospective analyses. This is a problematic approach for several reasons. First, patients may be quite ill and their physiological state may affect the degree of drug-associated ototoxicity. It is virtually impossible to control for variables such as age, sex, genetic susceptibility, nutritional status, noise exposure, or preexisting auditory or vestibular dysfunction. Furthermore, this patient population is frequently being treated for other conditions, making it difficult to assign the primary cause of adverse effects. Second, the debilitated state of some patients may preclude the testing necessary to establish the presence or extent of inner ear dysfunction. Prospective studies can avoid some of these complications. The often narrow therapeutic margin between toxic and therapeutic levels of many medications, and the often permanent nature of the toxic effects, make it essential to test potentially ototoxic drugs in animal models.

Animal models allow for control of age, developmental stage, sex, dosage administered, concurrent exposure to other drugs or noise, genetic makeup, and other factors that influence the pharmacokinetics or ototoxic potential of a drug. Experimental and clinical data often correlate in details of type, extent, and relative ototoxicities of agents tested. Animal studies involving interactions of multiple drugs or of drugs with environmental factors show marked similarities to effects seen with humans under similar conditions. Other advantages of using animal models include the ability to study teratogenic potential and the immediate availability of tissue for histological analysis. Most importantly, perhaps, only animal models permit the rational investigation of the mechanisms involved in insults to the auditory and vestibular systems. Ultimately, this information can enable more accurate testing for ototoxic effects of drugs, the development of safer drugs to replace those currently in use, and ways of protecting patients from the often devastating consequences of drug-induced ototoxicity.

2.2 Animal Models

Several factors must be weighed when choosing an animal model for the ototoxic effects of drugs. Chief among these are similarity of ototoxic responses to those seen in humans, anatomical factors, developmental considerations, ease in handling, and cost to obtain and maintain the animals.

2.2.1 Primates

Intuitively, primates would seem an obvious choice for the study of ototoxicity. Because of their ontogenetic closeness to humans, one might expect similar responses to pharmacological manipulations. While generally true, this is not always the case, as we will discuss later. In addition, the encasement in the petrous bone makes their inner ears difficult to study. Primates are also expensive, increasingly difficult to obtain, and prone to a large variety of illnesses. Despite these numerous drawbacks, the intellectual capabilities of these animals makes them ideal candidates for a variety of behaviorally based studies.

2.2.2 Guinea Pigs and Chinchillas

One of the most commonly used animals for studies of ototoxicity is the guinea pig. The guinea pig cochlea is easily accessible as it lies within a large air space inside the bulla instead of being encased in dense bone. This greatly simplifies placing electrodes within the cochlea, perfusing the fluid spaces directly, or obtaining samples of cochlear fluids or tissues. Moreover, guinea pigs develop auditory damage in response to drugs in a manner resembling that seen in humans. Chinchillas possess many of these same characteristics, although they are more sensitive to noise-induced hearing loss.

Guinea pigs are not well suited for investigations into dysfunction of the vestibular system. They do not readily display outward manifestations of balance disturbances, even when the inner ear is destroyed (Brummett and Fox 1982); however, reliable measures of vestibular function in these animals can be obtained by electronystagmography.

2.2.3 Rats and Mice

Rats and mice are easy to handle, inexpensive, and readily available. In contrast to many other mammals, the onset of cochlear function does not occur until the second postnatal week (Lenoir, Marat, and Uziel 1983), making rodents especially useful in the study of the relation between developmental influences and ototoxicity. Furthermore, rodents lend themselves to the study of ototoxic changes with aging since populations of aged rats or mice are easier to obtain than those of most other animals. The availability of many varieties of genetically homogeneous inbred mice may also prove useful in clarifying the issue of genetic susceptibility to drug-induced ototoxicity.

2.2.4 Cats and Rabbits

Cats provide a good model system in which to test the effects of ototoxic agents on the vestibular system, as they are primarily dependent on it for control of balance. Even mild vestibular derangements are manifested by readily apparent ataxia. Electronystagmography can be used as an additional measure of vestibular damage (Hawkins 1973b). Rabbits are similar to cats in many of these respects.

2.2.5 Other Animal Models

The lateral line organ of fishes and frogs and the avian lagena have all served as models to study specific aspects of ototoxicity. The tissues are easily accessible to pharmacological manipulations, electrophysiological testing, and morphological assessment. Extrapolation from these organs to the mammalian inner ear may be limited. On the other hand, avian models add a unique perspective on trauma and repair because their auditory hair cells can regenerate after ototoxic insults (Corwin and Cotanche 1988; Rubel, Oesterle, and Weisleder 1991). Recent evidence, however, suggests that mammalian hair cells, at least in the vestibular system, may have similar regenerative capacities (Forge et al. 1993; Warchol et al. 1993).

2.3 In Vitro Systems

Alternatives to in vivo animal experimentation are organ and cell cultures or cell-free systems. Organ cultures of different developmental stages can be studied under controlled environmental conditions, and the effects of drugs on development and morphology of the various cell types can be observed. Both early gestational (otocyst) and later stage embryonic ears will survive for several days in vitro (Anniko, Takada, and Schacht 1982). Short-term cultures of the postnatal cochlea (Richardson and Russell 1991) or isolated outer hair cells from mature animals (Dulon et al. 1989; Huang and Schacht 1990) also represent sensitive model systems to assess ototoxic influences. Such in vitro cultures not only allow for controlled drug application but also for direct access to morphological, biochemical, and physiological parameters. Thus, these preparations lend themselves to investigations of the extent of drug-induced ototoxicity, as well as mechanisms involved in cell death.

Cell-free systems are useful tools for the study of drug action at the biochemical and molecular level. Such biochemical or physicochemical systems serve two important purposes in a rational strategy of therapy: elucidation of the basic mechanisms of drug action or side effects and development of improved drugs or cotherapy to improve efficacy and minimize adverse actions. Before a cell-free system becomes a viable tool, some insight into potential mechanisms should be deduced from in vivo or in situ studies. Conversely, any results obtained from cell-free systems need to be transposed and verified in the intact system. As far as mechanisms of ototoxicity are concerned, cell-free model systems have been applied primarily to the study of aminoglycoside actions where they have yielded rational hypotheses of adverse drug action and potential therapeutic intervention (Wang et al. 1984; Schacht and Weiner 1986).

3. Pharmacokinetics and Drug Administration

3.1 Pharmacokinetics

The actions of a drug—both therapeutic and adverse—are largely determined by the concentration achieved at the target site and the duration of exposure. Most routes of drug administration allow limited control over these parameters as only the administered dose is known. In systemic application, for example, renal handling, hepatic metabolism, and uptake into body tissues affect concentrations of drugs in the plasma and inner ear fluids and tissues. While plasma levels can be monitored conveniently and adjusted by changing the dosage, concentrations in the inner ear are less amenable to measurement. The latter concentrations, however, are important for an understanding of drug action on cochlear and vestibular processes. Extrapolations from the injected amount, or the plasma levels, to concentrations in the inner ear are not possible as different drugs have different rates of penetration into and clearance from the inner ear. Therefore, concentrations in the tissues of the inner ear need to be analyzed for an exact assessment of drug pharmacokinetics. Alternatively, nonsystemic routes of drug administration may avoid some of the problems associated with a systemic dosing regimen.

3.2 Routes of Drug Administration

3.2.1 Systemic Application

Despite their limitations, the most common routes of drug application are systemic because of the relative ease of intramuscular, intraperitoneal, subcutaneous, or intravenous injection. The route selected depends on the absorptive properties of a drug and its potential side effects. The pharmacokinetics achieved also differ significantly and range from a slow rise of plasma concentrations after intramuscular administration to a rapid jump after bolus delivery by intravenous injection. Implantable osmotic pumps are an alternative to injections if sustained chronic administration of a drug is desired. Systemic applications also subject the drugs to potential metabolism by the liver, which can be important in the expression of toxicity or in detoxication. First-pass metabolism by the liver can be avoided if drugs are injected directly into the jugular vein.

Oral administration in food or water generally should not be considered a method of choice because the amount of drug consumed is not under direct experimental control. However, oral dosing is useful if the purpose is to mimic treatment in patients and if drug serum levels are monitored. Gavage by gastric tube controls the amount of medication administered but may stress the animal.

3.2.2 Topical Applications

This technique is one of several that allow targeting the inner ear specifically. Through topical applications, drugs can be applied to the tympanic membrane, to the middle ear space, or to the round window. Application to the tympanic membrane can also be combined with experimental conditions of relevance to clinical problems, for example, the presence of tympanic membrane perforations, tympanotomy tubes, or altered permeability of the tympanic membrane secondary to infection. It should be noted, however, that the kinetics of drug penetration and the concentrations on-site are frequently difficult to determine.

3.2.3 Cochlear Perfusion

In this experimental technique, drugs are infused directly into the perilymph or endolymph. Cochlear perfusion bypasses sites of systemic drug metabolism and gives direct control of time course and concentrations of the investigated agents in the inner ear fluids. Perfusions, however, may result in a different tissue distribution of the drugs as in the physiological case where the inner ear is reached by way of the vascular system. Nevertheless, perfusion studies combined with electrophysiological measurements are the most direct experimental measure of the ototoxic potential of different classes of drugs. As a form of drug administration, however, perfusions lack a clinical parallel and therefore remain a research method.

3.2.4 Arterial Perfusion

Pharmaceuticals can also be distributed into the inner ear via the vascular system. The basilar artery and its branch, the anterior inferior cerebellar artery, supply the cochlea. Although extensively used in the study of blood flow regulation, this technically difficult approach is rarely used for the study of ototoxic drugs.

4. Evaluation of Ototoxicity

A variety of qualitative and quantitative measures are available to test auditory performance. These measures not only provide a convenient armamentarium for the monitoring of toxicity in acute and chronic experiments but also aid in defining the site of damage. Appropriately selected and interpreted, specific tests can assess the integrity of individual components of the auditory system and can provide information from hair cell function to the integrated response of the trained and behaving animal.

While important in understanding the evaluation of ototoxicity, we will only provide a summary of experimental procedures and their theoretical basis. A detailed discussion would go beyond the scope of this review. The interested reader is referred to Dallos (1973; cochlear potentials), Moore (1983; auditory brain stem evoked responses), Fay (1988; comparative evaluation of methods), and Lonsbury-Martin (Chapter 6; otoacoustic emissions) for further information.

4.1 Cochlear Homeostasis

Blood flow provides oxygen and nutrients necessary for normal function of all components of the inner ear. It is a potential target for both ototoxic and therapeutic drugs and can be quantitated by several indirect means. These include the partitioning of circulating microspheres into the vessels of the inner ear, laser-Doppler flowmetry, and intravital microscopy.

Endocochlear (or endolymphatic) potential (EP) is measured in the scala media and represents the DC polarization of endolymph (compared to perilymph). It provides quantitative information about the physiological state of the cochlea and reflects the functionality of homeostatic mechanisms in the inner ear. Since the endocochlear potential is primarily maintained by stria vascularis, it is the measure of choice to assess the function of this tissue.

4.2 Auditory Function

Cochlear microphonic potential (CM) is an evoked potential and the electrical equivalent of the applied acoustic stimulus. It is believed to originate primarily from the outer hair cells.

Summating potential (SP) is an evoked DC response. It originates from the hyperpolarizing and depolarizing DC receptor potentials of cochlear hair cells.

Otoacoustic distortion products can be measured in the ear canal in a noninvasive manner and are, therefore, suitable for chronic studies. The origin and interpretation of this response is still debated, but there is evidence that it relies on the integrity of outer hair cells.

Cochlear whole nerve action potential (N1 or compound action potential, CAP) is another way to assess peripheral processing in a rather noninvasive manner. It depends on both hair cell transduction and synaptic activity and is the result of the synchronous action potentials of primary auditory afferents.

Single unit recordings assess the activity of individual neurons of the cochlear and vestibular nerves and their corresponding pathways in the brainstem.

Evoked auditory brainstem responses are electrical field potentials measured noninvasively by scalp electrodes. Their individual components reflect synchronous neuronal activity along the auditory pathways in response to sound stimuli.

Preyer reflex (or "pinna reflex") is a flicking response of the pinna in response to loud sound. It is a rather crude behavioral measure in rodents because it is observer dependent, difficult to quantify, and insensitive to subtle insults. It is noninvasive, however, and may serve as a rough indicator of auditory function or dysfunction.

Behavioral testing involves long-term training of animals using operant and other conditioning methods to obtain responses to specific auditory stimuli. Behavioral measurements of auditory thresholds reflect the entire peripheral auditory system as well as central processing. As a unique advantage, behavioral tasks can also include specific questions of auditory processing, such as pitch and loudness discrimination, which are difficult to address with electrophysiological measures.

4.3 Vestibular Function

Electronystagmography is an indirect measure of vestibular function. Hair cells of the ampullar cristae of the semicircular canals are stimulated and ultimately cause movement of the extraocular muscles via a central reflex arc. The electrophysiological measure of this nystagmus is influenced by many factors, such as the alertness of the animal and changes in the corneal-retinal potential (e.g., by exposure of the eyes to light).

Observational or behavioral tests rely on an animal's innate ability to correct for disturbances of its equilibrium. Some vestibular deficits are expressed in the unstimulated animal by tremor, difficulty in head or body righting, ataxia, circling, or waltzing. In addition, the normal response to a vestibular stimulus may be absent or diminished. This may be seen in the absence of nystagmus in response to rotation, impairment of righting reflexes, or the inability to maintain equilibrium in response to a vestibular challenge.

4.4 Morphology

Morphological studies assess the deviation from normal morphology in the inner ear after ototoxic trauma. Classic histological methods, such as surface preparations, can determine the integrity of tissues, in particular the presence or absence of hair cells in the organ of Corti. Transmission and scanning electron microscopy are needed to detect subtle effects at the ultrastructural level. Morphological evaluations achieve specificity for specific cell components when they are combined with cytochemical or fluorescent techniques.

Isolated hair cells have recently been introduced as models to study acute ototoxic actions. Noxious influences on cell viability can be assessed by morphological criteria (Zajic and Schacht 1987) or with viability dyes (Dulon et al. 1989). An advantage of this procedure is the potential for quick screening of drug effects on a defined cell population. As with in vitro studies, a correlation must eventually be drawn between these observations and the expression of ototoxic trauma in vivo.

5. Individual Drugs

Observations in humans and studies in animals have provided us with a wealth of information about potentially ototoxic drugs, their mechanisms of action, the relative ototoxicity of drugs within given classes, drug/drug interactions, and the role of exogenous variables on the extent of auditory or vestibular destruction. Knowledge of how drugs are metabolized, what determines their affinity for inner ear, renal and other tissues, how they are transported into cells, and how they influence intracellular or molecular events is systematically being gleaned. We will now focus on a description of the ototoxic effects of individual drugs based on both in vivo and in vitro studies.

Ototoxic drugs can be classified into several broad groups by the nature of their effects and their sites and mechanisms of action. There are agents that primarily produce permanent damage to the sensory cells (e.g., aminoglycoside antibiotics and cisplatin). Secondly, there are agents that may transiently affect hair cell function and other physiological parameters (e.g., salicylates and quinine). Thirdly, there are agents that primarily appear to affect stria vascularis and the maintenance of homeostatic processes such as the endolymphatic potential (e.g., diuretics and arsenicals). Such classifications are not always clear-cut since occasional cases of permanent hearing loss have been reported for drugs normally causing reversible damage. Nevertheless, for obvious reasons, drugs that consistently produce deficits of a predominantly permanent nature, such as the aminoglycosides and cisplatin, are of special clinical concern.

5.2 Aminoglycoside Antibiotics

5.2.1 Therapeutic Considerations

The aminoglycosides are a large group of antibiotics used primarily to treat serious enterococcal and gram-negative infections. Since their introduction in the 1940s after Waksman's discovery of streptomycin, a series of structurally related aminoglycosides have entered clinical practice: these include kanamycin, gentamicin, amikacin, tobramycin, netilmicin, and, outside the U.S.A., dibekacin and sisomicin. Aminoglycosides were the first effective drugs against tuberculosis but are no longer considered a first line treatment for that purpose. They are also employed infrequently in the treatment of uncommon diseases such as tularemia and brucellosis. As clinical tools they can be used to reduce vestibular responsiveness in patients with Ménière's syndrome. Their antibacterial efficacy is based on a not yet completely resolved multistep mechanism of damage to bacterial cell integrity and protein synthesis (Davis 1987).

5.2.2 Clinical Observations

The primary adverse effects of aminoglycoside antibiotics are nephrotoxicity and ototoxicity. While renal dysfunction is reversible, the auditory and vestibular effects can be permanent. Symptoms include vestibular dysfunc tion, aural fullness, tinnitus, and irreversible hearing loss. The hearing loss generally begins in the base of the cochlea, affecting high frequencies, and can proceed apically to the speech range. Hearing loss may even progress after cessation of therapy (Harrison 1954; Erlandson and Lundgren 1964; Ballantyne 1973). A few observations suggest that limited reversibility of hearing loss is possible, particularly if therapy is stopped immediately after the first indications of auditory damage. Monitoring of auditory function, therefore, is not only important in order to prevent further damage but also to allow for potential recovery of hearing function.

The type, incidence, and extent of ototoxicity varies with the individual aminoglycosides. The incidence of ototoxicity can be as high as 18% for gentamicin and as low as 2% for netilmicin (Fee 1980; Matz 1986). Furthermore, some aminoglycosides affect primarily the auditory system (neomycin, kanamycin, amikacin), whereas others are more prone to cause vestibular dysfunction (dibekacin) or to damage both organs to a significant extent (tobramycin, gentamicin). A striking example of selective ototoxicity is the pair streptomycin/dihydrostreptomycin. Although structurally differing by only the substitution of an aldehyde by an alcohol group, the former is preferentially vestibulotoxic, while the latter is more likely to injure the cochlea.

5.2.3 Pathology

The organ of Corti is considered the primary target of aminoglycoside damage in humans, with the outer hair cells the first and most severely affected. Hair cell damage is initially seen in the basal turns of the cochlea and then proceeds apically. Other cochlear structures including inner hair cells may remain normal (Keene et al. 1982). However, indicative of complex injury processes triggered by these drugs, degeneration of supporting cells, nerve fibers, and ganglion cells, as well as changes in stria vascularis, have been reported. It is possible that these alterations are not primary effects of the drugs but represent later changes following initial destruction of hair cells (Benitez, Schuknecht, and Bradenburg 1962; Johnsson and Hawkins 1972; Huizing and de-Groot 1987). In the case of vestibulotoxic aminoglycosides, damage in the vestibular apparatus includes loss of sensory cells and vacuolization in the cristae ampullaris and the maculae of the utricle and saccule. Type I hair cells of the cristae ampullaris appear to be affected before type II hair cells, while supporting cells remain intact. Degenerative changes in the utricle and saccule follow those in the crista (Wersäll and Hawkins 1962; Lindeman 1969; Hawkins 1973b; Johnsson et al. 1981; Keene et al. 1982).

The histopathology of animals treated with aminoglycosides parallels that seen in humans. In the guinea pig cochlea, the outer hair cells consistently are an early site of damage. Loss of these cells starts in the innermost row of the basal turns and progresses to the outer rows.

Subsequently, apical outer hair cells and inner hair cells starting from the base may be destroyed (Hawkins and Lurie 1952; Hawkins, Johnsson, and Aran 1969; Forge 1985). On the subcellular level, vesiculations of the mitochondrial cristae, clumping of nuclear chromatin, nuclear swelling, and loss of ribosomes occur shortly after treatment. Morphological alterations also include changes in the structure of the plasma membrane (Kaneko, Nakogawa, and Tanaka 1970; Theopold 1977; Forge 1985).

While destruction of outer hair cells figures prominently in experimental aminoglycoside toxicity, damage to the stria vascularis frequently accompanies the effects on the hair cells. Shrinkage of marginal cells and changes in their plasma membrane appear to be early and permanent structural effects (Forge and Fradis 1985; Forge, Wright, and Davies 1987). This underscores the complexity of drug action and may imply multiple sites and mechanisms involved in the final expression of toxicity.

These same changes are seen in virtually all mammalian species (cats: McCormick et al. 1985; rabbits: Hodges et al. 1985; squirrel monkeys: Igarashi 1973), as well as other animals evolutionarily more distinct from humans. The basilar papilla of lizards is susceptible to aminoglycoside damage (Bagger-Sjöbäck and Wersäll 1978), as are the hair cells of the basilar papilla (Cruz, Lambert, and Rubel 1987) and the dark cells of the ampulla (Park and Cohen 1982) of the chick. There seems to be one notable exception, however. Macaque monkeys are sensitive to the ototoxic effects of kanamycin and neomycin (Stebbins et al. 1969), but their auditory systems remain virtually unaffected by prolonged dihydrostreptomycin treatment. In contrast, patas monkeys show severe hair cell degeneration after dihydrostreptomycin (Hawkins et al. 1977). This unique intraspecies variation in susceptibility has yet to be explained.

The pathophysiological effects of aminoglycosides can also be induced by means other than systemic administration. Topical application of neomycin-containing solutions to the tympanic membranes of chinchillas with tympanostomy tubes destroys all cochlear hair cells and induces strial damage and degeneration of the vestibular apparatus (Wright and Meyerhoff 1984). Neomycin, when applied to the round window, very rapidly damages the cochlea, most noticeably the outer hair cells of the lower turns (Harada et al. 1986). Single transtympanic injections of aminoglycosides can also cause morphological changes to the sensory cells of both the auditory and vestibular systems of guinea pigs (Bareggi et al. 1990).

The effect of aminoglycosides on tissue cultures varies somewhat with the developmental age of the preparation. Normal embryogenesis continues for several days when inner ears are explanted late in development and exposed to aminoglycosides (Anniko and Nordemar 1979). In contrast, explanted inner ears of 16th day gestational mice begin to show morphological derangements after one day of exposure to the drug. Changes include inhibition of hair cell development with occasional fusing and complete destruction of these cells. Cytoplasmic swelling, degeneration of organelles, and disintegration of the nucleus occurs as the damage progresses, as well as severe damage to the epithelium of the cristae (Nordemar and Anniko 1983; Anniko, Takada, and Schacht 1982). In otocysts from earlier stages of development, low doses of aminoglycosides produce a more generalized dysmorphogenesis (Anniko 1981). Cochlear cultures from postnatal mice are also susceptible to aminoglycoside damage, which includes changes in the stiffness of their stereocilia (Kössl, Richardson, and Russell 1990) and membrane blebbing in the apical surfaces of the hair cells (Richardson and Russell 1991). In marked contrast to exposure in vivo or in culture, isolated cochlear outer hair cells do not die or show morphological changes upon exposure to high concentrations of gentamicin (Dulon et al. 1989; Williams, Zenner, and Schacht 1987). This apparent paradox points to the importance of drug/tissue interactions in the expression of aminoglycoside toxicity (see 'Mechanism of Action', Section 5.2.4) and illustrates both the usefulness and limitations of model systems.

The pattern of auditory pathophysiology sustained after aminoglycoside dosing in animals is quite similar to that seen in humans. Consistent with morphological observations, permanent threshold shifts or the absence of Preyer reflexes are documented for a variety of species (McGee and Olszewski 1962; Alleva and Balasz 1978; Brummett et al. 1978; Arpini et al. 1979). Behavioral audiometry has also shown progressive high, frequency hearing loss in macaque monkeys treated with kanamycin and neomycin and in patas monkeys treated with dihydrostreptomycin (Hawkins et al. 1977). Due to the destruction of hair cells, alteration of the CM and N1 occur first in the higher frequencies and later in the lower frequencies (Hawkins 1950; Davis et al. 1958; Logen et al. 1974; Brummett et al. 1978; Arpini et al. 1979). Thresholds of evoked potentials are decreased (Schwent, Williston, and Jewett 1980; Shepard and Clark 1985) and changes in acoustic distortion products appear shortly after drug administration (Brown, McDowell, and Forge 1989). Perilymphatic or endolymphatic infusion of aminoglycosides also decreases CM (Nuttall, Marques, and Lawrence 1977; Konishi 1979; Anniko, Takada, and Schacht 1982). In addition, endolymphatic perfusion of neomycin, kanamycin, or dihydrostreptomycin affects the endolymphatic potential (Konishi 1979).

Several lines of evidence indicate at least two distinctly different, possibly sequential, effects of aminoglycosides on hair cell systems. When aminoglycosides are applied locally (by perfusion or in culture), their initial actions are reversible, but after increasing drug concentrations or exposure times, irreversible damage sets in. This has been observed in both the fish lateral line organ and the mammalian cochlea, where suppression of the microphonic potential is initially reversible by solutions of high ionic strength or high calcium (Wersäll and Flock 1964; Takada and Schacht 1982). This effect may be due to a reversible block of transduction channels, which has been documented for streptomycin in the hair cells of the bullfrog saccule (Kroese, Das, and Hudspeth 1989) and for neomycin in cultures of the early postnatal mouse cochlea (Kössl, Richardson, and Russell 1990). In the latter system the reversible block of transduction channels is followed by irreversible changes in stiffness of the stereocilia bundle and hair cell morphology. The demonstration of dual effects again emphasizes that the actions of aminoglycosides are not limited to a singular site but probably encompass several sequential or perhaps concomitant mechanisms.

Vestibular disorders follow aminoglycoside administration in a variety of mammals including cats (Wersäll and Hawkins 1962; Duvall and Wersäll 1964; Waitz, Moss, and Weinstein 1971; McCormick et al. 1985), guinea pigs (Duvall and Wersäll 1964), rats (Alleva and Balasz 1978), mice (Caussé, Gondet, and Vallancien 1948), squirrel monkeys (Igarashi et al. 1971; Igarashi 1973), and chicks (Park and Cohen 1982). The short latency vestibular-evoked response and the vestibulocular reflex are markedly affected (Pettorossi et al. 1986; Elidan, Lin, and Honrubia 1987).

5.2.4 Mechanism of Action

The site of pathological damage by aminoglycosides, as well as the electrophysiological and perceptual consequences, have been known for decades. In contrast, the cellular and molecular mechanisms underlying the irreversible hair cell loss are only beginning to be unraveled. This is not due to a lack of data on biochemical effects of aminoglycosides but rather due to their abundance. At least in vitro, aminoglycosides affect the metabolism of proteins, lipids, and carbohydrates and can interfere with a plethora of other reactions. The difficulty has been to link such effects causally to the ototoxic actions of the drugs. An early appealing hypothesis was that these drugs accumulate in inner ear tissues to levels higher than in serum or other tissues, which then would account for their site-specific toxicity (Stupp 1970). This concept, however, has been invalidated by subsequent studies showing a lack of true accumulation and a lack of correlation between tissue concentrations and the ototoxic potential of several aminoglycoside antibiotics (reviewed by Henley and Schacht 1988).

Based on combinations of in vitro and in vivo studies, several extra- and intracellular reactions have been closely associated with the ototoxic effects of aminoglycosides (Schacht 1986). The acute and reversible actions of the drugs can be explained by a calcium antagonism at the external cell membrane and a block of transduction channels or voltage-gated calcium channels of hair cells (Dulon et al. 1989). Phosphoinositide lipids, which play an important role in the generation of second messengers, maintenance of normal cell physiology, and outer hair cell motility (Schacht and Zenner 1987), are a specific intracellular target (Stockhorst and Schacht 1977; Schacht 1979; Schacht 1986). Metabolism of these lipids is inhibited by aminoglycosides in inner ear tissues, and, furthermore, binding of aminoglycosides to phosphoinositides alters the structure and permeability of membranes (Schacht and Weiner 1986). Either one of these actions on membrane permeability or second messenger production could severely compromise cellular homeostasis and function.

These actions, however, cannot explain all features of chronic aminoglycoside ototoxicity. Most puzzling in the progression of aminoglycoside ototoxicity is the dissociation of the presence of the drug in the inner ear and its effects on function. Within hours after systemic injection, gentamicin enters cochlear tissues (Tran Ba Huy, Bernard, and Schacht 1986) and can be detected in hair cells (Hiel et al. 1992). Electrophysiological or morphological evidence of toxicity, however, is delayed for days or weeks. The paradox that cell structure and function can remain unaffected by the drug is most strikingly seen with isolated outer hair cells (Dulon et al. 1989). While being the primary targets of aminoglycosides in vivo, outer hair cells remain viable in the presence of gentamicin in vitro. These phenomena can be explained if gentamicin itself is not the toxic drug species but requires a metabolic conversion before its toxicity is expressed on hair cells. The possible existence of toxic metabolites was first inferred after it was shown that gentamicin toxicity in cochlear perfusions required an active metabolic process (Takada, Bledsoe, and Schacht 1985). Another argument for the existence of a metabolic process was seen in the potential involvement of glutathione in detoxication processes related to ototoxicity (Hoffman et al. 1988). In general, however, this possibility had been discounted as a mechanism of toxicity since the renal elimination of apparently unmetabolized aminoglycosides was essentially complete (Schentag and Jusko 1977). The existence of a metabolite was first confirmed by exposing isolated outer hair cells to gentamicin that had been incubated with a drug-metabolizing enzyme fraction from liver (Huang and Schacht 1990; Crann et al. 1992). In contrast to the parent drug, the resulting metabolite significantly decreased the viability of the hair cells. This finding should finally provide insight into the exact mechanisms of cell destruction by aminoglycoside antibiotics.

Furthermore, the hypothesis of aminoglycoside metabolism may also provide new approaches to the prevention of ototoxicity. For example, the toxic metabolite can be rendered inactive against outer hair cells in vitro in a glutathione-dependent reaction (Garetz, Rhee, and Schacht 1994). This may explain reports (Hoffman et al. 1987) that lowered hepatic glutathione levels potentiate the combined toxicity of ethacrynic acid and kanamycin. Glutathione, a low-molecular-weight sulfhydryl compound, is ubiquitous in mammalian cells and aids in the detoxication of xenobiotics and carcinogens. When glutathione was added to the dietary intake of guinea pigs in vivo, hearing loss induced by gentamicin was significantly attenuated (Garetz, Altschuler, and Schacht 1994), while serum levels of gentamicin remained unchanged. Expression of aminoglycoside toxicity may then be dependent on the balance of two processes, the potential of a tissue to synthesize and its capacity to detoxify the metabolite.

5.3 Antineoplastic Agents (Cisplatin)

5.3.1 Therapeutic Considerations

Several classes of chemotherapeutic agents possess ototoxic potential. Currently, attention focuses on a very effective new antineoplastic drug, cisplatin, but both nitrogen mustards and vinca alkaloids have also been implicated in ototoxic damage.

Cis-diamminedichloroplatinum II (cisplatin) is a cytotoxin effective against a variety of solid tumors. It exerts its cytotoxic effect by intercalating with DNA and suppressing cell growth and primarily the proliferation of fast-growing tumor cells. Other platinum-containing antineoplastic compounds such as carboplatin have been synthesized and tested, but cisplatin remains a mainstay of therapy despite its pronounced nephrotoxicity and ototoxicity. Furthermore, carboplatin, although often considered of lesser toxicity, also has ototoxic potential (Van der Hulst, Dreschler, and Urbanus 1988; Kennedy et al. 1990).

5.3.2 Clinical Observations

The ototoxicity of cisplatin is most commonly associated with high-frequency hearing loss and tinnitus. Vestibular dysfunction may also occur, particularly in patients with preexisting vestibular anomalies (Schafer et al. 1981; Black et al. 1982; Moroso and Blair 1983). The hearing loss is often slow in onset and subclinical because it begins at frequencies above the speech range. This may explain in part the large variation in the reported incidence from a few to one hundred percent: if the appropriate high frequencies are measured, all treated patients may be diagnosed with hearing loss (Kopelman et al. 1988). The auditory deficits are generally permanent, although limited recovery may be observed in the early stages of intoxication (Moroso and Blair 1983; Kobayashi et al. 1987). Cisplatin rarely causes acute, profound deafness (Guthrie and Gynther 1985).

It appears uncertain to what extent patient age, cumulative dosage, and prior hearing loss influence both incidence and severity of the ototoxic effects of cisplatin. Despite a relatively large number of reports on these questions, no clear picture seems to have emerged. Further clinical studies and animal experimentations are needed to resolve these issues. Until then, it may be appropriate to consider those parameters as potential complications of cisplatin therapy. Some consensus, however, seems to exist that children may be at risk at lower cumulative doses because of slower elimination of the drug (Murakami et al. 1990).

5.3.3 Pathology

There are relatively few descriptions of the histological changes associated with cisplatin ototoxicity in humans. Nevertheless, the obvious targets of the drug appear to be the outer hair cells (Wright and Schaefer 1982). Abnormalities can also be found in the spiral ganglia and cochlear nerve (Strauss et al. 1983). In the vestibular system, the maculae and cristae may be damaged by the administration of cisplatin (Wright and Schaefer 1982).

In animals treated with cisplatin, hair cells seem consistently affected, although other inner ear structures may also be damaged by this aggressive drug. The destruction of outer hair cells begins in the first and second rows of the basal turn of the cochlea. Outer hair cells of the upper turns and of the inner hair cells remain relatively well preserved (Estrem et al. 1981; Komune, Asakuma, and Snow 1981; Boheim and Bichler 1985; Marco-Algarra, Basterra, and Marco 1985). Outer hair cells of treated animals may show dilatation of internal membranes, alterations in the cuticular plate, increased vacuolization, and an increased number of lysosomal bodies (Estrem et al. 1981). The pattern of inner ear injury then becomes progressively more complex and expands to other cells and structures. Hensen's cells and Deiters' cells may show damage followed by collapse of Reissner's membrane and the entire organ of Corti (Laurell and Bagger-Sjöbäck 1991). Finally, the marginal cells of stria vascularis are affected (Nakai et al. 1982; Tange and Vuzevski 1984; Kohn et al. 1988; Laurell and Bagger-Sjöbäck 1991).

There is good agreement between the histopathology after cisplatin treatment and the behavioral and physiological effects of the drug. At high doses guinea pigs lose their Preyer reflex (Fleischman et al. 1975; Estrem et al. 1981; Barron and Daigneault 1987) and show dose-related increases in high-frequency thresholds as measured by evoked auditory brainstem response. There is, however, a fair amount of variability in the degree of threshold shift (Kohn et al. 1988) and in the permanence of the effect (Nakai et al. 1982). A decrease in CM (Komune et al. 1981a) and a dose-related suppression of N1 amplitude (Barron and Daigneault 1987), as well as reduction in the generation of distortion products (McAlpine and Johnstone 1991), all appear to correlate with a loss of neural response consistent with outer hair cell pathology. A decrease in the EP at lower doses of the drug would seem indicative of strial dysfunction. However, this effect has been dissociated from the observed hearing loss (Laurell and Engström 1989). No correlation between histopathology and functional deficits was seen in rhesus monkeys. While all of the cisplatin-treated animals displayed a loss of outer hair cells in the lower turns, the hearing loss was transient and only seen in half of the animals (Stadnicki et al. 1975).

5.3.4 Mechanism of Action

Attempts to explain the mechanism of cisplatin ototoxicity on a biochemical level have been largely unsuccessful. The observed decrease in endolymphatic potential has led to the speculation that alterations in the ionic composition of the endolymph or perilymph cause secondary damage to the sensory cells; however, the concentrations of sodium, potassium, and chloride in the cochlear fluids remain unchanged (Konishi, Gupta, and Prazma 1983). Therefore, the observed inhibition of adenylate cyclase activity (Bagger-Sjöbäck, Filipek, and Schacht 1980) or ATPases and membrane phosphatases (Aggarwal and Niroomand-Rad 1983; Tay et al. 1988) by cisplatin may be limited to in vitro systems. Furthermore, Na^+-K^+-ATPase activity in the cochlear lateral wall does not change with cisplatin administration (Barron and Daigneault 1987).

An immediate and site-specific loss of neural thresholds when cisplatin is injected into the scala media suggests that a blockade of the transduction channels of hair cells may be responsible for resulting hearing loss (McAlpine and Johnstone 1991). A block of depolarization-induced calcium entry in isolated outer hair cells (Saito, Moataz, and Dulon 1991) supports the notion of a cisplatin/ion channel interaction. This could also explain the changes in intracellular calcium levels observed after administration of cisplatin (Comis et al. 1986; Crespo et al. 1990), which correlate with shifts in auditory thresholds. However, competition with calcium and blockage of transduction channels does not explain the preferential destruction of outer hair cells since it should affect both inner and outer hair cells equally. Therefore, an antagonistic interaction with calcium may be part of an acute but not chronic mechanism of hearing loss. This would be similar to the acute mechanism proposed for aminoglycoside ototoxicity (Schacht 1986; see Section 5.2).

5.4 Salicylates

5.4.1 Therapeutic Considerations

For a century, salicylates have been among the most commonly used medications, their analgesic and anti-inflammatory actions in many respects unsurpassed by modern replacements. Aspirin, the *N*-acetyl ester of salicylic acid, is hydrolyzed in the body to the pharmacologically active salicylate. The therapeutic action is based on the inhibition of cyclooxygenase, an enzyme catalyzing one of the early steps in prostaglandin synthesis.

5.4.2 Clinical Observations

Although salicylates generally have a low incidence of severe side effects, they can produce tinnitus and transient hearing loss. Other nonsteroidal anti-inflammatory agents appear to be devoid of such ototoxic effects. The concentrations of aspirin necessary to produce the symptoms are high (2 to 5 g/d) and, therefore, usually outside normal therapeutic range except in treatment of rheumatoid arthritis. High-frequency tinnitus appears to be the first symptom, and associated hearing losses are generally mild to moderate, seldom exceeding 40 dB. They are typically bilateral and either flat or more pronounced at the higher frequencies, and recovery is usually complete within one to three days after the cessation of treatment. Changes in suprathreshold hearing characteristics, such as temporal integration, temporal resolution, frequency selectivity, and complex sound perception, may also accompany the salicylate-induced hearing losses (McCabe and Dey 1965; Pederson 1974; Young and Wilson 1982; McFadden, Plattsmier, and Pasanen 1984; Boettcher, Bancraft, and Salvi 1989). Only rarely has permanent deafness from salicylate administration been reported (Kapur 1965; Jarvis 1966), and it may be that these cases were compromised by confounding factors (Kisiel and Bobbin 1981).

5.4.3 Pathology

Analysis of temporal bones from patients suffering from salicylate ototoxicity have failed to reveal a clear anatomical correlate for the observed symptoms. The usual finding has been normal anatomy without undue hair cell loss for the age of the patient (Bernstein and Weiss 1967). Strial atrophy has been reported in one case, as well as loss of spiral ganglion cells, but it is unclear whether these changes were secondary to aging or due to salicylate administration (DeMoura and Hayden 1968).

The same lack of histologically observable causes for ototoxicity is generally seen in animals treated with salicylates. There were no significant changes in sensory epithelium, spiral ganglia, or stria vascularis of guinea pigs and squirrel monkeys (Myers and Bernstein 1965; Deer and Hunter-Duvar 1982). However, guinea pigs showed vacuolization of the endoplasmic reticulum and bending of their outer hair cell stereocilia (Douek, Dodson, and Bannister 1983). Scattered outer hair cell loss after salicylate treatment in chinchillas (Woodford, Henderson, and Hamernick 1978) has not been seen consistently (Spongr et al. 1992).

Behavioral studies confirm a pattern of hearing loss similar to humans in several species of animals given salicylates. Guinea pigs and chinchillas develop a moderate temporary threshold shift after single or multiple doses (Gold and Wilpizeski 1966; Woodford, Henderson, and Hamernick 1978; Cazals et al. 1988). Responses of higher primates to salicylate, however, are somewhat contradictory. Squirrel monkeys showed a moderate loss of hearing twenty-four hours after a single injection as assessed by shock-avoidance audiometry (Myers and Bernstein 1965). This is in contrast to most human and animal studies in which the observed threshold shifts were much more transient. Higher single doses produced a similar shift in monkeys but with virtually complete recovery after several hours (Stebbins 1970).

Although the pattern of electrophysiological changes is not consistent, it nevertheless points to outer hair cells as a possible target of salicylates. Salicylates may decrease the amplitude of the CM and CAP in cats

(Silverstein, Bernstein, and Davies 1967) and guinea pigs (McPherson and Miller 1974); decrease CAP with normal CM or SP (summating potential) in guinea pigs (Puel, Bobbin, and Fallon 1988); or increase both CAP and SP with a corresponding decrease in CM (Stypulkowski 1990). Furthermore, evoked responses are inhibited at stimulation of low intensities but not of high intensities (Puel, Bobbin, and Fallon 1988; Stypulkowski 1990), implying an effect on the amplifier properties of outer hair cells. This view is further supported by the fact that both spontaneous and evoked otoacoustic emissions are temporarily reduced after salicylate use (McFadden and Plattsmier 1984; Long and Tubis 1988).

A question that has been difficult to resolve is whether animals experience tinnitus with salicylate treatment. Spontaneous firing rates were elevated after salicylate administration in single neurons and the auditory nerve of cats (Evans and Kline 1982) and in the neurons of the inferior colliculus of guinea pigs (Jastreboff and Sasaki 1986), indicating a tinnitus-like activity. A behavioral model interpreted to be consistent with the existence of salicylate-induced tinnitus is discussed later in this book (Penner and Jastreboff, Chapter 7).

5.4.4 Mechanism of Action

Prominent among several biochemical theories of salicylate action has been the suggestion that the well-established inhibitory effect on prostaglandin synthesis is also the cause of the observed auditory deficits. Prostaglandin levels indeed decrease transiently in cochlear tissues and fluids after systemic (Escoubet et al. 1985) or local (Jung et al. 1988) application of salicylates. In contrast, other known inhibitors of prostaglandin synthesis have no effect on cochlear physiology after local instillation (Puel, Bobbin, and Fallon 1990), arguing against prostaglandins as an ototoxic mediator. The drug is also a known inhibitor of several enzymatic reactions and will affect energy metabolism by uncoupling oxidative phosphorylation. Whether any such actions occur at drug concentrations associated with hearing deficits has not been established.

An effect on cochlear blood flow, either directly or via an alteration in prostaglandin levels, has been suggested to underlie the auditory derangements. Vasoconstriction of capillaries in several cochlear structures is a possible effect in vivo (Hawkins 1976). Consistent with this theory is the decreased cochlear blood flow after both systemic and local salicylate application (Didier, Miller, and Nuttall 1993). Furthermore, concurrent administration of a vasodilating agent decreases the salicylate-induced threshold shift (Cazals et al. 1988). However, changes in vasculature have been difficult to evaluate because they do not appear to encompass the entire cochlea. Moreover, it is questionable whether the generally observed small reductions in blood flow are sufficient to explain the pathophysiological responses.

Finally, a different site of action was deduced from the effect of salicylates on outer hair cells in vitro. Turgor and contractile capability of these cells were decreased when they were incubated in millimolar concentrations of salicylate (Shehata, Brownell, and Dieler 1991). Such impairment of outer hair cell function could account for a temporary threshold shift but would not explain the generation of tinnitus. The precise cellular mechanism of salicylate ototoxicity, thus, remains unknown.

5.5 Diuretics

5.5.1 Therapeutic Considerations

Loop diuretics are a potent group of drugs used to treat states of fluid overload. Commonly prescribed loop diuretics include furosemide, bumetanide, and ethacrynic acid. Furosemide, bumetanide, and piretanide inhibit a luminal $Na^+/Cl^-/K^+$ cotransporter resulting in salt and water reabsorption in the ascending limb of the loop of Henle in the renal tubules. The action of ethacrynic acid may differ somewhat and involve an inhibition of mitochondrial ATP synthesis (Greger and Wangemann 1987).

5.5.2 Clinical Observations

Furosemide, ethacrynic acid, and related diuretics can produce transient hearing loss (Maher and Schreiner 1965; Schneider and Becker 1966; Schwartz et al. 1970; Venkateswaran 1971). High intravenous doses of these drugs and impaired renal function enhance the risk of ototoxicity, and permanent hearing deficits do occur albeit much less frequently (Pillay et al. 1969; Lloyd-Mostyn and Lord 1971; Brown et al. 1974; Quick and Hoppe 1975; Keefe 1978; Rifkin et al. 1978; Gallagher and Jones 1979; Rybak 1988). Tinnitus is also associated with the administration of diuretics (Maher and Schreiner 1965; Schneider and Becker 1966; Pillay et al. 1969; Schwartz et al. 1970), while instances of vestibular disturbances are rare.

5.5.3 Pathology

Histological changes in human temporal bones obtained from patients with documented diuretic-induced ototoxicity mainly involve the stria vascularis (Matz and Hinojosa 1973). Damage may include edema, thickening, and degeneration of the stria vascularis especially in the basal region (Arnold, Nadol, and Weidauer 1981). A more generalized endolymphatic hydrops has also been seen, and hair cells may be affected as well. Effects on these cells range from a granularity in cochlear hair cells and type I and II hair cells of the cristae and maculae to a scattered loss of basal outer hair cells (Matz, Beal, and Krames 1969).

Animal experiments concur with the human data in that the stria vascularis is the primary or at least initial site of action of loop diuretics in the cochlea. Within minutes after injection of either ethacrynic acid or furosemide, there is a drop of the endolymphatic potential and consequently a loss of microphonic and action potentials (Brummett et al. 1977; Bosher 1980a; Pike and Bosher 1980). Suppression of the endolymphatic potential seems to be a property of most loop diuretics including ethacrynic acid, furosemide, piretanide, bumetanide, and indacrinone (Rybak, Whitworth, and Scott 1991), although the individual kinetics appear to differ. Endolymphatic potential loss is rapid in onset for furosemide but gradual for ethacrynic acid. A temporary, high-frequency hearing loss has been documented in guinea pigs treated with these agents (Ernstson 1972; Brown 1975), as would be expected from the greater severity of effects seen in the basilar portion of the cochlea.

Concomitant with the reduction of the endolymphatic potential, morphological changes occur in the stria vascularis. These changes begin in the marginal cells and develop into edema throughout the cell layers of the stria (Quick and Hoppe 1975; Brummet et al. 1977; Bosher 1980a, 1980b; Pike and Bosher 1980). Changes in the tight junctions of the marginal cells (Rarey and Ross 1982) raise the possibility that these drugs could cause alterations in the endolymph/perilymph barrier. Intermediate cells may also be involved because they atrophy in treated guinea pigs (Forge 1976). After cessation of the drug, the cochlear potentials and the morphology of the tissues recover.

There is disagreement as to the extent of hair cell damage associated with loop diuretics. Reversible swelling or other damage to hair cells was seen in some studies; (Mathog, Thomas, and Hudson 1970; Johnsson and Hawkins 1972; Crifo 1973; Akuyoshi 1981) while others report no damage to or loss of hair cells (Federspil and Mausen 1973; Brown et al. 1979).

5.5.4 Mechanism of Action

Attempts to elucidate the biochemical pathways responsible for derangements associated with loop diuretics have met with limited success. Alteration of strial function is indicated by the changed concentration of endolymphatic potassium (Rybak and Morizono 1982), chloride (Rybak and Whitworth 1986), and calcium (Ninoyu and Meyer zum Gottesberge 1986). Therefore, it was initially thought that inhibition of ion transport processes play a major role in the cochlear actions of these drugs. Inhibition of Na^+-K^+-ATPase was proposed but subsequently ruled out as a cause of the ototoxicity (Kuijpers and Wilberts 1976; Thalmann et al. 1977). Similarly, interference with fluid and ion transport by adenylate cyclase was discussed, but actions of diuretics on this enzyme can be dissociated from their ototoxic potential (Marks and Schacht 1981). In an analogy to the pharmacological action on the kidney, an inhibition of a luminal $Na^+/CL^-/K^+$ cotransporter could be postulated. Loop diuretics block transport of potassium chloride out of strial marginal cells (Santi and

Lakhani 1983), but whether such a mechanism is responsible for strial edema and the suppression of endolymphatic potential remains unknown.

Although most evidence points to a strial site of action for the loop diuretics, the question remains as to whether such primary damage and its sequelae are responsible for the hair cell damage and permanent deficits sometimes seen. Respiratory enzymatic activity in outer hair cells is depressed with administration of loop diuretics (Kaku, Farmer, and Hudson 1973; Comis, Pratt, and Hayward 1981), but nonototoxic drugs cause a similar effect (Koide, Hata, and Hando 1966). Furosemide, however, may indeed affect active processes in the cochlea. These are possibly associated with outer hair cell activity, resulting in altered responses to acoustic stimuli (Ruggero and Rich 1991).

There is good evidence that an organic acid transport system participates in the overall mechanism of diuretic action. The drugs enter the cochlea via a carrier-mediated transport, and prevention of entry will also prevent the ototoxic actions of these drugs. Consistent with their differential action of the kidney, furosemide-like diuretics and ethacrynic acid behave somewhat differently: organic acids including probenecid, salicylate, and penicillin G are effective competitors of furosemide penetration into the inner ear (Rybak and Whitworth 1987; Rybak, Whitworth, and Scott 1990), while quinine reduces the effects of both drugs (Rybak and Whitworth 1988). Since the therapeutic effects of diuretics remain unaltered by the coadministration of organic acids (Rybak et al. 1984), this is an exciting approach to the attenuation of the ototoxic effects of these drugs.

5.6 Miscellaneous Drugs

The drugs discussed so far constitute the most important clinically used ototoxic agents. Furthermore, they are drugs for which clinical pathology and potential mechanisms have been established. A variety of other agents, however, also show distinct ototoxic effects. For a good number of these, evidence is based on anecdotal observations without controlled experimentation. Others are no longer in common use and are therefore of more historic interest, such as the arsenical- and heavy-metal based therapeutics. A few ototoxic drugs, however, have recently received renewed attention.

5.6.1

Quinine has been in use for centuries as a treatment for fever-causing diseases and malaria in particular. Tinnitus and temporary threshold shift are two of many well-established side effects of quinine in humans and experimental animals (Hawkins 1976; Alván, Karlsson, and Villén 1989; Alván et al. 1991). Vasoconstriction of strial vessels (Hawkins, Johnsson, and Preston 1972) and effects on outer hair cell mechanics (Karlsson et al. 1991) are among the possible bases for its action. Since the 1940s, quinine has largely been replaced by the less toxic chloroquine, which, nevertheless, also induces reversible auditory deficits (Bernard 1985). Since some malarial parasites are developing resistance against chloroquine, the use of quinine (or other analogues) may again become widespread. Although both quinine and chloroquine have been reported to cause permanent hearing loss, rigorously controlled studies on this important question are lacking.

5.6.2

Erythromycin has become the drug of choice for the treatment of *Legionella pneumophilia* and related pathogens. It was long considered to be devoid of ototoxic side effects but has now been established to damage both the cochlea and the vestibular system. The symptoms are threshold shift, tinnitus, and vertigo. These effects are generally reversible (Cramer 1986; Brummett and Fox 1989), although one case of permanent hearing loss has been documented (Dylewski 1988).

5.6.3

Polypeptide antibiotics have both nephrotoxic and ototoxic potential. Vancomycin, rarely used since the 1950s, has recently been reintroduced in the treatment of the increasing number of *Staphylococcus aureus* infections that are resistant to methicillin (Watanakunakorn 1982). Although patients have experienced hearing loss and reversible tinnitus after vancomycin administration, they were usually receiving combination therapy. It is not clear from the clinical literature that vancomycin alone can produce auditory deficits. In experimental animals, this drug does not have ototoxic potential, even at near-lethal doses (Brummett and Fox 1989).

5.6.4

α-Difluoromethylornithine is a novel antitumor and antiprotozoal agent. Sensorineural hearing loss developed in a number of patients treated for cancer with this drug (Abeloff et al. 1984, 1986). Auditory function generally recovered once treatment was stopped. The pathology of difluoromethylornithine-induced hearing loss in guinea pigs points to a cochlear site of damage (Jansen et al. 1989). This is corroborated by effects of the drug on acoustic distortion products in the developing rat (Henley et al. 1990). α-Difluoromethylornithine has a known primary biochemical site of action as an irreversible inhibitor of ornithine decarboxylase, a regulatory enzyme in polyamine biosynthesis. Polyamines play a major role in cellular growth, development, maintenance, and regeneration, and cochlear ornithine decarboxylase is inhibited by the drug in vitro (Henley, Gerhardt, and Schacht 1987). Speculations that this action is responsible for the ototoxic side effects are intriguing but remain to be confirmed.

6. Factors Influencing Ototoxicity

6.1 The State of the Animal

The above discussions of drug actions emphasize the general nature of observed effects. Extrapolations from one model to another can indeed be made, with few exceptions. It needs to be emphasized, however, that sensitivity of the auditory and vestibular systems to drugs and other traumatic insults is influenced by a number of variables, such as developmental stage and age of the animal. For example, a sensitive period exists for aminoglycosides, difluoromethylornithine, and perhaps other drugs, which coincides with the latest stages of cochlear development (Pujol 1986; Henley et al. 1990). This may have some bearing on the increased ototoxicity sometimes reported in children. Conversely, elderly patients may show an increased susceptibility to ototoxic trauma, a phenomenon that still needs to be explored in detail in appropriate animal models.

The nutritional status of an animal influences body defenses and the biochemical makeup of tissues. Food deprivation can increase susceptibility to drug-induced hearing loss (Hoffman et al. 1987). Finally, sex, pigmentation, diurnal rhythm, and the presence of anesthesia may all influence drug disposition and the expression of ototoxicity (Wästerström et al. 1986; Jastreboff et al. 1988; Yonovitz and Fisch 1991).

One other confounding variable is usually absent from animal experimentation, namely, a disease under treatment. In contrast, drug therapy in patients may be complicated by the effects of the disease itself and other preexisting conditions. Some risk factors may vary with the type of drug, but two factors are of general concern: liver and kidney dysfunction. The liver is the major organ for drug metabolism either to metabolites with altered pharmacological properties or to conjugation products to facilitate excretion. Kidney function determines urinary elimination. Dysfunctions of either or both organs potentially change drug metabolism and clearance and thus the effective concentrations of the ototoxic agent achieved in serum and tissues.

Concomitant or prior treatment with other drugs may influence the effects of ototoxic agents as discussed in detail below. Perhaps more unexpectedly, even seemingly unrelated therapeutic intervention may affect susceptibility to drugs. For example, prior central nervous system radiation therapy increases the incidence of hearing loss sustained in cisplatin treatment in patients (Schell et al. 1989; Weatherly 1991) and in animals (Baranak, Wetmore, and Packer 1988).

6.2 Drug–Drug Interactions

Many patients receiving potentially ototoxic drugs in clinical settings are being treated with multiple medications. When combining two drugs, it is to be expected that effects and side effects will be additive, but, in reality, the interaction of such combinations is not predictable. Drugs may act synergistically (e.g., antimicrobial combination therapy) or independently if given for different therapeutic targets. In some cases, a second drug may inactivate the first (e.g., aminoglycoside and penicillin). However, if certain specific combinations of ototoxic agents are administered, the results can be devastating to the auditory system, although treatment with either drug alone would be safe.

6.2.1 Potentiation of Adverse Effects

The observation that aminoglycoside therapy in combination with diuretics may lead to unexpected hearing loss was first made in patients and subsequently replicated in animal models (Mathog and Klein 1969; West, Brummett, and Himes 1973). Experimental conditions can be defined where neither the treatment with the aminoglycoside antibiotic alone nor the diuretic alone will cause a hearing deficit, and the combination will lead to complete deafness.

Ototoxic potentiation is not limited to the interaction between aminoglycosides and diuretics, although these two classes of drugs appear to be most frequently associated with this phenomenon. Aminoglycosides (Schweitzer et al. 1984) and loop diuretics (Laurell and Engström 1989) also potentiate the ototoxicity of cisplatin in experimental models. Furthermore, combinations of aminoglycosides with vancomycin or aminooxyacetic acid, among others, are more harmful than either treatment alone would predict. Likewise, loop diuretics interact with viomycin or polymixin B to enhance their ototoxic effects (Brummett 1981).

The damage induced by such drug combinations is generally irreversible, the time course much faster, and the extent of damage much larger than with the individual drugs. The mechanisms of synergistic ototoxicity are unclear. One drug may alter the clearance or uptake of the other, allowing a prolonged or increased exposure to susceptible organs. Alternately, one drug may inhibit detoxication mechanisms needed to minimize ototoxicity of the other drug. Thirdly, sublethal attacks at two different target sites may combine to produce greater damage to the end organ.

6.2.2 Amelioration of Adverse Effects

A desired result of combination therapy would be an amelioration of the adverse effects. Such strategies have been designed since ototoxicity of drugs became known. Few if any have been successful in clinical application. Examples of potentially beneficial approaches have been mentioned in previous sections for aminoglycosides (coadministration of glutathione) and furosemide (coadministration of organic acids).

Diuretics as well as probenecid (Jacobs et al. 1991), diethyldithiocarbonate (Gratton et al. 1990), and a variety of clinically used sulfur-containing compounds (Jones, Basinger and Holscher 1992) can ameliorate the nephrotoxicity associated with cisplatin. Whether these interventions also reduce or prevent the associated ototoxicity remains to be established. Concurrent administration of fosfomycin or selective antioxidants, however, appears to inhibit the elevation of hearing thresholds following cisplatin administration (McGinness et al. 1978; Schweitzer et al. 1986; Otto 1988). Other treatments apparently effective against aminoglycoside nephrotoxicity still need to be tested for their efficacy against the ototoxic side effects. These include protection by poly-L-aspartic acid (Kishore et al. 1992) and the acceleration of tissue repair by growth factors (Morin et al. 1992). The latter concept seems particularly interesting in view of the suggestions that mammalian hair cells have the capacity to regenerate (Forge et al. 1993; Warchol et al. 1993).

6.3 Combination of Drugs with Noise Trauma

Concurrent exposure to noise is another risk factor in therapy with potentially ototoxic drugs. Primarily aspirin and aminoglycosides (Dayal and Barok 1975; McFadden and Plattsmier 1983) interact with noise, but diuretics, cisplatin (Gratton et al. 1990), and industrial chemicals (Fechter 1989) have also been implicated in such interactions. In contrast to the drug-drug combinations, the enhanced damage is less apparent and frequently associated with very specific conditions of noise exposure. Such a potential interaction should, however, always be considered if chemotherapy is combined with noise exposure in the workplace.

7. Summary

A number of drugs from widely disparate therapeutic classes and with totally unrelated structures can cause damage to the auditory and vestibular systems. Most of these agents continue to be widely used despite their potential ototoxicity because their beneficial effects outweigh possible risks. Moreover, less toxic substitutes are often not available to treat some life-threatening diseases. We should not forget, however, that improvements in monitoring of serum drug levels have drastically diminished the risk of side effects. Early detection of ototoxicity and termination of drug administration may further reduce the severity of the ototoxicity. Unfortunately, the sometimes moribund state of the recipients can preclude the attainment of baseline measurements and the monitoring of auditory function during treatment.

The use of a variety of in vivo and in vitro models has enabled researchers to glean a wealth of information about ototoxic drugs. The site of action is most often clearly localized in humans and animals, and the exact cellular and biochemical mechanisms of ototoxicity are beginning to be unraveled.

Unfortunately, it is not yet possible to predict the potential ototoxicity of new drugs or the effects of combined drug therapy. Thus, further animal studies are required to provide more in-depth information as to the precise molecular derangements caused by these ototoxic agents. The recently introduced use of isolated cochlear hair cells as in vitro models and noninvasive physiological measurements such as otoacoustic emissions may prove invaluable in such investigations. With increased focus on these problems, prevention of damage from ototoxic drugs may be perfected and safe but effective therapeutic modalities developed.

Acknowledgments. The writing of this chapter was concluded in November 1992. Thereafter, references were updated, but no new material was included. The authors wish to thank their colleagues S. Bledsoe, A. Forge, J. E. Hawkins, Jr., and A. Nuttall for critical and constructive comments. S.G. was supported by training grant DC-00024; research on ototoxicity is supported by research grant DC-00124 to J.S.

References

Abeloff M, Rosen S, Luk G, Baylin S, Zeltzman M, Sjoerdsma A (1986) Phase II trials of α-difluoromethylornithine, an inhibitor of polyamine synthesis, in advanced small cell lung cancer and colon cancer. Cancer Treat Rep 70:843–845.

Abeloff MD, Slavik M, Luk GD, Griffin CA, Hermann J, Blanc O, Sjoerdsma A, Baylin SB (1984) Phase I trial and pharmacokinetic studies of α-difluoromethylornithine — an inhibitor of polyamine biosynthesis. J Clin Oncol 2:124–130.

Aggarwal SK, Niroomand-Rad I (1983) Effect of cisplatin on the plasma membrane phosphatase activities in ascites sarcoma-180 cells: a cytochemical study. J Histochem Cytochem 31:307–317.

Akuyoshi M (1981) Effect of loop-diuretics on hair cells of the cochlea in guinea pigs. Histological and histochemical study. Scand Audiol Suppl 14:185–199.

Alleva FR, Balazs T (1978) Toxic effects of postnatal administration of streptomycin sulfate to rats. Toxicol Appl Pharmacol 45:855–859.

Alván G, Karlsson KK, Villén T (1989) Reversible hearing impairment related to quinine blood concentrations in guinea pigs. Life Sci 45:751–755.

Alván G, Karlsson KK, Hellgren U, Villén T (1991) Hearing impairment related to plasma quinine concentration in healthy volunteers. Br J Clin Pharmacol 31:409–412.

Anniko M (1981) Elemental composition of the developing inner-ear. Ann Otol Rhinol Laryngol 90:25–32.

Anniko M, Nordemar H (1979) Embryogenesis of the inner ear. I. Development and differentiation of the mammalian insta ampullaris in vivo and in vitro. Arch Otorhinolaryngol 224:285–299.

Anniko M, Takada A, Schacht J (1982) Comparative ototoxicities of gentamicin and netilmicin in three model systems. Am J Otolaryngol 3:422–433.

Arnold W, Nadol JB, Weidauer H (1981) Ultrastructural histopathology in a case of human ototoxicity due to loop diuretics. Acta Otolaryngol 91:399–414.

Arpini A, Cornacchia L, Albiero L, Bamonte F, Parravicini L (1979) Auditory function in guinea pigs treated with netilmicin and other aminoglycoside antibiotics. Arch Otorhinolaryngol 224:137–142.

Bagger-Sjöbäck D, Filipek CS, Schacht J (1980) Characteristics and drug responses of cochlear and vestibular adenylate cyclase. Arch Otorhinolaryngol 228:217–222.

Bagger-Sjöbäck D, Wersäll J (1978) Gentamicin induced mitochondrial damage in inner ear sensory cells of the lizard *Calotes versicolor*. Acta Otolaryngol 86:35–51.

Ballantyne JC (1973) Ototoxicity: A clinical review. Audiology 12:325–336.

Baranak CC, Wetmore RF, Packer RJ (1988) *Cis*-platinum ototoxicity after radiation treatment; an animal model. J Neurooncol 6:261–267.

Bareggi R, Grill V, Narducci P, Zweyer M, Tesei L, Russolo M (1990) Gentamicin ototoxicity: Histological and ultrastructural alterations after transtympanic administration. Pharmacol Res Commun 22:635–644.

Barron SE, Daigneault EA (1987) Effect of cisplatin on hair cell morphology and lateral wall Na, K-ATPase activity. Hear Res 26:131–137.

Benitez JT, Schuknecht HF, Bradenburg JH (1962) Pathologic changes in human ear after kanamycin. Arch Otolaryngol 75:192–197.

Bernard P (1985) Alterations of auditory evoked potentials during the course of chloroquine treatment. Acta Otolaryngol (Stockh) 99:387–392.

Bernstein JM, Weiss AD (1967) Further observations on salicylate ototoxicity. J Laryngol Otol 81:915–925.

Black FO, Myers EN, Schramm VL, Johnson J, Sigler B, Thearle PB, Burns DS (1982) Cisplatin vestibular ototoxicity: A preliminary report. Laryngoscope 92:1363–1368.

Boheim K, Bichler E (1985) Cisplatin-induced ototoxicity andrometric findings and exper cochlear pathology. Arch Otorhinolaryngol 242:1–6.

Bosher SK (1980a) The nature of the ototoxic actions of ethacrynic acid upon the mammalian endolymph system. I. Functional aspects. Acta Otolaryngol 89:407–418.

Bosher SK (1980b) The nature of the ototoxic actions of ethacrynic acid upon the mammalian endolymph system. II. Structural-functional correlates in the stria vascularis. Acta Otolaryngol 90:40–54.

Brown AM, Mcdowell B, Forge A (1989) Acoustic distortion products can be used to monitor the effects of chronic gentamicin treatment. Hear Res 42:143–156.

Brown CG, Ogg CS, Cameron JS (1974) High dose furosemide in acute fevers. Intrinsic renal failure. Scott Med J Suppl 19:35–38.

Brown RD (1975) Ethacrynic acid and furosemide: Possible cochlear sites and mechanisms of ototoxic action. Medikon 4:33–40.

Brown RD, Manno JE, Daigneault EA, Manno BR (1979) Comparative acute ototoxicity of intravenous bumetanide and furosemide in the purebred beagle. Toxicol Appl Pharmacol 48:157–169.

Brummett RE (1981) Effects of antibiotic-diuretic interactions in the guinea pig model of ototoxicity. Rev Infect Dis Suppl 3:216–223.

Brummett RE, Fox KE (1982) Studies of aminoglycoside ototoxicity in animal models. In: Whelton A & Neu HC (eds) The Aminoglycosides. Microbiology, Clinical Use and Toxicology. New York: Dekker, pp. 419–451.

Brummett RE, Fox KE (1989) Vancomycin- and erythromycin-induced hearing loss

in humans. Antimicrob Agents Chemother 33:791-796.
Brummett RE, Fox KE, Bendrick TW, Himes DL (1978) Ototoxicity of tobramycin, gentamicin, amikacin and sisomicin in the guinea pig. J Antimicrob Chemother Suppl A 4:78-83.
Brummett R, Smith CA, Ueno Y, Cameron S, Richter R (1977) The delayed effects of ethacrynic acid on the stria vascularis of the guinea pig. Acta Otolaryngol 83:98-112.
Caussé R, Gondet I, Vallancien B (1948) Action vestibulaire de la streptomycine chez la souris. Compt Rend Soc Biol 142:747-749.
Cazals Y, Li S, Aurousse C, Didier A (1988) Acute effects of noradrenalin related vasoactive agents on the ototoxicity of asprin – an experimental study in the guinea pig. Hear Res 36:89-96.
Comis SD, Pratt SR, Hayward TL (1981) The effect of furosemide, phetanide and bumetanide on cochlear succinic dehydragenase. Neuropharmacology 20:405-407.
Comis SD, Rhys-Evans PM, Osborne MP, Pickles JO, Jeffries DJ, Pearse HA (1986) Early morphological and chemical changes induced by cisplatin in the guinea pig organ of Corti. J Laryngol Otol 100:1375-1383.
Corwin JT, Cotanche DA (1988) Regeneration of sensory hair cells after acoustic trauma. Science 240:1772-1774.
Cramer R (1986) Erythromycin ototoxicity. Drug Intell Clin Pharm 20:764-765.
Crann SA, Huang MY, McLaren JD, Schacht J (1992) Formation of a toxic metabolite from gentamicin by a hepatic cytosolic fraction. Biochem Pharmacol 43:1835-1839.
Crespo PV, Fernandez F, Ciges M, Campos A (1990) X-ray microanalysis of cisplatin ototoxicity in the cochlea. Adv Otorhinolaryngol 45:129-132.
Crifo S (1973) Ototoxicity of sodium ethacrynate in the guinea pig. Arch Otorhinolaryngol 206:27-38.
Cruz RM, Lambert PR, Rubel EW (1987) Light microscopic evidence of hair cell regeneration after gentamicin toxicity in chick cochlea. Arch Otolaryngol Head Neck Surg 113:1058-1062.
Dallos P (1973) The Auditory Periphery. New York: Academic Press.
Davis BD (1987) Mechanism of bactericidal action of aminoglycosides. Microbiol Rev 51:341-350.
Davis H, Deatheridge BH, Rosenblut B, Fernandez C, Kemura R, Smith CA (1958) Modification of cochlear potentials produced by streptomycin poisoning and extensive venous obstruction. Laryngoscope 68:596-627.
Dayal VS, Barok WG (1975) Cochlear changes from noise, kanamycin, and aging. Laryngoscopy Suppl 85:1-18.
Deer BC, Hunter-Duvar I (1982) Salicylate ototoxicity in the chinchilla: A behavioral and electron microscope study. J Otolaryngol 11:260-264.
DeMoura LEP, Hayden RC (1968) Salicylate ototoxicity. A human temporal bone report. Arch Otolaryngol 87:368-372.
Didier A, Miller JM, Nuttall AL (1993) The vascular component of sodium salicylate ototoxicity in the guinea pig. Hear Res 69:199-206.
Douek EE, Dodson HC, Bannister LH (1983) The effects of sodium-salicylate on the cochlea of guinea pigs. J Laryngol Otol 97:793-799.
Dulon D, Zajic G, Aran JM, Schacht J (1989) Aminoglycoside antibiotics impair calcium entry but not viability and motility in isolated cochlear outer hair cells. J

Neurosci Res 24:338-346.
Duvall AJ, Wersäll J (1964) Site of action of streptomycin upon inner ear sensory cell. Acta Otolaryngol 57:581-598.
Dylewski J (1988) Irreversible sensorineural hearing loss due to erythromycin. Can Med Assoc J 139:230-231.
Elidan J, Lin J, Honrubia V (1987) Vestibular ototoxicity of gentamicin assessed by the recording of a short-latency vestibular-evoked response in cats. Laryngoscope 97:865-870.
Erlandson P, Lundgren A (1964) Ototoxic side effects following treatment with streptomycin, dihydrostreptomycin and kanamycin. Acta Med Scand 176:147-163.
Ernstson S (1972) Ethacrynic acid-induced hearing loss in guinea pig. Acta Otolaryngol 73:476-483.
Escoubet B, Amsallem P, Ferrary E, Tran Ba Huy P (1985) Prostaglandin synthesis by the cochlea of the guinea pig. Influence of aspirin, gentamicin, and acoustic stimulation. Prostaglandins 29:589-599.
Estrem SA, Babin RW, Ryu JM, Moore KC (1981) *Cis*-diamminechloroplatinum (II) ototoxicity in the guinea pig. Otolaryngol Head Neck Surg 89:638-645.
Evans EF, Kline R (1982) The effects of intracochlear and systemic furosemide on the properties of single cochlear nerve fibers in the cat. J Physiol 331:409-427.
Fay RR (1988) Comparative psychoacoustics. Hear Res 34:295-305.
Fechter LD (1989) A mechanistic basis for interactions between noise and chemical exposure. Arch Compl Environ Studies 1:23-28.
Federspil P, Mausen H (1973) Experimentelle Untersuchungen zur Ototoxizität des Furosemids. Res Exp Med 161:175-184.
Fee WE Jr (1980) Aminoglycoside ototoxicity in the human. Laryngoscope Suppl 24:1-19.
Fleischman RW, Stadnicki SW, Ethier MF, Schaeppi U (1975) Ototoxicity of *cis*-dichlorodiammine platinum (II) in the guinea pig. Toxicol Appl Pharmacol 33:320-332.
Forge A (1976) Observations on the stria vascularis of the guinea pig cochlea and changes resulting from the administration of the diuretic furosemide. Clin Otolaryngol 1:211-219.
Forge A (1985) Outer hair cell loss and supporting cell expansion following chronic gentamicin treatment. Hear Res 35:39-46.
Forge A, Fradis M (1985) Structural abnormalities in the stria vascularis following chronic gentamicin treatment. Hear Res 20:233-244.
Forge A, Wright A, Davies SJ (1987) Analysis of structural changes in the stria vascularis following chronic gentamicin treatment. Hear Res 31:253-266.
Forge A, Li I, Corwin JT, Nevill G (1993) Ultrastructural evidence for hair cell regeneration in the mammalian inner ear. Science 259:1616-1619.
Gallagher KI, Jones JK (1979) Furosemide-induced ototoxicity. Ann Intern Med 91:744-745.
Garetz SL, Altschuler RA, Schacht J (1994) Attenuation of gentamicin ototoxicity by glutathione in the guinea pig in vivo. Hear Res 77:81-87.
Garetz SL, Rhee DJ, Schacht J (1994) Sulfhydryl compounds and antioxidants inhibit cytotoxicity to outer hair cells of a gentamicin metabolite in vitro. Hear Res 77:75-80.
Gold A, Wilpizeski CR (1966) Studies in auditory adaption. The effects of sodium salicylate on evoked auditory potentials in cats. Laryngoscope 76:674-685.

Gratton MA, Salvi RJ, Kamen BA, Saunders SS (1990) Interaction of cisplatin and noise on the peripheral auditory system. Hear Res 50:211-224.

Guthrie TM Jr, Gynther L (1985) Acute deafness. A complication of high-dose cisplatin. Arch Otolaryngol 111:344-345.

Harada T, Iwamori M, Nagai Y, Nomura Y (1986) Ototoxicity of neomycin and its penetration through the round window membrane into the perilymph. Ann Otol Rhinol Laryngol 95:404-408.

Harrison WH (1954) Ototoxicity of dihydrostreptomycin. Q Bull Northwest Univ Med Sch 28:271.

Hawkins JE Jr (1950) Cochlear signs of streptomycin intoxication. J Pharmacol Exp Ther 100:38-44.

Hawkins JE Jr (1973a) Comparative otopathology: Aging, noise and ototoxic drugs. Adv Otorhinolaryngol 20:125-141.

Hawkins JE Jr (1973b) Ototoxic mechanisms – a working hypothesis. Audiology 12:383-393.

Hawkins JE Jr (1976) Drug ototoxicity. In: Keidel WD, Neff WD (eds) Handbook of Sensory Physiology Vol. 5, part 3, Berlin: Springer-Verlag, pp. 707-748.

Hawkins JE Jr, Lurie MH (1952) The ototoxicity of streptomycin. Ann Otol 61:789-806.

Hawkins JE Jr, Johnsson LG, Aran JM (1969) Comparative tests of gentamicin ototoxicity. J Infect Dis 119:417-426.

Hawkins JE Jr, Johnsson LG, Preston RE (1972) Cochlear microvasculature in normal and damaged ears. Laryngoscope 82:1364-1374.

Hawkins JE Jr, Stebbins WC, Johnsson LG, Moody DB, Maraski A (1977) The patas monkey as a model for dihydrostreptomycin ototoxicity. Acta Otolaryngol (Stockh) 83:123-129.

Henley CM, Schacht J (1988) Pharmacokinetics of aminoglycoside antibiotics in blood, inner-ear fluids and tissues and their relationship to ototoxicity. Audiology 27:137-146.

Henley CM, Gerhardt H-J, Schacht J (1987) Inhibition of inner ear ornithine decarboxylase by neomycin in-vitro. Brain Res Bull 19:695-698.

Henley C, Atkins J, Martin G, Lonsbury-Martin B (1990) Critical period for alpha-difluoromethylornithine (DFMO) ototoxicity in the developing pigmented rat. J Cell Biochem Suppl 14F:22.

Hiel H, Bennani H, Erre J-P, Aurousseau C, Aran J-M (1992) Kinetics of gentamicin in cochlear hair cells after chronic treatment. Acta Otolaryngol 112:272-277.

Hodges GR, Watanabe IS, Singer P, Rengachary S, Brummett RE, Reeves D, Justesen DR, Worley SE, Gephardt EP Jr (1985) Ototoxicity of intravenously administered gentamicin in adult rabbits. Res Comm Chem Pathol Pharmacol 50:337-347.

Hoeffding V, Fechter LD (1991) Trimethyltin disrupts auditory function and cochlear morphology in pigmented rats. Neurotoxicol Teratol 13:135-145.

Hoffman DW, Whitworth CA, Jones KL, Rybak LP (1987) Nutritional status, glutathione levels, and ototoxicity of loop diuretics and aminoglycoside antibiotics. Hear Res 31:217-222.

Hoffman DW, Whitworth CA, Jones-King KL, Rybak LP (1988) Potentiation of ototoxicity by glutathione depletion. Ann Otol Rhinol Laryngol 97:36-41.

Huang MY, Schacht J (1989) Drug-induced ototoxicity. Pathogenesis and prevention. Med Toxicol Adverse Drug Exp 4:452–467.

Huang MY, Schacht J (1990) Formation of a cytotoxic metabolite from gentamicin by liver. Biochem Pharmacol 40:R11–R14.

Huizing EH, de-Groot JC (1987) Human cochlear pathology in aminoglycoside ototoxicity—a review. Acta Otolaryngol S436:117–125.

Igarashi M (1973) Vestibular ototoxicity in primates. Audiology 12:337–349.

Jacobs C, Kaubisch S, Halsey J, Lum BL, Gosland M, Coleman CN, Sikic BI (1991) The use of probenecid as a chemoprotector against cisplatin nephrotoxicity. Cancer 67:1518–1524.

Jansen C, Mattox DE, Miller KD, Brownell WE (1989) An animal model of hearing loss from α-difluoromethylornithine. Arch Otolaryngol Head Neck Surg 115:1234–1237.

Jarvis JF (1966) A case of unilateral permanent deafness following acetylsalicylic acid. J Laryngol Otol 80:318–320.

Jastreboff PJ, Issing W, Brennan JF, Sasaki CT (1988) Pigmentation, anesthesia, behavioral factors, and salicylate uptake. Arch Otolaryngol Head Neck Surg 114:186–191.

Jastreboff PJ, Sasaki CT (1986) Salicylate-induced changes in spontaneous activity of single units in the inferior colliculus of the guinea pig. J Acoust Soc Am 80:1384–1391.

Johnsson LG, Hawkins JE Jr, Kingsley TC, Black FO, Matz GJ (1981) Aminoglycoside-induced cochlear pathology in man. Acta Otolaryngol Suppl 383:1–19.

Johnsson LG, Hawkins JE Jr, (1972) Strial atrophy in clinical and experimental deafness. Laryngoscope 82:1105–1125.

Jones MM, Basinger MA, Holscher MA (1992) Control of the nephrotoxicity of cisplatin by clinically used sulfur-containing compounds. Fundam Appl Toxicol 18:181–188.

Jung TTK, Woo HY, Baer W, Miller S, Juhn SK (1988) Effect of non-steroidal anti-inflammatory drugs on the hearing and prostaglandin levels in the perilymph. Otolaryngol Head Neck Surg 99:154.

Kaku Y, Farmer JC Jr, Hudson WR (1973) Ototoxic drug effects on cochlear histochemistry. Arch Otolaryngol 98:282–286.

Kaneko Y, Nakogawa T, Tanaka K (1970) Reissner's membrane after kanamycin administration. Arch Otolaryngol 92:457–462.

Kapur YP (1965) Ototoxicity of acetylsalicylic acid. Arch Otolaryngol 81:134–138.

Karlsson KK, Ulfendahl M, Khanna SM, Flock A (1991) The effects of quinine on the cochlear mechanics in the isolated temporal bone preparation. Hear Res 53:95–100.

Keefe PE (1978) Ototoxicity from oral furosemide. Drug Intell Clin Pharm 12:428.

Keene M, Hawke M, Barber HO, Farkashidy J (1982) Histopathological findings in clinical gentamicin ototoxicity. Arch Otolaryngol 108:65–70.

Kennedy ICS, Fitzharris BM, Colls BM, Atkinson CH (1990) Carboplatin is ototoxic. Cancer Chemother Pharmacol 26:232–234.

Kishore BK, Ibrahim S, Lambricht P, Laurent G, Maldague P, Tulkens PM (1992) Comparative assessment of poly-L-aspartic and poly-L-glutamic acids as protectants against gentamicin-induced renal lysosomal phospholipidosis, phospholipiduria and cell proliferation in rats. J Pharmacol Exp Ther 262:424–432.

Kisiel DL, Bobbin RP (1981) Miscellaneous Ototoxic Agents. In: Brown RD (ed) Pharmacology of Hearing: Experimental Clinical Bases. New York: John Wiley & Sons, Inc., pp. 231–270.

Kobayashi H, Ohashi N, Watanabe Y, Mizukoshi K (1987) Clinical features of cisplatin vestibulotoxicity and hearing loss. ORL J Otorhinolaryngol Relat Spec 49:67-72.

Kohn S, Fradis M, Pratt H, Zidan J, Podoshin L, Robinson E, Nir I (1988) Cisplatin ototoxicity in guinea pigs with special reference to toxic effects in the stria vascularis. Laryngoscope 98:865-871.

Koide Y, Hata A, Hando R (1966) Vulnerability of the organ of Corti in poisoning. Acta Otolaryngol 61:332-344.

Komune S, Asakuma S, Snow JB Jr (1981) Pathophysiology of the ototoxicity of cis-diamminedichloroplatinum. Otolaryngol Head Neck Surg 89:275-282.

Konishi T (1979) Effects of local application of ototoxic antibiotics on cochlear potentials in guinea pigs. Acta Otolaryngol 88:41-46.

Konishi T, Gupta BN, Prazma J (1983) Ototoxicity of cis-dichlorodiammine platinum (II) in guinea pigs. Am J Otolaryngol 4:18-26.

Kopelman J, Budnick AS, Sessions RB, Kramer MB, Wong GY (1988) Ototoxicity of high-dose cisplatin by bolus administration in patients with advanced cancers and normal hearing. Laryngoscope 98:858-864.

Kössl M, Richardson GP, Russell IJ (1990) Stereocilia bundle stiffness—effects of neomycin sulphate, A23187 and concanavalin-A. Hear Res 44:217-229.

Kroese ABA, Das A, Hudspeth AJ (1989) Blockage of the transduction channels of hair cells in the bullfrog's sacculus by aminoglycoside antibiotics. Hear Res 37:203-217.

Kuijpers W, Wilberts DPC (1976) The effect of ouabain and ethacrynic acid on the ATPase activities in the inner ear of the rat and guinea pig. ORL 38:321-327.

Laurell G, Bagger-Sjöbäck D (1991) Degeneration of the organ of Corti following intravenous administration of cisplatin. Acta Otolaryngol 111:891-898.

Laurell G, Borg E (1986) Cis-platin ototoxicity in previously noise-exposed guinea pigs. Acta Otolaryngol 101:66-74.

Laurell G, Engström B (1989) The combined effect of cisplatin and furosemide on hearing function in guinea pigs. Hear Res 38:19-26.

Lenoir M, Marat M, Uziel A (1983) Comparative ototoxicity of four aminoglycoside antibiotics during the critical period of cochlear development in the rat. Acta Otolaryngol Suppl 405:3-16.

Lindeman HH (1969) Regional differences in sensitivity of the vestibular sensory epithelia to ototoxic antibiotics. Acta Otolaryngol 67:177-189.

Lloyd-Mostyn RM, Lord IJ (1971) Ototoxicity of IV furosemide. Lancet 2:1156.

Logen TB, Prazma J, Thomas WG, Fischer ND (1974) Tobramycin ototoxicity. Arch Otolaryngol 99:190-193.

Long GR, Tubis A (1988) Modification of spontaneous and evoked otoacoustic emissions and associated psychoacoustic microstructure by aspirin consumption. J Acoust Soc Am 84:1343-1353.

Maher JE, Schreiner GE (1965) Studies on ethacrynic acid in patients with refractory edema. Ann Intern Med 62:15-29.

Marco-Algarra J, Basterra J, Marco J (1985) Cis-diaminedichloroplatinum ototoxicity. An experimental study. Acta Otolaryngol 99:343-347.

Marks SC, Schacht J (1981) Effects of ototoxic diuretics on cochlear Na, K-ATPase and adenylate cyclase. Scand Audiol Suppl 14:131-138.

Mathog RH, Klein WJ Jr (1969) Ototoxicity of ethacrynic acid and aminoglycoside antibiotics in uremia. N Engl J Med 280:1223-1224.

Mathog RH, Thomas WG, Hudson WR (1970) Ototoxicity of new and potent diuretics. Arch Otolaryngol 90:152-155.

Matz GJ (1986) Aminoglycoside ototoxicity. Am J Otolaryngol 7:117-119.

Matz GJ, Beal DD, Krames L (1969) Ototoxicity of ethacrynic acid demonstrated in a human temporal bone. Arch Otolaryngol 90:152-159.

Matz GJ, Hinojosa R (1973) Histopathology following use of ethacrynic acid. Surg Forum 24:488.

McAlpine D, Johnstone BM (1991) The ototoxic mechanism of cisplatin. Hear Res 47:191-204.

McCabe PA, Dey FL (1965) The effect of aspirin on auditory sensitivity. Ann Otol Rhinol Laryngol 74:312-325.

McCormick GC, Weinberg E, Szot R, Schwartz E (1985) Comparative ototoxicity of netilmicin, gentamicin, tobramycin in cats. Toxicol Appl Pharmacol 77:479-489.

McFadden D, Plattsmier HS (1984) Aspirin abolishes spontaneous otoacoustic emissions. J Acoust Soc Am 76:443-448.

McFadden D, Plattsmier HS, Pasanen EG (1984) Aspirin-induced hearing loss as a model of sensorineural hearing loss. Hear Res 16:251-260.

McGee TM, Olszewski J (1962) Streptomycin sulfate and dihydrostreptomycin toxicity. Behavioral and histopathologic studies. Arch Otolaryngol 75:295-311.

McGinness JE, Proctor PH, Demopoulos HB, Hokanson JA, Kirkpatrick DS (1978) Amelioration of *cis*-platinum nephrotoxicity by orgotein (superoxide dismutase). Physiol Chem Phys 10:267-277.

McPherson DL, Miller JF (1974) Choline salicylate effects on cochlear function. Arch Otolaryngol 99:304-308.

Moore EJ, ed (1983) Bases of Auditory Brain-Stem Evoked Responses. New York: Grune & Stratton.

Morin NJ, Laurent G, Nonclercq D, Toubeau G, Heuson-Stiennon J-A, Bergeron MG, Beauchamp D (1992) Epidermal growth factor accelerates renal tissue repair in a model of gentamicin nephrotoxicity in rats. Am J Physiol 263 (Renal Fluid Electrolyte Physiol 32): F806-F811.

Moroso MJ, Blair RL (1983) A review of *cis*-platinum ototoxicity. J Otolaryngol 12:365-369.

Murakami T, Inoue S, Sasaki K, Fujimoto T (1990) Studies on age-dependent platinum pharmacokinetics and ototoxicity of cisplatin. Sci Cancer Ther 6:145-151.

Myers EN, Bernstein JM (1965) Salicylate ototoxicity. Arch Otolaryngol 82:483-493.

Nakai Y, Konishi K, Chang KC, Ohashi K, Morisaki N, Minowa Y, Morimoto A (1982) Ototoxicity of the anti cancer drug cisplatin. An experimental study. Acta Otolaryngol 93:227-232.

Ninoyu O, Meyer zum Gottesberge AM (1986) Calcium transport in the endolymphatic space of cochlea and vestibular organ. Acta Otolaryngol 102:222-227.

Nordemar H, Anniko M (1983) Organ culture of the late embryonic inner ear as a model for of ototoxicity studies. Acta Otolaryngol 96:457-466.

Nuttall AL, Marques DM, Lawrence M (1977) Effects of perilymphatic perfusion with neomycin on the cochlear microphonic potential in the guinea pig. Acta Otolaryngol 83:393-400.

Otto WC, Brown RD, Gage-White L, Kupetz S, Anniko M, Penny JE, Henley CM (1988) Effects of cisplatin and thiosulfate upon auditory brainstem responses of guinea pigs. Hear Res 35:79-86.

Park JC, Cohen GM (1982) Vestibular ototoxicity in the chick: Effects of streptomycin on equilibrium and on ampullary dark cells. Am J Otolaryngol 3:117–127.

Pederson CP (1974) Brief-tone audiometry in persons treated with salicylate. Audiology 13:311–319.

Pettorossi VE, Bamonte F, Errico P, Ongini E, Draicchio F, Sabetta F (1986) Vestibulo-ocular reflex (VOR) in guinea pigs. Impairment induced by aminoglycoside antibiotics. Acta Otolaryngol 101:378–388.

Pike DA, Bosher SK (1980) The time course of the strial changes produced by furosemide. Hear Res 3:79–89.

Pillay VKG, Schwartz FD, Aimi K, Kark RM (1969) Transient and permanent deafness following treatment with ethacrynic acid in renal failure. Lancet 1:77–79.

Puel JL, Bobbin RP, Fallon M (1988) The active process is affected first by intense sound exposure. Hear Res 37:53–64.

Puel JL, Bobbin RP, Fallon M (1990) Salicylate, mefenamate, meclofenamate, and quinine on cochlear potentials. Otolaryngol Head Neck Surg 102:66–73.

Pujol R (1986) Periods of sensitivity to antibiotic treatment. Acta Otolaryngol Suppl 429:29–33.

Quick CA, Hoppe W (1975) Permanent deafness associated with furosemide administration. Ann Otol 84:94–101.

Rarey KE, Ross MD (1982) A survey of the effects of loop diuretics on the zonulae occudentes of the perilymph-endolymph barrier by freeze fracture. Acta Otolaryngol 94:307–316.

Richardson GP, Russell IJ (1991) Cochlear cultures as a model system for studying aminoglycoside induced ototoxicity. Hear Res 53:293–311.

Rifkin SI, de Quesada AM, Pickering MJ, Shires DL Jr (1978) Deafness associated with oral furosemide. South Med J 71:86–88.

Rubel EW, Oesterle E, Weisleder P (1991) Hair cell regeneration in the avian inner ear. CIBA Found Symp 160:77–102.

Ruggero MA and Rich NC (1991) Furosemide alters organ of Corti mechanics: Evidence for feedback of outer hair cells upon the basilar membrane. J Neurosci 11:1057–1067.

Rybak LP (1988) Ototoxicity of ethacrynic acid (a persistant clinical problem). J Laryngol Otol 102:518–520.

Rybak LP, Morizono T (1982) Effect of furosemide upon endolymph potassium concentration. Hear Res 7:223–231.

Rybak LP, Whitworth C (1986) Changes in endolymph chloride concentration following furosemide injection. Hear Res 24:133–136.

Rybak LP, Whitworth C (1987) Some organic acids attenuate the effects of furosemide on the endocochlear potential. Hear Res 26:89–93.

Rybak LP, Whitworth C (1988) Quinine reduces noxious cochlear effects of furosemide and ethacrynic acid. Am J Otolaryngol 9:238–243.

Rybak PR, Green TP, Juhn SK, Morizono T (1984) Probenecid reduces cochlear effects and perilymph penetration of furosemide in chinchilla. J Pharm Exp Ther 230:706–709.

Rybak LP, Whitworth C, Scott V (1990) Organic acids do not alter the cochlear effects of ethacrynic acid. Hear Res 46:95–100.

Rybak LP, Whitworth C, Scott V (1991) Comparative acute ototoxicity of loop diuretic compounds. Eur Arch Otorhinolaryngol 434:1–5.

Saito T, Moataz R, Dulon D (1991) Cisplatin blocks depolarization-induced calcium entry in isolated cochlear outer hair cells. Hear Res 56:143-147.

Santi PA, Lakhani BN (1983) The effect of bumetanide on the stria vascularis: A stereological analysis of cell volume density. Hear Res 12:151-165.

Schacht J (1979) Isolation of an aminoglycoside receptor from guinea pig inner ear tissues and kidney. Arch Otorhinolaryngol 224:129-134.

Schacht J (1986) Molecular mechanisms of drug-induced hearing loss. Hear Res 22:297-304.

Schacht J, Weiner N (1986) Aminoglycoside-induced hearing loss: A molecular hypothesis. ORL J Otorhinolaryngol Relat Spec 48:116-123.

Schacht J, Zenner HP (1987) Evidence that phosphoinositides mediate motility in cochlear outer hair cells. Hear Res 31:155-159.

Schafer SD, Wright CG, Post JD, Frenkel EP (1981) *Cis*-platinum vestibular toxicity. Cancer 47:857-859.

Schell MJ, McHaney VA, Green AA, Kun LE, Hayes FA, Horowitz M, Meyer WH (1989) Hearing loss in children and young adults receiving cisplatin with or without prior cranial irradiation. J Clin Oncol 7:754-760.

Schentag JJ, Jusko WJ (1977) Renal clearance and tissue accumulation of gentamicin. Clin Pharmacol Ther 22:364-370.

Schneider WJ, Becker EL (1966) Acute transient hearing loss after ethacrynic acid therapy. Arch Intern Med 117:715-717.

Schwartz GH, David DS, Riggio RR, Stenzel KH, Rubin AL (1970) Ototoxicity induced by furosemide. N Engl J Med 282:1413-1414.

Schweitzer VG, Dolan DF, Davidson T, Abrams GE, Snyder R (1986) Amelioration of cisplatin-induced ototoxicity by fosfomycin. Laryngoscope 96:948-958.

Schweitzer VG, Hawkins JE, Lilly DJ, Litterst CJ, Abrams G, Davis JA, Christy M (1984) Ototoxic and nephrotoxic effects of combined treatment with *cis*-diamminedichloroplatinum and kanamycin in the guinea pig. Otolaryngol Head Neck Surg 92:38-49.

Schwent VL, Williston JS, Jewett DL (1980) The effects of ototoxicity on the auditory brain stem response and the scalp recorded cochlear microphonic in guinea pigs. Laryngoscope 90:1350-1359.

Shehata WE, Brownell WE, Dieler R (1991) Effects of salicylate on shape, electromotility and membrane characteristics of isolated outer hair cells from guinea pig cochlea. Acta Otolaryngol 111:707-718.

Shepard RK, Clark GM (1985) Progressive ototoxicity of neomycin monitored using derived brainstem response audiometry. Hear Res 18:105-110.

Silverstein H, Bernstein JM, Davies DG (1967) Salicylate ototoxicity: a biochemical and electrophysiological study. Ann Otol Rhinol Laryngol 76:118-128.

Silverstein H, Yules RB (1971) The effect of diuretics on cochlear potentials and inner ear fluids. Laryngoscope 81:873-888.

Spandow O, Anniko M, Møller AR (1988) The round window as access route for agents injurious to the inner ear. Am J Otolaryngol 9:327-335.

Spongr VP, Boettcher FA, Saunders SS, Salvi RJ (1992) Effects of noise and salicylate on hair cell loss in the chinchilla cochlea. Arch Otolaryngol Head Neck Surg 118:157-164.

Stadnicki SW, Fleischman RW, Schaeppi U, Merriman P (1975) *Cis*-dichlorodiammineplatinum (II) (NSC-11987S): Hearing loss and other toxic effects in rhesus monkeys. Cancer Chemother Rep 591:467-480.

Stebbins WC (1970) Studies of hearing and hearing loss in the monkey. In: Stebbins WC (ed) Animal Psychophysics. New York: Appleton Century-Crafts, pp. 41-66.

Stebbins WC, Miller JM, Johnsson LG, Hawkins JE Jr (1969) Ototoxic hearing loss and cochlear pathology in the monkey. Ann Otol 78:1007-1025.

Stebbins WC, Moody DB, Hawkins JE Jr, Johnsson LG, Norat MA (1987) The species-specific nature of the ototoxicity of dihydrostreptomycin in the patas monkey. Neurotoxicology 8:33-44.

Stockhorst E, Schacht J (1977) Radioactive labeling of phospholipids and proteins by cochlear perfusion in the guinea pig and the effects of neomycin. Acta Otolaryngol 83:401-409.

Strauss M, Towfighi J, Lord S, Lipton A, Harvey HA, Brown B (1983) Cis-platinum ototoxicity: Clinical experience and temporal bone histopathology. Laryngoscope 93:1554-1559.

Stupp HF (1970) Untersuchung der Antibiotikaspiegel in den Innenohrflüssigkeiten und ihre Bedeutung für die spezifische Ototoxizität der Aminoglykosidantibiotika. Acta Otolaryngol Suppl 262:1-85.

Stypulkowski PM (1990) Mechanisms of salicylate ototoxicity. Hear Res 46;113-145.

Takada A, Schacht J (1982) Calcium antagonism and reversibility of gentamicin-induced loss of cochlear microphonics in the guinea pig. Hear Res 8:179-186.

Takada A, Bledsoe S, Schacht J (1985) An energy-dependent step in aminoglycoside ototoxicity: Prevention of gentamicin ototoxicity during reduced endolymphatic potential. Hear Res 19:245-251.

Tange RA, Vuzevski VD (1984) Changes in the stria vascularis of the guinea pig due to cis-platinum. Arch Otorhinolaryngol 239:41-47.

Tay LK, Bregman CL, Masters BA, Williams PD (1988) Effects of cis-diaminodichloroplatinum (II) on rabbit kidney in vivo and on rabbit renal proximal cells in culture. Cancer Res 48:2538-2543.

Thalmann R, Ise I, Bohne BA, Thalman I (1977) Actions of loop diuretics and mercurials upon the cochlea. Acta Otolaryngol 83:221-232.

Theopold HM (1977) Schädigung von Kochlea und Hörkernen nach Aminoglykosidantibiotika-Therapie. Laryngol Rhinol Otol 56:40-49.

Tran Ba Huy P, Bernard P, Schacht J (1986) Kinetics of gentamicin uptake and release in the rat: Comparison of inner ear tissues and fluids with other organs. J Clin Invest 77:1492-1500.

Van der Hulst RJ, Dreschler WA, Urbanus NA (1988) High frequency audiometry in prospective clinical research of ototoxicity due to platinum derivatives. Ann Otol Rhinol Laryngol 97:133-137.

Venkateswaran PS (1971) Transient deafness from high doses of furosemide. Br Med J 3:113-114.

Waitz JA, Moss EL Jr, Weinstein MJ (1971) Aspects of the chronic toxicity of gentamicin sulfate in cats. J Infect Dis 124S:125-129.

Wang BM, Weiner ND, Takada A, Schacht J (1984) Characterization of aminoglycoside-lipid interactions and development of a refined model for ototoxicity testing. Biochem Pharmacol 33:3257-3262.

Warchol ME, Lambert PR, Goldstein BJ, Forge A, Corwin JT (1993) Regenerative proliferation in inner ear sensory epithelia from adult guinea pigs and humans. Science 259:1619-1622.

Wästerström SA, Bredberg G, Linquis NG, Lyttkens L, Raskande H (1986) Ototoxicity of kanamycin in albino and pigmented guinea-pigs. 1. A morphologic and electrophysiologic study. Am J Otol 7:11-18.

Watanakunakorn C (1982) Treatment of infections due to methicillin-resistant

Staphylococcus aureus. Ann Intern Med 97:376-378.

Weatherly RA, Owens JJ, Catlin FI, Mahoney DM (1991) *Cis*-platinum ototoxicity in children. Laryngoscope 101:917-924.

Wersäll J, Flock A (1964) Suppression and restoration of the microphonic output from the lateral line organ after local application of streptomycin. Life Sci 3:1151-1155.

Wersäll J, Hawkins JE Jr (1962) The vestibular sensory epithelia in the cat labyrinth and their reactions in chronic streptomycin intoxication. Acta Otolaryngol 54:1-23.

West BA, Brummett RE, Himes DL (1973) Interaction of kanamycin and ethacrynic acid. Arch Otolaryngol 98:32-37.

Williams SE, Zenner HP, Schacht J (1987) Three molecular steps of aminoglycoside ototoxicity demonstrated in outer hair cells. Hear Res 30:11-18.

Woodford DM, Henderson D, Hamernick RP (1978) Effects of combinations of sodium salicylate and noise on the auditory threshold. Hear Res 42:129-142.

Wright CG, Meyerhoff WL (1984) Ototoxicity of otic drops applied to the middle ear in the chinchilla. Am J Otolaryngol 5:166-176.

Wright CG, Schaefer SD (1982) Inner ear histopathology in patients treated with *cis*-platinum. Laryngoscope 92:1408-1413.

Yonovitz A, Fisch JE (1991) Circadian rhythm dependent kanamycin-induced hearing loss in rodents assessed by auditory brainstem responses. Acta Otolaryngol 111:1006-1012.

Young LL, Wilson KA (1982) Effects of acetylsalicylic acid on speech discrimination. Audiology 21:342-349.

Zajic G, Schacht J (1987) Comparison of isolated outer hair cells from five mammalian species. Hear Res 26:249-256.

6
The Role of Viral Infection in the Development of Otopathology: Labyrinthitis and Autoimmune Disease

Nigel K. Woolf

1. Introduction

The development of the field of virology as a science has been accompanied by an increased awareness of the potential for viruses to influence virtually all areas of clinical pathology, including disorders of the inner ear. As attention has focused increasingly on the role of viral infections in the pathogenesis of different diseases of the labyrinth, the body of clinical and experimental literature in this area has grown rapidly. In spite of the relatively large body of information that is now available concerning the etiology of labyrinthine disorders, our understanding about the adverse influences of viral infections on the inner ear is still relatively incomplete. While congenital, perinatal, and postnatally acquired viral infections have long been assumed to directly or indirectly contribute to the pathogenesis of various inner ear disorders, the clinical diagnosis of a viral infection frequently has not been confirmed by sensitive and specific laboratory tests. The goal of this chapter is to review and update earlier surveys (e.g., Strauss and Davis 1973; Jaffe 1978; Davis and Johnsson 1983; Schuknecht and Donovan 1986; Davis 1990, 1993) of the role of viruses in development of hearing loss and vertigo. As part of this review, the current evidence implicating a viral etiology in the development of autoimmune diseases of the inner ear also will be examined.

2. Background

Viral infections of the labyrinth are considered to be a major source of auditory and vestibular system pathology. However, the actual etiology of hearing disorders is frequently unknown, and accurate estimates are difficult to achieve. Each year thousands of children are born with congenital deafness or perinatally acquired auditory deficits. Davis and Wood (1992) reported that the incidence of hearing impairment was 1 in 174 neonatal intensive care unit (NICU) graduates and 1 in 1278 non-NICU babies. Postnatally, the incidence of idiopathic sudden sensorineural hearing loss has been estimated at between 5 and 20 per 100,000 population per year (Byl 1984). In addition to these auditory pathologies, thousands more patients each year experience vestibular neuritis, accompanied by acute vestibular dysfunction (Schuknecht and Kitamura 1981). Jaffe (1973) found that over 40% of viral induced sensorineural hearing losses are accompanied by vestibular disturbances. Conservative estimates indicate that acute, persistent, and reactivated latent viral infections are responsible for more than 40,000 cases of hearing loss and vestibular dysfunction per year (Davis, Johnsson, and Hornfeld 1981). However, because of difficulties imposed by the location of the cochlea within the temporal bone, the evidence supporting a clinical diagnosis of viral etiology for many of these disorders of the labyrinth has been limited to circumstantial evidence. With notable exceptions, the role of viruses in the etiology of auditory and vestibular disorders has most frequently not been confirmed by direct examination of the cochlea.

Clinically, it can be extremely difficult to establish a viral etiology for cases of deafness and vertigo. Otologic exams can demonstrate the existence of an auditory or vestibular pathophysiology, but they generally are not able to confirm the cause of the disorder. Advances in computerized imaging, including CT and MRI, can now provide general information about the shape of the middle ear and bony labyrinth, but they provide little information about the structural and functional condition of the membranous labyrinth. Postmortem cytopathological exams frequently have limited value when the onset of the hearing loss or vertigo occurred years preceding death. In early studies, the reliance upon clinical diagnoses in relating viral infections with the development of auditory and vestibular deficits has been of questionable value. Rashes and other physical signs produced by one virus are usually indistinguishable from those resulting from other viral infections. Conversely, although there may be no overt symptomology during viral infection, in many cases significant auditory and vestibular system pathologies may still ensue. In the absence of laboratory isolation of a virus, serologic studies can be used to confirm that an individual has been infected with a particular virus. However, such a finding does not prove that a specific virus was the cause of the labyrinthine pathology. Furthermore, viral persistence and latency can make serological findings difficult to interpret, particularly in immunosuppressed patients.

While sensitive, specific, and reliable immunological and molecular biologic laboratory tests are now available to confirm the existence of a viral infection, prior to the 1970's the standard criteria for including patient cases in most studies was based solely upon clinical observations. It is now well recognized that the temporal association of hearing loss and vertigo with clinical signs of viral infection is not sufficient to attribute a viral etiology to an auditory or vestibular pathology. Instead, the establishment of a viral etiology for a labyrinthine disorder should be based on satisfying the conditions of Koch's postulates, as modified for infectious agents (Rivers 1937; Johnson and Gibbs 1974; Cole and Jahrsdoerfer 1988). These postulates require not only that the virus has to be associated with a disease, but that the virus has to be shown to be the actual cause of the resulting pathology (Rivers 1937).

As described by Davis and Johnsson (1983), the requirements to be satisfied for viral infections include: identification (i.e., isolation) of the virus from within the inner ear; a close association of the virus with a specific clinical syndrome of deafness and vertigo; the ability to transfer the viral infection to an experimental animal model; and the capacity to recreate a homologous disease in the animal model following viral transfer. Those viruses that have been associated clinically with diseases of the labyrinth are listed in Table 6.1. At present, only mumps, cytomegalovirus (CMV), and rubeola have been shown to unambiguously satisfy all of the modified Koch's postulates.

A major limitation to satisfying Koch's postulates has been the frequent inability to transfer human viruses to experimental animal models. For some viruses, such as cytomegalovirus, the different viral isotypes have been shown to be species-specific both in vivo and in vitro. Consequently, while the different members of the CMV family cause related diseases in their homologous species, human strains of CMV do not infect experimental animals. For other classes of virus in Table 6.1, such as human immunodeficiency virus (HIV), the virology is not as well established, and a failure to establish experimental animal models by transfer of the virus to animal species may simply reflect the choice of an inappropriate species or inadequate experimental techniques. Future studies with animal models, including transgenic species, may eventually lead to more of the viruses in Table 6.1 satisfying all four of the modified Koch's postulates.

The isolation of a virus from within the inner ear is considered to hold

TABLE 6.1 Viruses Associated with Hearing Loss and Vertigo

Adenovirus	Mumps*
Columbia SK virus	Parainfluenza
Coxsackievirus	Polio
Cytomegalovirus*	Rubella
Encephalomyocarditis	Rubeola*
Epstein-Barr virus	St. Louis encephalitis
Hepatitis virus	Tickborne encephalitis
Herpes (simplex, zoster, hominis)	Varicella-zoster (chickenpox)
Human immunodeficiency virus (HIV)	Variola
Influenza	Western equine encephalitis
Mononucleosis	Yellow fever

*Viruses satisfying Koch's postulates for causing hearing loss

significant etiological importance, since it is generally assumed viruses are not normally present in the cochlea. However, some caution is required concerning this assumption, since it has been demonstrated that guinea pigs frequently can harbor viruses in their inner ears without any apparent histological or functional signs of pathology (Craft et al. 1973; Lohle et al. 1982). A variety of approaches are currently available for isolating viruses within the temporal bone. First, infectious virus can be cultivated in vitro from perilymph, endolymph, or from inner ear tissues. Second, viral antigens can be localized in the inner ear utilizing immunohistochemistry. Third, nucleic acid sequences can be detected with in situ hybridization or polymerase chain reaction (PCR) assays. Fourth, viral particles can be identified directly by electron microscopy. Finally, morphological evidence of virus-specific cytopathology with cellular-specific tropisms has frequently been accepted as evidence of viral infection. In comparison to the first three criteria, however, cytopathological patterns of evidence are less specific, since different viruses can produce homologous cellular pathologies. Similarly, caution is required in interpreting electron microscopical results, since artifacts can be misinterpreted as viral particles.

3. Routes of Inner Ear Infection

There are four main routes through which viruses can reach the inner ear. First, following systemic infection, viruses can enter the inner ear through the blood supply to the labyrinth (i.e., hematogenous route). Second, viruses can spread centrifugally from the central nervous system by direct extension along the meninges and auditory nerve (VIIIth cranial nerve). Third, viruses can travel through fluid pathways from the central nervous system, either from the cerebrospinal fluid into the perilymphatic compartment along the cochlear aqueduct or alternatively into the endolymphatic compartment through the endolymphatic duct. Finally, viruses can infect and traverse the round or oval windows after establishing a middle ear infection.

4. Consequences of Inner Ear Infection

Once a virus has reached the inner ear, four possible pathological outcomes can develop (Strauss and Davis 1973). First, direct effects can result from acute infection of the cells of the inner ear, followed by altered cellular homeostasis or cell lysis. Second, indirect effects can arise from infections of the vascular endothelial cells resulting in diffuse or focal ischemia. Third, persistent infection may result in a continuous low-level state of viral replication, which spreads slowly and affects only a few cells at any given time. Finally, latent infections can develop when infectious virus is harbored in cells without active viral replication. While persistent and latent viral infections may not directly damage the inner ear, both can stimulate host immune responses, which in turn can lead to pathological changes via immunopathological reactions or by triggering an autoimmune reaction (Hughes et al. 1984; Rarey, Davis and Deshmukh 1984).

5. Laboratory Diagnosis of Viral Infection

Historically, the confirmation of a viral infection has been dependent upon clinical diagnosis of symptomatic disease. A clinical diagnosis was further supported when there was epidemiological evidence of a viral epidemic present in the community. Since there are no overt manifestations of infection in many cases [e.g., up to 50% of cases of mumps meningitis (Azimi, Cramblett and Haynes 1969)], laboratory confirmation is required to establish a viral etiology for a hearing loss. Viral infection can be confirmed by 1) isolating the virus by cultivating specimens from the throat, urine, blood or CSF; 2) demonstrating a significant rise (i.e., greater than fourfold increase) in the viral antibody titer of antibodies in comparisons between acute and convalescent serum samples; 3) establishing virus-specific cytopathology in the inner ear (e.g., viral inclusion bodies, giant cells); 4) identifying viral particles by electron microscopy; 5) recovering viral antigen by enzyme immunoassay for soluble antigens or immunostaining of infected epithelial cells obtained by swabbing the pharynx, conjunctiva, cervix or buccal cavity; and 6) documenting viral nucleic acids by either gel electrophoresis or dot-blot hybridization following PCR amplification of viral DNA.

The demonstration of positive viral antibody titers in patients, secondary to auditory or vestibular pathology, is not considered a sufficient proof of a viral etiology for an inner ear disorder. While the presence of a positive antibody titer can be used to demonstrate prior exposure to a virus, it does not prove the virus was the specific cause of a given inner ear pathology. The lowest standard to which the diagnosis of a viral etiology for an inner ear pathology can be held is evidence of seroconversion temporally coincident with the onset of the hearing loss and/or vestibular dysfunction. Stronger evidence of a viral etiology for an inner ear pathology is provided by isolation of the virus from the inner ear fluids or tissues. To date, only mumps (Westmore, Pickard, and Stern 1979) and CMV (Davis et al. 1979) have been cultured from human perilymph. However, as described below, through the use of modern, more sensitive immunological and molecular biological techniques, CMV, mumps, rubeola, and rubella have all been identified within the labyrinth of patients with inner ear pathologies.

6. Viral Infections Associated with Hearing Loss and Vestibular Pathologies

6.1 Mumps

Mumps was historically one of the first viruses to be associated with deafness. The first reported case of mumps associated deafness was reported by Toynbee in 1860 (cited in Evenberg 1957). The incidence of hearing loss secondary to mumps is considered to be very low, at a rate of only approximately 1 per 2300 cases (Evenberg 1957). In a study of laboratory confirmed mumps meningoencephalitis, where hearing loss can be assumed to be more common than in uncomplicated mumps cases, only 2% developed hearing losses (Koskiniemi, Donner, and Pettay 1983).

6.1.1 Clinical Findings

Mumps infection occurs most frequently in children and young adults. Hearing loss usually appears within a few days to two weeks following inflammation of the parotid gland, but may occur without parotitis (Dawes 1963). The hearing loss is usually rapid in onset, most frequently profound, and usually permanent (Vuori, Lahkainen, and Peltonen 1962). For unknown reasons the loss is monaural in approximately 80% of cases (Lindsay 1973; Smith and Gussen 1976; Jaffe 1978). The hearing loss tends to be maximal in the higher frequencies. Reports of tinnitus and a sensation of fullness are common in involved ears. Vomiting, nausea, and vertigo are also experienced by some patients. The sensation of disequilibrium usually resolves within a few weeks, although some patients are left with diminished or absent caloric responses. Absent caloric responses have occasionally been reported for mumps patients with deafness, but no history of dizziness (Hyden, Odkvist, and Kylen 1979).

Mumps meningitis occurs in up to 50% of infected patients (Bang and Bang 1943). While it has been suggested that mumps meningitis should precipitate more labyrinthine pathology than uncomplicated mumps, Azimi, Cramblett, and Haynes (1969) detected no hearing loss in their series of fifty-one children with mumps meningitis. Koskiniemi, Donner, and Pettay (1983), in a study of patients with confirmed mumps meningoencephalitis, found only 2% experienced hearing loss, while 37% had balance impairments and 13% had vertigo. In one reported case, mumps virus was isolated in perilymph from a patient with unilateral deafness associated with parotitis (Westmore, Pikcard, and Stern 1979).

6.1.2 Temporal Bone Histopathology

No studies have been reported for patients who died after recent acquisition of mumps associated deafness. However, the temporal bones from three patients with a long standing deafness that followed infection with mumps have been reported (Lindsay, Davey, and Ward 1960; Smith and Gussen 1976; Oliveira and Schnuknecht 1990). Chronic degenerative pathologies in endolymphatic structures were observed. There was relative sparing of the organ of Corti, except in the cochlear base where there was atrophy of the organ of Corti and stria vascularis. Secondary degeneration of some of the spiral ganglion cells and their distal processes was also noted. In two of these cases, the tectorial membrane was either missing, spherically deformed and encapsulated, or displaced from the organ of Corti (Smith and Gussen 1976; Oliveira and Schnuknecht 1990). In one case the sensory epithelia of the vestibular organs were degenerated and no hair cells remained (Oliveira and Schnuknecht 1990). This latter case also showed endolymphatic hydrops in both the cochlear duct and the vestibular labyrinth.

6.1.3 Comparative Animal Studies

Mumps labyrinthitis has been investigated in hamsters (Davis and Johnson 1976), guinea pigs (Fukuda 1986; Katsuhiko et al. 1988) and monkeys (*Macaca irus*) (Tanaka et al. 1988). In each of these species, the virus principally affected the endolymphatic structures, with evident infection in the nonneuroepithelial cells of the cochlear and vestibular membranous labyrinth. Immunohistochemical studies of these tissues detected viral antigen in the supporting cells of the organ of Corti, cells of both layers of Reissner's membrane, stria vascularis, and the membranous walls of the saccule and utricle. Following infection with the neurotropic wild strain of the mumps virus, neurons in the spiral and vestibular ganglia were also found to be infected (Davis and Johnson 1976). Viral antigen was not observed in the perilymphatic structures. The patterns of infection were independent of whether the infection was initiated by inoculation directly into the inner ear perilymph or by intracerebral injection. Thus, experimental findings support the hypothesis that, in addition to viremic spread to the inner ear, mumps virus can reach the cochlea via centrifugal spread from the CSF into the perilymph, or along the auditory nerve from the CNS. Davis and Johnsson (1983) have suggested that a route of infection from the CNS is more likely to occur in children, since their cochlear aqueduct is wider than in adults (Palva and Dammert 1969).

6.2 Measles (Rubeola)

Before the introduction of the rubeola vaccination, measles was a leading infectious cause of hearing loss in children. In the prevaccination era, up to 10% of cases of acquired deafness were reported to occur secondary to measles infection (Kinney 1953). However, the incidence of hearing loss induced by rubeola is very low, occurring in less than 1 case per 1000

(Miller, Stanton, and Gibbons 1956). Since the widespread implementation of childhood measles vaccination programs, the incidence of measles has been reduced significantly, and clinical cases of measles induced hearing loss are now rare.

6.2.1 Clinical Findings

Hearing loss in infected children usually develops abruptly, coincident with the appearance of the measles rash. The loss is usually bilateral, although unilateral deafness has been reported (Kinney 1953). Shambaugh, Hagens, and Holderman (1928) reported that among children who developed hearing loss secondary to measles, approximately 45% became deaf, while 55% experienced mild to moderate hearing losses. The audiograms for this group were characteristically asymmetric, with a downward slope towards the higher frequencies (Kinney 1953; Davis and Johnsson 1983). Lower frequencies (i.e., the cochlear apical turns) apparently are less likely to be impacted. The hearing loss induced by measles is typically permanent (Shambaugh, Hagens, and Holderman 1928). Tinnitus and vertigo have also been reported in conjunction with the hearing loss. Additionally, absent or diminished caloric responses have been found unilaterally or bilaterally in up to 72% of afflicted patients (Shambaugh, Hagens, and Holderman 1928).

6.2.2 Temporal Bone Histopathology

In the small number of temporal bone cases reported from individuals with deafness secondary to measles, histopathology has been observed in both the cochlear and vestibular divisions of the membranous labyrinth (Schuknecht 1974; Suboti 1976). Rubeola primarily causes an endolymphatic labyrinthitis, with the pathology confined almost entirely to the cochlear duct, saccule, and utricle. In the cochlea, degenerative changes were generally focal in the organ of Corti, stria vascularis, and tectorial membrane, and tended to be more severe in the base of the cochlea. Subacute inflammatory processes with granulomas and multinucleated Warthin-Finkeldy giant cells have also been identified in stria vascularis (Lindsay and Hemenway 1954). Pathology in the saccule and utricle consisted primarily of collapse of the membranous wall and macular neuroepithelial degeneration; however, the macula utriculi and crista ampularis usually exhibited only minimal changes.

6.2.3 Comparative Animal Studies

Experimental rubeola infection of the membranous labyrinth has been induced in hamsters (Davis and Johnson 1976). Following inoculation of the virus into either the inner ear or brain, measles infected both perilymphatic and endolymphatic structures. Viral antigen was seen in perilymphatic mesenchymal cells lining the cochlea and the cochlear aqueduct. Viral antigen was also present in the sensory cells of the organ of Corti, utricle, saccule, and the cristae of the semicircular canals. Rubeola antigen was not observed in the supporting cells of the endolymphatic membranes. Cochlear and vestibular ganglions cells also frequently exhibited viral antigen. Multinucleated giant cells, characteristic of rubeola infection, were identified within the organ of Corti and the cochlear and vestibular ganglia (Davis and Johnson 1976).

6.3 Cytomegalovirus

Cytomegalovirus is a separate genus within the family Herpesviridae. While different strains of CMV exist, the clinical significance of these variants remains uncertain. Antigenic diversity among CMVs is so great that it has not been possible to sort them into serotypic groupings. However, the different human CMV strains share at least 80% of their DNA's genetic information, which has permitted the development of accurate laboratory tests for the identification of CMV. In immunocompetent adults and infants, CMV infection has little clinical importance. The groups primarily at risk for significant sequelae from CMV infection include the developing fetus and immunodeficient individuals.

CMV is the leading cause of human congenital viral infection, occurring in up to 2.5% of all live births (Hanshaw and Dudgeon 1978). Epidemiological studies indicate that the impact of congenital CMV infection today is comparable to that of congenital rubella in the era before world-wide implementation of rubella immunization (Davis 1979). Congenital CMV is now considered such a major public health problem that Yow (1989) has proposed mandatory screening of all newborn infants for infection. The consequences of CMV infection to the developing fetus range from severe generalized illness to more subtle deficiencies that are not manifest until later in childhood. Approximately 15% of CMV congenitally infected infants manifest varying degrees of hearing loss and central nervous system damage at birth or later in childhood (Stagno et al. 1977a; Davis 1979). It has been conservatively estimated that the sequelae of congenital CMV infection, including both auditory deficits apparent at birth and late appearing hearing losses following persistent infection or reactivation of latent congenital infection, may account for 40,000 new cases of sensorineural hearing loss per year (Davis, Johnsson, and Kornfeld 1981). Consequently, CMV is now recognized as the leading nonhereditary cause of human congenital auditory impairment. At present there is no effective vaccine available for the prevention of congenital CMV infection.

Unlike rubella and toxoplasmosis, prior maternal infection with CMV does not prevent future congenital CMV infection of the fetus (Stagno et al. 1977b). However, while maternal immunity does not prevent the occurrence of congenital infection, it does generally protect against the more serious pathologies associated with congenital CMV infection (Stagno et al. 1982). Primary CMV infection during pregnancy can be detected by sequential serological testing, but at present there are no techniques for identifying women with reactivated CMV infections during pregnancy. In neonates, the presence of serum immunoglobulin M (IgM) has been accepted as evidence of congenital infection, since maternal IgM does not cross the placenta. However, the value of serological evidence remains limited, since not all infected individuals develop significant serum titers and for some viruses (e.g., herpes simplex, Buddingh et al. 1953), serum titers vary intermittently.

A significant new development in the clinical perspective of CMV infection is the impact of AIDS. Cytomegalovirus (CMV) is by far the most common opportunistic infection in AIDS patients, occurring in over 95% of all cases (Reichert et al. 1983; Quinan et al. 1984). The most widely recognized sensory system pathology in AIDS is CMV retinitis, which occurs in up to 45% of all cases (Stulting, Hendricks, and Manais 1987). However, recent data suggest that greater than 45% of asymptomatic HIV-positive patients, and 57% of symptomatic AIDS patients, also experience peripheral and central auditory and vestibular pathologies (Hausler et al. 1991). To date, more than 100 reports have appeared in the literature documenting significant otologic disease in AIDS patients, including hearing loss, vertigo, tinnitus, otalgia, and CMV infection of the temporal bone (e.g., Levy, Bredesen, and Rosenblum 1985; Kohan, Rothstein, and Cohen 1988; Kwartler et al. 1991; Hausler et al. 1991; Chandrasekhar, Siverls, and Sekhar 1992; Michaels, Soucek, and Liang 1994). These sensory system pathologies arise primarily from CMV and other opportunistic coinfections that develop in immunocompromised AIDS patients, not from direct infection with HIV (Reichert et al. 1983; Quinan et al. 1984). Given that CMV is the most common opportunistic infectious pathogen in AIDS patients, and the established propensity for CMV to infect the labyrinth (Meyers and Stool 1968; Stagno et al. 1977a; Davis 1979; Davis, Johnsson, and Kornfeld 1981; Woolf et al. 1985, 1989), it is probable that, as with AIDS-related retinitis, CMV is a primary cause of AIDS-associated inner ear pathologies.

6.3.1 Clinical Findings

Of the estimated 33,000 infants born with congenital CMV infection in the United States each year, approximately 10% develop the symptomatic form of congenital CMV: cytomegalic inclusion disease (CID) (Stagno et al. 1983). Approximately 25% of CID infants die and up to 90% of the survivors experience significant and permanent auditory (50%), intellectual (39%), or visual (22%) sequelae as a direct result of their intrauterine exposure to CMV (Stagno et al. 1977a; Hanshaw and Dudgeon 1978; Davis 1979). The remaining 90% of infants congenitally infected with CMV are asymptomatic at birth. However, as many as 20% of these infants who appear normal at birth later develop sensorineural hearing loss, neurologic abnormalities, mental retardation, expressive-language delays and learning disabilities (Dahle et al. 1979a; Hanshaw et al. 1976). The single most important late appearing sequelae of congenital CMV is sensorineural hearing loss. In over 25% of these cases, the auditory deficits initially appear only after the first year of life (Stagno et al. 1977a).

The most prevalent clinical pathologies associated with congenital CID are sensorineural hearing loss, chorioretinitis, inflammatory lymphocytic proliferation in reticuloendothelial organs, generalized teratogenic effects, and most serious, involvement of the central nervous system (CNS) with resultant severe brain damage (Stagno et al. 1977a; Hanshaw and Dudgeon 1978; Davis 1979). The cytopathologic hallmark of congenital CMV is the presence of greatly enlarged, cytomegalic cells with both intranuclear and cytoplasmic inclusions (Meyers and Stool 1968). However, inflammatory reactions secondary to the viral infection are frequently found in affected individuals (Stagno et al. 1977a), suggesting that host immunity may also play an etiologic role in congenital CMV pathogenesis.

The otologic manifestations of congenital CMV infection range from subtle auditory sensorineural and neurologic deficiencies to congenital deafness (Hanshaw and Dudgeon 1978). The hearing losses are bilateral in nearly half of the affected patients. Auditory deficits tend to be greater at higher frequencies, and they are substantial enough to interfere with learning and verbal communication. The consensus of numerous prospective and retrospective epidemiological studies is that approximately 20% of CMV congenitally infected children suffer hearing loss and/or neurological pathologies by the time they are four years old (Hanshaw et al. 1976; Stagno et al. 1977a, 1983; Dahle et al. 1979a,b; Pass et al. 1980).

CMV can be reactivated frequently during childbearing years and transmitted to the fetus in spite of the presence of substantial maternal humoral immunity (Stagno et al. 1977b, 1982). Recurrent maternal CMV infections are mainly due to reactivation of latent virus. Like rubella, an innate barrier limits CMV vertical transmission: only about 50% of primary maternal CMV infections lead to fetal congenital infection (Stagno et al. 1983). It is unclear why some congenitally infected infants develop symptomatic disease while others remain free of symptoms. While maternal immunity does not protect the fetus from congenital CMV infection, it does exert a beneficial effect by reducing the virulence of fetal infection (Stagno et al. 1982). It now appears that the stage of gestation when CMV is acquired is not correlated with the severity of fetal infection (Stagno et al. 1986).

6.3.2 Temporal Bone Histopathology

A 1981 review article (Davis, Johnsson, and Kornfeld 1981) reported finding only seven examples of temporal bone histopathology for human neonates with confirmed congenital CMV infections. Since that time only two additional temporal bone reports from CMV congenitally infected neonates have been presented (Rarey and Davis 1985; Davis et al. 1987), and a third temporal bone report on a 14-year-old who died of sequelae of congenital CID was recently reported (Rarey and Davis 1993). The cytopathologic hallmark of CMV infection is the presence of greatly enlarged cytomegalic cells, which are further characterized by the presence of both intranuclear and cytoplasmic inclusions. In CMV congenitally infected neonates with temporal bone involvement who have been studied, cytomegalic inclusion bearing cells were stereotypically identified within the marginal cells lining scala media (e.g., stria vascularis, Reissner's membrane, spiral limbus), saccule, utricle, and semicircular canals (Meyers and Stool 1968; Stagno et al. 1977a; Davis 1979; Davis, Johnsson, and Kornfeld 1981). Cochlear and saccular hydrops, as well as the collapse of Reissner's membrane and the saccular wall have also been observed. In the majority of cases, no signs of inflammation were observed and there was no evidence of teratogenesis in any of the temporal bones (Strauss 1990). In comparison with the cochlear signs, vestibular labyrinthitis was more severe, particularly in the saccule and utricle. Inclusion bearing cells were not observed in the organ of Corti, the cristae or the ganglia. Following 14 years survival after congenital CID, more extensive signs of endolymphatic as well as perilymphatic cytopathology were observed than that reported in neonates (Rarey and Davis 1993).

The method of immunofluorescent staining for CMV antigen, a more sensitive approach than gross histopathological examination, has also been applied in three of the nine reported neonatal temporal bone cases (Stagno et al. 1977a; Davis et al. 1979; Davis, Johnsson, and Kornfeld 1981): CMV antigen was detected in cells lining the endolymphatic surface of the cochlear duct, in the organ of Corti, and in spiral ganglion cells. Stagno et al. (1977a) concluded that CMV antigen production appears to be far more common in the inner ear than viral cytopathology (i.e., cytomegaly). Since inflammatory responses were more common than signs of direct viral cytopathology, it has been hypothesized that immunological responses in congenitally infected neonates may be a more significant source of cochlear pathology than direct effects of viral infection (Stagno et al. 1977b).

6.3.3 Comparative Animal Studies

Because of the species-specific nature of the viruses in the CMV family, it has not been possible to infect experimental animals with human CMV. However, reproducible in vitro and in vivo models of CMV labyrinthitis have been established in guinea pigs (e.g., Harris et al. 1984; Woolf et al. 1985, 1988, 1989; Woolf 1991; Keithley et al. 1988; Keithley, Woolf, and Harris 1989; Woolf 1990) and mice (Davis and Strauss 1973; Davis and Hawrisiak 1977; Davis 1981) following infection with their species-specific strains of CMV. In contrast to the regular sequence of progressive labyrinthitis and hearing loss observed after primary guinea pig CMV (GPCMV) infection of the inner ear (Harris et al. 1984; Woolf et al. 1985, 1988, Woolf 1990; Keithley, Woolf, and Harris 1989), inoculation of GPCMV into the scala tympani of seropositive guinea pigs (i.e., secondary infection) did not induce either GPCMV labyrinthitis or hearing loss (Woolf et al. 1985). Unlike the CMV endolabyrinthitis most typically found in humans, primary CMV infection in guinea pigs and mice, induced by inoculation of species-specific CMV into either the inner ear or brain, has consisted predominately of a cochlear and vestibular perilabyrinthitis.

The histopathological signs of murine CMV (MCMV) and GPCMV labyrinthitis have typically consisted of perilabyrinthine inflammation, fibrosis, hemorrhage, and cytomegaly (i.e., swelling, rounding, and formation of intranuclear inclusions (Fig. 6.1) in cells lining the perilymphatic spaces (i.e., scala tympani, scala vestibuli, and Reissner's membrane) and in spiral ganglion cells. Cytopathology typically consisted of partial degeneration of the organ of Corti and the loss of spiral ganglion cells. At the

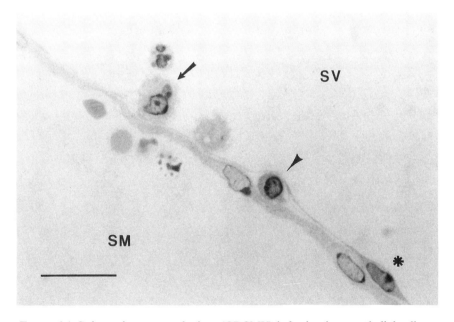

FIGURE 6.1 Guinea pig cytomegalovirus (GPCMV) infection in mesothelial cells on the scala vestibuli side of Reissner's membrane during GPCMV labyrinthitis. Within the space of three cells, the cells changed from a normal flat morphology (star), to a slightly abnormal profile (arrowhead), to a typical cytomegalic profile (arrow), with characteristic cell rounding, loss of adherence, and inclusion formation in the nucleus. Epithelial cells on the scala media side of Reissner's are normal. Infiltrating neutrophils and platlets are present in scala media (SM) and scala vestibuli (SV) adjatent to Reissner's membrane. Scale bar = 10 μm.

electron microscopical level, GPCMV nucleocapsids (i.e., the complete viral DNA without its surrounding membrane) have been identified within the nucleus of inflammatory cells (Fig. 6.2) infiltrating the cochlea during GPCMV labyrinthitis. Using immunohistochemical staining for GPCMV antigens (Fig. 6.3), CMV infection has been observed in spiral ganglion cells and mesothelial cells lining the perilymphatic compartment of GPCMV infected cochleas (Keithley et al. 1988; Keithley, Woolf, and Harris 1989; Woolf et al. 1989). Cells stained with GPCMV viral antigen were much more common than cells showing virus-specific cytopathology. More recent molecular biological investigations of GPCMV labyrinthitis, including in situ hybridization studies localizing GPCMV mRNA expression (Woolf 1994) and PCR analysis of GPCMV DNA (Woolf unpublished), have also confirmed viral infection within the cochlea. With respect to the absolute sensitivity of these measures, the smallest number of GPCMV infected cells was observed by cytopathology, significantly more infected cells were detected by immunohistochemical staining, and the greatest number of infected cells was found with in situ hybridization assays. Thus, as the sensitivity of the investigative approach has increased, there has been a corresponding significant increase in not only the number of cells, but also in the categories of cells showing infection. Similar trends are to be expected

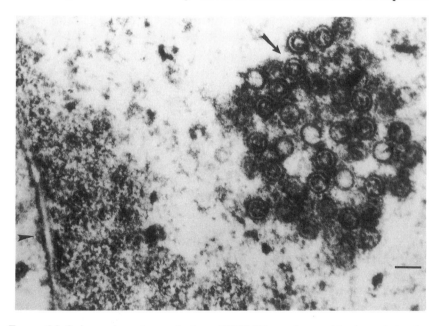

FIGURE 6.2 Guinea pig cytomegalovirus (GPCMV) nucleocapsids (arrow) in the nucleus adjacent to the nuclear membrane (arrowhead) of a polymorphonuclear neutrophil infiltrating scala tympani during GPCMV labyrinthitis. Scale bar = 100 nm.

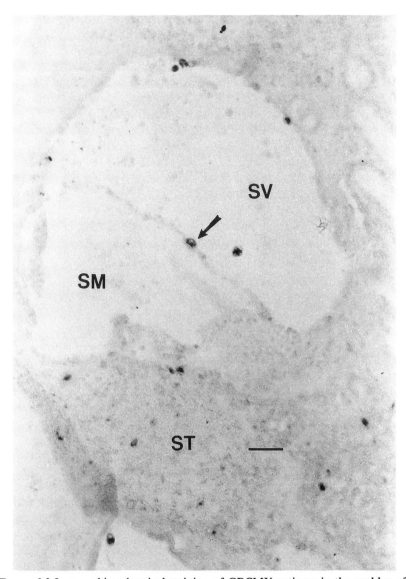

FIGURE 6.3 Immunohistochemical staining of GPCMV antigens in the cochlea of a guinea pig at 8 d postinoculation of scala tympani (ST) with $10^{2.5}$ Median tissue culture infective dose $TCID_{50}$ GPCMV. Immunostained cells (e.g., mesothelial cell in Reissner's membrane (arrow)) appeared darkly stained in this lightly counterstained section. GPCMV antigen immunostaining was observed in mesothelial cells lining ST and scala vestibuli (SV), in inflammatory infiltrating cells in the peripymphatic compartments (SV and ST), and in spiral ganglion cells. Immunostaining was not observed in the endolymphatic compartment (i.e., scala media: SM). Scale bar = 50 μm.

in future studies of human temporal bones using the more sensitive molecular biological techniques (e.g., in situ hybridization, PCR).

Unlike the situation in the murine CMV model (Johnson 1969), GPCMV crosses the placenta and infects the fetus in utero in the guinea pig model (Choi and Hsiung 1978). Previous studies have demonstrated that the auditory system is susceptible to GPCMV congenital infection (Woolf et al. 1989; Woolf 1990, 1991). In a striking example, Nomura, Hara, and Kurata (1988) have shown congenital GPCMV infection can also cause significant cochlear teratologies (Fig. 6.4).

Using immunohistochemical staining for GPCMV antigens, 45% of congenitally infected neonates had GPCMV isolated within the inner ear at birth, primarily in the spiral ganglion cells (21%), vestibular ganglion cells (24%), endothelial cells of the modiolar vein (7%), and inflammatory infiltrates within the perilymphatic scalae (7%) (Woolf et al. 1989). In 28% of the GPCMV congenitally infected neonates, GPCMV was also localized in the temporal bone marrow (Woolf et al. 1989; Woolf 1990, 1991). In

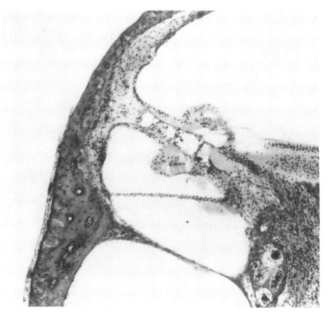

FIGURE 6.4 Teratology induced following congenital GPCMV infection in the apex of the cochlea of a guinea pig neonate. The mother became pregnant shortly after she was infected by GPCMV inoculation ($2 \times 10^3 TCID_{50}$ GPCMV) into scala tympani through the round window. The number of turns in the offspring's cochlea was normal, but there was hyperplasia of the cochlear duct and the organ of Corti. The organ of Corti did not end normally at the apex, but continued to grow basalwards forming a mirror image (i.e., double) organ of Corti. Note also the supernumerary (i.e., 7 instead of 3) outer hair cells in the abnormal (i.e., basal) organ of Corti in the pair. From Nomura et al. (1988), reprinted by permission.

comparison to GPCMV labyrinthitis in adult guinea pigs, where extensive inflammatory reactions were a hallmark of primary infection (Harris et al. 1984; Woolf et al. 1985), inflammation was relatively infrequent in congenitally infected animals (Woolf et al. 1989). The high prevalence of GPCMV antigens in the temporal bone of these congenitally infected neonates (i.e., 66%) suggested that the developing ear is a specific target tissue for congenital infection. Of the guinea pigs shown to be congenitally infected, 28% also exhibited significant auditory deficits, as measured by cochlear microphonic, N1 compound action potential, and auditory brainstem evoked response thresholds (Woolf et al 1989; Woolf 1990, 1991). GPCMV congenitally infected neonates also exhibited significant increases in their central auditory system conduction times, as indicated by prolongation of their auditory brainstem responses (ABR) Wave I-IV intervals (Woolf 1991). No significant differences were noted in the cochlear infection patterns and hearing losses emanating from the first two trimesters of pregnancy. However, the consequences of congenital infection on third trimester maternal infection group offspring could not be assessed, since all of the mothers inoculated with GPCMV in the middle of the third trimester of pregnancy delivered dead babies (13/13) at term (Woolf et al. 1989).

The pattern of auditory system neuropathology in congenital CMV infected guinea pigs is similar to that reported in humans (reviewed above). Notably, congenital CMV infection of the auditory nerve in both humans (Stagno et al. 1977a) and guinea pigs (Woolf et al. 1989) has been confirmed through the use of immunohistochemical staining techniques. However, cytomegalic inclusion bodies have not been observed in the auditory nerve spiral ganglion cells in any of the congenitally infected species examined to date. While cytopathology is not necessarily a characteristic of congenital CMV infections in the peripheral auditory system, congenital CMV infection has been associated with auditory nerve pathophysiology in both guinea pigs (Woolf et al. 1989) and humans (Stagno et al. 1977a). Thus, the common pattern of auditory nerve infection and sensorineural hearing loss in both guinea pigs and humans implies that the neuropathogenesis of congenital CMV induced auditory impairment in both species is similar (Woolf et al. 1989; Woolf 1990, 1991).

In contrast to the similar pattern of congenital CMV infection in cochlear neural elements for humans and guinea pigs, a distinct species difference occurs in nonneural regions of the temporal bone. Both in vivo and in vitro, animal CMV replicates in cells lining the perilymphatic compartments (Davis 1979, 1990; Harris et al. 1984; Woolf et al. 1985, 1988; Fukuda, Keithley, and Harris 1988; Keithley et al. 1988), while human CMV replicates primarily in cells lining the endolymphatic compartment (Meyers and Stool 1968; Stagno et al. 1977a; Davis 1979; Davis et al. 1981). The significance of this species difference in the cellular tropisms for CMV infection in nonneural regions of the cochlea is currently unknown.

In experimental studies, 72% of congenitally infected guinea pig neonates had normal hearing at birth (Woolf et al. 1989; Woolf 1991). These animals were comparable to humans with so called "silent" congenital CMV infections (i.e., confirmed congenital CMV infection in the absence of overt anatomical or functional pathology) (Hanshaw et al. 1976). In children, silent congenital CMV infections are associated with progressive sensorineural hearing loss (Hanshaw et al. 1976; Dahle et al. 1979b). The experimental data suggest that similar progressive hearing losses also occurs in guinea pigs (Woolf unpublished).

6.4 Rubella

Postnatal infection with rubella virus, a member of the Togaviridae, usually causes a mild disease (German measles) of short duration, generally without complications, and induces a permanent immunity. In contrast, the ear, eye, and heart of the fetus are particularly vulnerable to congenital infection during the first trimester of pregnancy (Gregg 1941; Swan et al. 1943). The correlation between congenital rubella infection and hearing loss was first established during the Australian epidemics of 1939–41 (Swan et al. 1943), however, it was only during the period of the last major rubella epidemic in the United States, between 1963–65, that the full significance of congenital rubella emerged. During this epidemic, congenital rubella infection was responsible for the development of hearing loss in over 12,000 infants (Trybus et al. 1980). Thus, prior to the introduction of widespread rubella vaccination, congenital rubella infection was a major source of acquired deafness. Following the introduction of childhood rubella vaccination, there has been a significant drop in the incidence of rubella infection (Davis and Johnsson 1983), although sporadic small-scale epidemics of congenital rubella infection and hearing loss in unvaccinated individuals continue to occur (Orenstein and Greaves 1982). In addition to the threat from primary rubella infections for uninoculated individuals, there is also an established risk for reinfection of mothers previously immunized against rubella. Severe deafness following congenital infection has been reported for mothers with preexisting immunity against rubella (Keith 1991).

The most serious consequences of maternal rubella infection are associated with infection during the first trimester of pregnancy, although hearing impairment can occur following maternal infection at any time during pregnancy (reviewed in Bordley, Brookhauser, and Worthington 1972). During the first trimester of pregnancy, rubella is most teratogenic; in the second trimester rubella teratologies are less frequent, but may still occur; and during the third trimester damage to the fetus is rare (Bordley and Hardy 1969). With respect to identifying rubella infection during pregnancy, it is now widely recognized that clinical diagnosis of rubella is unreliable. The rubella rash cannot be distinguished from the rashes of other viruses, and, consequently, rubella infection must be confirmed by laboratory testing. Furthermore, it has been demonstrated that the auditory system can be seriously damaged by rubella infection even in cases where there have been no obvious signs of viral infection (Karmody 1968). In prospective studies of maternal seroconversion during pregnancy, it has been reported that approximately 50% of the maternal infections were subclinical (Bordley and Hardy 1969).

6.4.1 Clinical Findings

Rubella, cytomegalovirus, and herpes hominis (simplex) are the three groups of viruses known to produce teratogenic effects in humans (Catalano and Sever 1971). The sequelae of congenital rubella can involve either malformation, pathophysiology (with or without concurrent cytopathology), or latent infection (Strauss and Davis 1973). Rubella is categorized as a noncytolytic virus, acting primarily through the inhibition of mitosis and increased incidence of chromosomal breakage (Plotkin et al. 1965). Because of the difficulties associated with testing hearing in young children, hearing losses are frequently not detected at birth, but are instead identified only later in childhood. Desmond et al. (1969) observed an age-related increase in the incidence of hearing loss in her laboratory confirmed cases of congenital rubella: 42% at 40 weeks postpartum, 45% at 18 months, and 72% at 3 years. While these increases in incidence may have reflected in part the technical demands of testing hearing in neonates, others have concluded that the observed increases in the incidence of hearing loss with postnatal age are the result of a true progressive hearing loss (Alford 1968; Bordley and Hardy 1969; Bordley and Alford 1970; Desmond et al. 1970).

Hearing losses secondary to congenital rubella are predominately bilateral (Borton and Stark 1970; Fitzgerald et al. 1970). However, unilateral hearing losses have also been reported (Hodgson, 1969). Approximately 55% of children with hearing loss secondary to congenital rubella had profound deficits (>91 dB ISO), 30% had severe deficits (71 to 90 dB ISO), and 15% mild to moderate losses (<70 dB ISO) (Bordley and Hardy 1969; Hodgson 1969; Borton and Stark 1970; Fitzgerald, Sillon, and McConnell 1970; Trybus et al. 1980). The audiograms associated with congenital rubella typically are flat across stimulus frequencies, but "belly-type curves" (i.e., flat with maximal loss in the 0.5 to 2.0 kHz range), and high frequency slopes (i.e., falling contour with increasing frequency) have also been reported (Hardy et al. 1969; Bordley and Alford 1970).

6.4.2 Temporal Bone Histopathology

The number of temporal bones describing rubella histopathology remains quite limited, and there has been an emphasis on material acquired from fetal, premature and young infants (Lindsay et al. 1953; Friedmann and Wright 1966; Alford 1968; Ward, Honrubia, and Moore 1968; Brookhouser and Bordley 1973). Congenital rubella infection in the first trimester of pregnancy typically produced cochleosaccular inner ear malformations (i.e., Scheibe inner ear anomaly) (Meyers and Stool 1968). Characteristically, there has been varying degrees of degeneration in the cochlear neuroepithelium (Brookhauser and Bordley 1973). There often was a lack of development of the hair cells and pillar cells, particularly in the apical end of the cochlea (Meyers and Stool 1968). The tectorial membrane was frequently displaced from the organ of Corti, rolled up against the limbus, in contact with Reissner's membrane, and enclosed in a sheath of flattened cells (Meyers and Stool 1968; Davis and Johnsson 1983). The stria vascularis was partly or totally atrophied (Lindsay et al. 1953; Alford 1968; Brookhouser and Bordley 1973). Partial collapse of the saccule, with saccular membranes collapsed and adherent to a degenerated macula, has also been reported (Lindsay and Hemenway 1954; Esterley and Oppenheimer 1973).

The earliest histopathological studies emphasized the presence of hemorrhage in the inner ear (Carruthers 1945; Keleman and Gottlieb 1959). While it is possible that these findings were related to artifacts of temporal bone processing, a connection between rubella and hemorrhage has also been noted in patients with rubella related thrombocytopenic purpura (Monif et al. 1965). While the significance of inner ear hemorrhage has been discounted by some investigators (Lindsay and Hemenway 1954), large aggregations of inflammatory cells forming small granulomas in stria vascularis have been reported (Friedmann and Wright 1966), pathologies that could in turn affect the production of endolymph and the overall condition of the epithelium of the cochlear duct (Lindsay et al. 1953). These inner ear histopathological findings are consistent with degeneration of neuroepithelial structures after they have matured, suggesting the rubella virus may damage the cochlea continuously during embryogenesis, and not merely arrest development of the inner ear at the time of infection (Lindsay et al. 1953; Lindsay 1973).

6.4.3 Comparative Animal Studies

Experimental evidence of rubella infection in animals is limited to a single in vitro study of chick embryo otocyst organ cultures (Shikani and Richtsmeier 1991). Morphological and microscopic changes in the otocyst following rubella infection consisted of individual cell death and delayed organ differentiation. These results suggested that the peripheral component of congenital deafness associated with rubella infection was a direct effect of the virus itself.

6.5 Herpes Zoster Oticus

Primary infection with varicella zoster virus (VZV), a DNA virus in the herpes virus group, causes varicella or chickenpox. In children, varicella is a self-limited infection, and only rarely is varicella associated with hearing loss (Jaffe 1978). However, in adults and immunocompromised individuals VZV infection may cause devastating effects, including auditory and vestibular pathologies (Byl and Adour 1977; Heathfield and Mee 1978). VZV is one of the most common infections of childhood, and almost everyone has been infected by the end of adolescence (Myers and Connelly 1992). VZV infection is transmitted by direct contact with skin lesions or exposure to infected respiratory secretions (Brunell 1989). Following primary infection, VZV establishes a cell associated viremia, vesicular cutaneous lesions and latent or persistent infection in neurons (Gilden et al. 1987) or supporting cells (Croen et al. 1988) of the dorsal root ganglia. As with other herpes viruses, currently available clinical treatment to control VZV infection does not prevent the establishment of persistent infections.

Herpes zoster oticus, or Ramsy Hunt (1907) syndrome, is induced by reactivation of latent VZV, frequently years after the primary infection. Reactivation of the virus in neural (e.g., geniculate) ganglia results in the development of vesicular eruptions in the outer ear and the external auditory canal, followed over the next few days by facial paralysis, and auditory and vestibular pathologies (Aleksic, Budzilovich, and Lieberman 1973). Since other viruses can mimic the clinical symptoms and facial paralysis induced by herpes zoster infection (e.g., mumps and cytomegalovirus), laboratory confirmation of herpes zoster infection is required (Najoo et al. 1988; Koropchak et al. 1991).

6.5.1 Clinical Findings

Auditory and/or vestibular findings develop in up to 50% of patients with herpes zoster oticus (Byl and Adour 1977; Heathfield and Mee 1978). Symptoms include hyperacusis, tinnitus, nystagmus, and disequilibrium. Severe vertigo can also occur in the absence of facial paralysis (Zajtchuk, Matz, and Lindsay 1972; Proctor et al. 1979) or hearing loss (Proctor et al. 1979). While ear complaints are relatively common with herpes zoster oticus infection, less than 7% of patients experience sensorineural hearing loss (Byl and Adour 1977). The facial paralysis, hearing loss, and vertigo usually improve over the course of several weeks following onset. However, patients occasionally experience permanent hearing loss and reduced caloric responses (Blackley, Friedman, and Wright 1967; Heathfield and Mee 1978).

6.5.2 Temporal Bone Histopathology

Temporal bones from patients with herpes zoster oticus have been collected from within a few weeks to years after the onset of acute disease (Denny-Brown, Adams, and Fitzgerald 1944; Rose, Brett, and Burston 1964; Blackley, Friedman, and Wright 1967; Zajtchuk, Matz, and Lindsay 1972; Proctor et al. 1979). The main histopathological findings in these temporal bone reports have been extensive perivascular, perineural and intraneural inflammatory lymphocytic infiltrations in the facial nerve, auditory nerve and cochlea. Edema, necrosis, and hemorrhage have also been observed in these temporal bones (Denny-Brown, Adams, and Fitzgerald 1944; Rose, Brett, and Burston 1964). Some specimens exhibited inflammatory cells in the saccule and utricle (Davis and Johnsson 1983). Extensive perivascular cuffing by lymphocytes was frequently apparent in the modiolus and in spiral and vestibular ganglia; however, for some cases inflammation was not observed in the ganglia. After prolonged survival, atrophy of some neurones in the spiral and vestibular ganglia was evident, although the majority of the neurones appeared to have survived (Blackley, Friedman, and Wright 1967; Proctor et al. 1979).

6.5.3 Comparative Animal Studies

VZV is highly species-specific, and until recently natural infection has only been demonstrated in humans and gorillas (Myers, Kramer, and Stanberry 1987). While the natural history of VZV in gorillas resembles that in humans (Marennikova et al. 1974), practical considerations have prevented their use as an experimental model system. African green monkeys, Patas monkeys, and pygmy marmosets have also been shown to be susceptible to VZV infection, but they do not develop symptomatic disease resembling that in humans (reviewed in Myers and Connelly 1992).

Guinea pigs inoculated with an attenuated strain of VZV developed a limited viremic infection, acquired antibody, and exhibited VZV-specific lymphoproliferative responses (Matsunaga et al. 1982; Myers, Stanberry, and Edmund 1985; Myers, Connelly, and Stanberry 1991). Rats also seroconverted after VZV inoculation, and virus persisted for months in rat dorsal root ganglion neurons (Sadzot-Delvaux et al. 1990). Isolation of VZV in rat trigeminal and thoracic dorsal root ganglia has been accomplished using both polymerase chain reaction and in situ hybridization techniques (Sadzot-Delvaux et al. 1990; Lowry et al. 1993). While guinea pigs, rats, or other small animal models may eventually prove to be useful for investigations of herpes zoster oticus, an experimental model of VZV latency and pathology in the temporal bone has not yet been established.

7. Other Viruses Associated with Hearing and Vestibular Pathology

The remaining viruses in Table 6.1 have been less well characterized with respect to their role in the etiopathogenesis of inner ear disorders. For most of these viruses, only a temporal association has been made between laboratory confirmation of viral infection and the development of inner ear pathophysiology, and thus they have not been shown to satisfy the rigorous requirements of Koch's modified postulates for viruses (Davis and Johnsson 1983). At the minimum level of proof, evidence of viral infection has rested upon serologic confirmation of a four-fold or greater rise in antibody titer for convalescent sera in comparison to sera obtained during or preceding acute infection (Jaffe 1978). In addition, the association between these viruses and inner ear pathology has frequently been further supported by epidemiological evidence. Notably, viruses that cause upper respiratory infections (e.g., influenza, parainfluenza, and adenovirus) have generally been assumed to play a role in the etiopathogenesis of subsequent hearing loss. Khetarpal et al. (1990), after reviewing the literature, concluded that there was a clinical history of preceding or concurrent viral infection of the upper respiratory tract in up to 33% of patients with idiopathic sudden hearing loss (Van Dishoeck and Bierman 1957; Jaffe 1973; Shaia and Sheehy 1976). Finally, in some cases, signs of virus-specific cochlear cytopathology in the organ of Corti, tectorial membrane, stria vascularis, cochlear and/or vestibular neurons have been accepted as evidence of viral labyrinthitis (Lindsay, Davey, and Ward 1960; Beal, Davey, and Lindsay 1967; Sando et al. 1977; Schuknecht and Donovan 1986).

7.1 Influenza Types A and B

Studies of the incidence of positive serologic conversions from paired acute versus convalescent sera have implicated influenza, along with adenovirus, mumps, and parainfluenza, as the viruses most frequently involved in sudden hearing loss (Khetarpal, Nadol, and Glynn 1990). In addition to hearing loss, influenza types A and B have also been associated with vestibular neuritis (Hirata et al. 1989). In the Veltri et al. (1981) study of patients with idiopathic sudden hearing loss, 26% had documented seroconversions to influenza types A and B viruses.

Experimental animal studies with influenza types A and B have shown infection occurred primarily in the mesenchymal cells lining the perilymphatic channels of the cochlear aqueduct and scala tympani of hamsters (Davis and Johnson 1976) and guinea pigs (Davis 1993). Neuroadaptive strains of influenza caused extensive infection of the spiral and vestibular ganglia of newborn hamsters, while nonadapted strains only rarely infected neurons in these ganglia (Davis and Johnson 1976). Rarey, David, and Deshmukh (1984) demonstrated that intranasal inoculation of influenza B in ferrets (a model of Reyes's syndrome) induced vacuolation of the epithelial cells in stria vascularis and the semicircular canals. Secondary loss of sensory hair cells in the cristae of semicircular canals was also observed. As in Reyes's syndrome in humans, the inner ear pathologies observed in the ferrets occurred in the absence of direct evidence of viral infection of the labyrinth (Rarey et al. 1984).

7.2 Parainfluenza Serotypes 1, 2 and 3

Based on serologic conversion data, parainfluenza is also considered to be among the most common viral agents associated with hearing loss (Shaia and Sheehy 1976; Veltri et al. 1981; Wilson et al. 1982). Laboratory based studies of Veltri et al. (1981) and Wilson et al. (1982) have also suggested that parainfluenza virus, possibly reactivated following acute infection by another viral strain, is one of a cluster of viruses frequently involved in idiopathic sudden hearing loss. Veltri et al. (1981) found parainfluenza to be the sole infectious agent in 1.3% of their patients with sudden hearing loss (confirmed by significant seroconversion) and as one of a cluster of viruses in 10% of patients with sudden hearing loss.

Experimental models using mice (Shimokata et al. 1977) and guinea pigs (Fukuda 1986) have identified parainfluenza virus type 1 (Sendai virus) infection in both perilymphatic and endolymphatic structures of the labyrinth. Sendai viral antigen was localized in stria vascularis, Reissner's membrane, and the organ of Corti (Fukuda 1986). Evidence of fibrosis in scala tympani, vacuolization in stria vascularis, and inflammatory infiltrates within scalae tympani and vestibuli were also observed in the latter study.

7.3 Adenovirus

Maassab (1973), in collaboration with Jaffe (1973), reported in their paired clinical and virologic studies that adenoviruses and *Mycoplasma pneumoniae* were the leading causes of idiopathic sudden hearing loss. Adenoviruses were identified as the primary infectious agent correlated with hearing loss in 28% of their patients diagnosed by the serologic criteria of a four-fold increase in antibody titer, and in 15% of the patients diagnosed by nasopharyngeal swabs. Veltri et al. (1981) and Wilson et al. (1982) also identified adenoviruses as being correlated with hearing losses, but only in cases where there were seroconversions to multiple viruses, suggesting that adenoviruses may become involved only after acute infection with another strain of virus reactivates latent adenovirus infection. Adenoviruses are nonneurotropic viruses and the mechanism(s) by which they cause cochlear pathology is unknown (Khetarpal, Nadol, and Glynn 1990). Davis (1993) did not connect the adenovirus infection with vestibular pathology in his review of viral vestibular neuritis. No experimental animal models of adenovirus infection of the inner ear have been reported.

7.4 Herpes Hominis (Simplex Types 1 and 2)

Veltri et al. (1981) found herpes simplex type 1 to be the sole specific infectious agent identified in 1% of his patients with sudden hearing loss and one of a cluster of multiple viruses exhibiting seroconversion in 9% of their cases. The incidence of herpes virus involvement was much more frequent for patients with multiple seroconversions (40%) than for those with only a single seroconversion (4%). While herpes simplex is also known to produce teratogenic effects in humans, a direct association with congenital deafness has not been established (Catalano and Sever 1971).

In rabbits (Kumagami 1972) and neonatal hamsters (Davis and Johnson 1976), herpes simplex has been shown to infect the perilymphatic, neuroepithelial and peripheral ganglion structures within the membranous labyrinth. Viral antigen immunostaining was found in both the spiral and vestibular ganglia, the organ of Corti, maculae of the saccule and utricle, and cristae of the semicircular canals (Davis and Johnson 1976). Herpes simplex antigen was also observed in the mesenchymal cells lining scala tympani and the cochlear aqueduct. In guinea pigs, herpes simplex viral antigens were detected within the cochlear duct in Reissner's membrane, stria vascularis, the spiral ligament, and the basilar membrane (Nomura, Hara, and Kurata 1985). Virus-like particles resembling herpes simplex virus have also been identified by electron microscopy in otherwise healthy guinea pigs, suggesting humans could also harbor latent herpes simplex infections in the inner ear (Lohle et al. 1982).

7.5 Variola

Variola, or smallpox, has been reported to infect the inner ear of humans. Temporal bones from patients who died from smallpox have exhibited a pattern of infection suggestive of an acute middle ear infection that spread secondarily to the inner ear via the round window (Bordley and Kapur 1972). An experimental model of variola infection in the inner ear has not been established, but vaccinia (cowpox), a related apoxvirus, has been investigated in neonatal hamsters (Davis and Johnson 1976). Vaccinia virus, inoculated into the inner ear or brain in hamsters, infected both perilymphatic and endolymphatic structures. Mesenchymal cells lining the perilymphatic scalae of the cochlea, the cochlear aqueduct, and the saccule were immunostained for viral antigen. Viral antigen was also present in the organ of Corti, limbus, and basilar membrane. No vaccinia virus antigen was observed in the spiral or vestibular ganglia.

7.6 HIV and AIDS

Hearing loss and central auditory system neuropathology are common features of patients with acquired immune deficiency syndrome (AIDS) (Petito 1988; Kohan, Rothstein, and Cohen 1988; Kwartler et al. 1991; Hausler et al. 1991; Chandrasekhar, Siverls, and Sekhar 1992). Auditory and vestibular complaints have been reported as the presenting symptoms for AIDS patients (Breda et al. 1988; Kohan, Rothstein, and Cohen 1988). Estimates indicate that over 90% of AIDS patients suffer CNS infections

(Petito 1988), 70% have ocular signs (Stulting, Hendricks, and Marais 1987), and up to 60% develop peripheral and central auditory and vestibular pathologies (Kohan, Rothstein, and Cohen 1988). Approximately 45% of AIDS related complex (ARC) patients, prior to the onset of full-blown AIDS, also exhibit audiological and vestibular neuro-otological disorders (Levy et al. 1985). Direct infection of the nervous system with HIV (human immunodeficiency virus) has been well documented, and tropisms for a specific neural cell population have been reported (Elder and Sever 1988). However, while a role for direct infection with HIV has also been proposed for the pathogenesis of AIDS related sensory system neuropathologies, most of the clinical pathologies in AIDS patients are not caused directly by HIV, but instead are the result of opportunistic coinfections (Nelson et al. 1988; Wiley and Nelson 1988). CMV is by far the leading cofactor in AIDS: more than 95% of AIDS patients are coinfected with CMV (Reichert et al. 1983; Quinnan et al. 1984). In immunocompetent individuals CMV infections are usually benign, but in immunocompromised patients CMV infections frequently produce significant morbidity and mortality (Nankervis and Kumar 1978).

While the prevalence of neuro-otologic disease in AIDS patients is still largely unrecognized at present by the medical community, more than 100 reports of significant auditory and vestibular pathophysiology in AIDS and ARC patients have appeared in the clinical literature (e.g., Kohan et al. 1988; Petito 1988; Hausler et al. 1991; Kwartler et al. 1991; Chandrasekhar, Siverls, and Sekhar 1992). Several imaging studies have demonstrated peripheral and central auditory system pathologies (Belman et al. 1986; Kohan, Hammerschlag, and Holliday 1990), and other recent reports have detected viral labyrinthitis in the temporal bones of AIDS patients (Brivio et al. 1991; Collier et al. 1992; Michaels et al. 1994). In only two of these temporal bone reports (Kwartler et al. 1991; Michaels et al. 1994), however, were audiometric background data available. Michaels et al. (1994) reported hearing losses greater than 20 dB, of the sensorineural type, in 69% (42/61) of their patients: 41 of the 68 affected ears showed hearing losses of 20 to 40 dB across the whole range of frequencies tested, 27 showed losses greater than 40 dB, and 2 had losses in excess of 95 dB. Opportunistic infections in these inner ears included cryptococcus, cytomegalovirus, and deposits of Kaposi's sarcoma. CMV infection was considered to be a particularly significant cause of the peripheral hearing loss in these cases, since 5 of 25 patients (20%) with AIDS showed CMV changes in their inner ears (i.e., cochlear CMV inclusion bearing cells or CMV genome by in situ hybridization).

In the Kwartler et al. (1991) case, audiometric thresholds were not presented because the patient was uncooperative, but acoustic reflex thresholds were elevated at 500 Hz and absent at 1000 Hz and 2000 Hz. The major finding in this study was marked invasion of the auditory nerve by *Cryptococcus neoformans*, early necrosis of the cochlear and vestibular nerves, and necrosis of the organ of Corti. The stria vascularis was edematous and mild endolymphatic hydrops was seen in the cochlea and saccule.

In the most detailed temporal bone investigation to date, Chandrasekhar, Silverls, and Sekhar (1992) reported the organ of Corti was normal in all seven AIDS temporal bones they examined, but pathological findings consisted of petromastoiditis, perilymphatic and endolymphatic precipitates, and neuroepithelial changes. The unusual finding of inflammatory endolymphatic precipitate in the semicircular canals and fibrillar perilymphatic precipitations suggested a viral infectious etiology. However, while auditory and vestibular pathologies are common findings in AIDS patients, because of the ubiquitous presence of opportunistic infections in these individuals, establishing a direct role for HIV in inner ear pathologies must depend upon identifying either viral antigens in the cochlea by immunohistochemistry, or viral nucleic acids by either in situ hybridization or PCR.

7.7 Arenavirus

Lassa fever, an acute febrile illness due to arenavirus infection, is pandemic throughout West Africa. In a prospective study of patients with Lassa fever, Cummins et al. (1990) found the incidence of sensorineural hearing loss in infected individuals exceeded that previously reported for any other postnatally acquired infection. Approximately one-third of patients hospitalized with Lassa fever, and none of the controls, developed acute sensorineural hearing losses. Fifty-three percent of patients who developed hearing losses were left with some degree of permanent hearing loss following recovery from acute infection: two-thirds of the losses were unilateral, one-third were bilateral. Clinical characteristics associated with the hearing losses included sudden onset, frequent tinnitus, and a close relationship to febrile illness. The severity of the hearing losses ranged from mild (<25 dB) to profound (>100 dB). Because of language difficulties, dizziness and vertigo were not discriminated, but no signs of nystagmus or ataxia were associated with hearing loss. The natural history and pathogenesis of hearing loss secondary to Lassa fever are poorly understood at this time and no animal models have been described.

7.8 Polio, Columbia SK, Yellow Fever and Epstein-Barr Virus

Single references to sudden deafness secondary to poliomyelitis (Mawson 1963), Columbia SK viral infection (Van Dishoeck and Bierman, 1957), and yellow fever virus infection (Moe 1947) have been reported in the literature. In addition, Hirata et al. (1989) and Silverstein, Steinberg and Nathanson (1972) have reported transient vertigo in patients with infectious mononucleosis from Epstein-Barr virus infections. Temporal bone findings were not reported for these cases and no animal models of these viral infections have been established.

8. Viral Infection and Autoimmune Diseases of the Inner Ear

8.1 Background

Idiopathic sensorineural hearing loss affects approximately 1 in 10,000 individuals each year (Byl and Adour 1977). The leading etiologic theories for the causes of these idiopathic sudden sensorineural hearing losses include viral infection, vascular insults, labyrinthine membrane rupture, and autoimmune disease (Cole and Jahrsdoerfer 1988). Autoimmune inner ear disease is perhaps the least well understood of these etiologic factors. The term "autoimmune sensorineural hearing loss" was first proposed by McCabe (1979) to describe a group of patients for which no other explanation for their hearing losses could be found. Subsequently, it was recognized that the term autoimmune sensorineural hearing loss was too restrictive, since the vestibular system is also frequently involved, and the expression "autoimmune inner ear disease" (AIED) has now become more widely accepted (McCabe 1989).

While the clinical profile for patients with AIED has been well defined (Hughes et al. 1988), the specific mechanism responsible for immune-mediated inner ear disease remains unknown. In practice, AIED has been diagnosed by a combination of clinical manifestations, positive responses to immunological laboratory tests, and symptomatic improvement after immunosuppressive therapy (McCabe 1979). However, to date, Wetebsky's postulates, a modified version of Koch's postulates for immunological phenomena (Cole and Jahrsdoerfer 1988), have not yet been satisfied to confirm a relationship between autoimmunity and AIED sensorineural hearing loss.

8.2 Clinical Findings and Laboratory Diagnosis

Clinically, AIED patients present with bilateral, asymmetric, progressive sensorineural hearing loss, with or without dizziness, and occasionally with symptoms of tinnitus and ear pressure. A characteristic of this disease is that pathologies develop over periods ranging from several days to months, and not suddenly in less than twenty-four hours or over a period of several years (McCabe 1979). In principle, the evidence for autoimmunity in these patients should depend upon positive immunological laboratory tests and findings of substantial improvement following intense steroid therapy. In practice, the diagnosis of AIED has primarily been dependent upon the results achieved by empirical treatment with immunosuppressive drugs and circumstantial evidence of other autoimmune conditions, such as Cogan's syndrome, rheumatoid arthritis, systemic lupus erythematosus, Sjogren's syndrome, polyarteritis nodosa, Wegener's granulomatosis and ulcerative colitis (Harris 1987, 1993). McCabe (1991) reported that 25% of his AIED patients also developed other autoimmune diseases.

Ideally, AIED should be confirmed by a sensitive and valid laboratory test. However, there has been some controversy over the value of available laboratory tests for the diagnosis of AIED. Hughes et al. (1987) have divided the available laboratory immune tests into antigen-specific and antigen-nonspecific classes. The antigen-specific tests include lymphocyte migration inhibition and transformation assays, ELISA, and immunohistochemical staining procedures. The antigen-specific tests currently used rely on crude inner ear preparations, since the antigen(s) that trigger the immune response(s) in AIED is (are) unknown. Furthermore, the antigen-specific tests with lymphocytes are currently limited by the requirement for whole fresh patient blood. Yoo et al. (1982) have utilized an ELISA approach to screen sera and demonstrate the presence of elevated antibodies to Type II collagen in patients with Meniere's disease and otosclerosis. Other investigators have attempted to diagnose AIED by correlating specific immunofluorescent staining patterns within inner ear tissues incubated with sera from suspected patients (Arnold, Pfaltz, and Altermatt 1985; Soliman 1988). While antigen-specific tests, particularly the lymphocyte migration inhibition and transformation assays, have been strongly endorsed by some laboratories for the diagnosis of AIED (McCabe 1979; Hughes et al. 1984, 1985, 1987; Berger et al. 1989), other investigators have not found these tests to be particularly helpful (Arnold et al. 1985; Harris 1987; Moscicki et al. 1988; Harris and Sharp 1990). Similarly, neither the ELISA antibody titer protocols nor the immunostaining of inner ear tissue approaches available have, as yet, received widespread general acceptance as clinical tests for AIED.

Antigen-nonspecific tests listed by Hughes et al. (1987) include erythrocyte sedimentation rate, C-reactive protein, Cl_q binding assays, total hemolytic complement assay (CH_{50}) for circulating immune complexes, and cryoprecipitation. These antigen-nonspecific tests, as well as tests for rheumatoid factor and anti-DNA antibodies, may be helpful when positive results are found, but they are less specific and sensitive than the antigen-specific tests (Hughes et al. 1984, 1986; Veldman et al. 1987). In reviewing these clinical tests for AIED, McCabe (1987) concluded that, at that time, there was no reliable, established laboratory test to identify AIED.

More recently, a promising new diagnostic approach is the application of Western blotting to detect antibodies in patient sera that identify specific proteins in extracts of inner ear tissues. A number of investigators have employed preparations of inner ear proteins in these immunoelectrophoretic assays for clinical investigations of AIED (McCabe 1979; Hughes et al. 1984; Arnold, Pfaltz, and Altermatt 1985; Veldman 1988; Harris and Sharp 1990; Harris 1993; Yamanobe and Harris 1993; Moscicki et al. 1994). At present at least five putative (isologous or heterologous) inner ear specific proteins (32kd, 33-35kd, 58kd, 60kd, and 220kd) have been identified which react with serum autoantibodies from putative AIED patients (Yamanobe and Harris 1993). In addition, at least six noncochlear-specific proteins (40-42kd, 47kd, 50kd, 55kd, 62kd and 68kd) from a variety of bovine (heterologous) tissue extracts (i.e., brain, kidney, liver, skeletal muscle, and small intestine) have been identified that react with serum autoantibodies from AIED patients (Harris and Sharp 1990; Harris 1993; Yamanobe and Harris 1993; Moscicki et al. 1994). In particular, Harris (1993) found that 33% of 138 patients with rapidly progressive sensorineural hearing loss were reactive with the 68kd protein by Western blot. When the patient population was restricted to those with actively progressing (i.e., evidence of hearing loss on serial audiograms less than three months apart), idiopathic, bilateral sensorineural hearing loss, Moscicki et al. (1994) found 89% had serum antibody reactive with the 68kd protein by Western blot. In contrast, none of the serum from AIED patients with inactive hearing loss in the latter study reacted with 68kd proteins.

While the immunoelectrophoretic assay approaches appear very promising, additional studies will be required to replicate and validate the exceptional specificity and sensitivity reported by Moscicki et al. (1994), and to characterize the identity of the antigen with which the patient serum autoantibodies react. The disadvantages inherent in using crude autologous or heterologous antigen preparations (e.g., inner ear or other tissues) in these immunoassays will only be circumvented once the specific antigen(s) associated with AIED have been identified, their amino acid sequences determined, and antigens synthesized for routine use in Western blot immunoassays (Yamanobe and Harris 1993). In a recent report, Billings et al. (1995) provided preliminary evidence linking the 68-kd antigen reactive with AIED patient sera to a specific member of the family of stress proteins. Their attempts to purify and characterize the 68-kd diagnostic antigen suggest that it is ubiquitous, rather than specific to the inner ear, and that it may represent the highly inducible heat shock protein 70 (hsp 70). Additional studies will be required to confirm the identity of the antigen(s) reactive with the sera of AIED patients, and to determine whether the autoantibodies in patient sera reflect an active immune process, or a secondary response following damage to the inner ear.

8.3 Possible Mechanisms for a Viral Etiology for AIED

What evidence is there that viruses play a role in the etiopathogenesis of AIED? Oldstone (1989) has identified three general lines of evidence suggesting possible relationships between viruses and autoimmune disease.

First, autoimmune responses are induced or enhanced by infection with a wide array of human DNA and RNA viruses (Fujinami and Oldstone 1989). Second, in animal models acute and persistent viral infections induce, accelerate, or enhance autoimmune responses and initiate autoimmune diseases (Oldstone 1972). Third, as a result of similar antigenic determinants between viruses and host components, viruses can potentiate autoimmune responses (Oldstone 1987; Atkinson et al. 1993). In the latter case, when similar structures are shared by molecules from dissimilar genes or their protein products, the relationship is defined as "molecular mimicry" (Oldstone 1987). Following viral infection, the shared epitopes produce cross-reacting immune responses that can initiate autoimmune disease (Fujinami and Oldstone 1985). Once an effector immune response (either B cell (humoral) or T cell (cytotoxic T cell)) is directed against a virus, it might possibly cross-react with self protein or autoantigen, evoking an autoimmune response. Among the viruses that are known to induce autoantibody production are hepatitis, chickenpox, measles, mumps, and infectious mononucleosis (reviewed in Fujinami and Oldstone 1989). Thus, as a result of molecular mimicry between viruses and inner ear tissues, humoral or cellular immune responses directed against the inner ear could cause or promote auditory and/or vestibular pathologies.

For several decades viruses have been associated with autoimmune-like diseases (Waksman 1962). The best established examples are to be found in postinfectious encephalopathies that develop following measles, mumps and vaccinia viral infection (Johnson et al. 1984). These encephalopathies (e.g., encephalomyelitis) share common histopathological features with the experimental autoimmune disease models denoted experimental allergic encephalitis (EAE) (Freund, Stern, and Pisani 1947). While the mechanisms by which viral infections can induce encephalopathies after the virus has been cleared remains unknown, it is intriguing that in many ways the characteristics of immunity in the central nervous system are similar to those that exist in the inner ear (Yamanobe and Harris 1992).

8.4 Comparative Animal Studies

Perhaps the best available evidence suggesting an immune mediated etiopathogenesis for AIED is the production of experimental sensorineural hearing loss in animal models. Lehnhardt (1958) first postulated an autoimmune mechanism to be responsible for what later was classified as AIED. The first reported experimental models of AIED followed shortly afterwards in the early 1960's (Beickert 1961; Terayama and Sasaki 1964). Since that time, a number of groups have made substantial progress in developing animal models of AIED. Yoo and his co-workers have suggested a role for Type II collagen in the development of immune-mediated auditory deficits and structural lesions in the cochlear membranous labyrinth and spiral ganglion of both patients and experimental animals (Yoo,

Tomoda, and Hernandez 1982, 1983a,b, 1984). Cochlear pathologies observed in these studies included spiral ganglion cell degeneration, atrophy of the organ of Corti, atrophy of stria vascularis, endolymphatic hydrops, and sensorineural hearing loss (i.e., auditory brainstem threshold shifts). While subsequent studies have partially supported Yoo's results (Huang, Yi, Abramson 1986; Ohashi, Tomada, and Yoshie 1989; Cruz et al. 1990; Soliman 1990), studies by other investigators were unable to replicate these findings (Harris, Woolf, and Ryan 1986; Solvesten-Sorensen et al. 1988).

The earliest experimental models of AIED immunized guinea pigs with isologous inner ear extracts (Beickert 1961; Terayama and Sasaki 1964). These studies reported that autoimmunization induced both histopathology and hearing loss (i.e., frequency-specific Pryer reflex). Since that time, and using more modern immunological and electrophysiological techniques, other investigators have immunized animals with isologous and/or heterologous inner ear extracts (Harris 1987; Soliman 1989; Orozco et al. 1990; Yamanobe and Harris 1992; Harris et al. 1993), or antibodies immunoreactive with specific inner ear structures (Cruz et al. 1990; Nair et al. 1993), in attempts to induce autoimmune-mediated inner ear pathologies. While these studies individually contain strong presumptive evidence for the induction of experimental AIED, collectively they exhibited inconsistencies that remain to be resolved.

Some of the differences between the various inner ear antigen immunization studies may be the result of methodological considerations (e.g., source of the immunizing inner ear antigen and variation in the immunization schedule). However, review of this literature has revealed a number of significant qualitative and quantitative differences with respect to the incidence of inner ear pathology, the extent of the cochlear histopathology, the presence or absence of cellular inflammatory reaction, and the level and permanence of the associated hearing losses. Most striking has been disparities in the relationships between cellular and humoral immune responses, inner ear histopathology, and the development of hearing loss. Notably, while several experimental inner ear antigen immunization studies reported a strong positive correlation between hearing losses and the presence of significant inner ear antigen-specific serum antibody titers (Orozco et al. 1990; Harris et al. 1993; Nair et al. 1993), other studies found a lack of correlation between hearing loss and antibody titer (Terayama and Sasaki 1964; Harris 1987; Yamanobe and Harris 1992).

Considerable variability has also been noted between the extent to which hearing loss was associated with cochlear inflammation in models of AIED. While several studies reported an association between hearing loss and cochlear inflammatory responses (i.e., vasculitis, lymphocyte infiltration, and inflammatory precipitates) in the cochlear scalae (Terayama and Sasaki 1964; Harris 1987; Soliman 1989; Yamanobe and Harris 1992), hearing loss has also been reported in the absence of cochlear inflammation (Orozco et al. 1990; Yamanobe and Harris 1992). Additional investigations with these experimental models of AIED will be required to resolve the apparent differences in these experimental results and to further elucidate the mechanisms responsible for the induction of autoimmune cochlear pathologies. Once specific inner ear antigens have been identified and purified, it will be possible to duplicate in experimental AIED studies the degree of control and reproducibility currently found in other autoimmune model systems, such as experimental autoimmune uveoretinitis (e.g., Peng, Yoshitoshi, and Shichi 1992; Sunil et al. 1993). Successful isolation and synthesis of specific cochleopathogenic antigens will also ultimately facilitate the development of new diagnostic procedures and therapeutic approaches for the management of AIED.

9. Summary

The development of effective treatments for viral labyrinthitis requires an increased understanding of the mechanisms by which viruses damage the inner ear. Clinical and experimental studies have clearly established that viruses can infect the inner ear and cause auditory and vestibular pathologies. However, while knowledge about the pathogenesis of viral induced labyrinthine histopathology and pathophysiology has increased, the etiologic mechanisms responsible for the development of these auditory and vestibular deficits are still largely unknown. Because of the technical difficulties associated with human clinical investigations, experimental animal studies offer unique opportunities to define and explain the etiopathogenesis of viral induced deafness and vertigo. However, while animal studies can contribute greatly to our understanding of inner ear disease progression, it will remain critically important to continue to collect temporal bones for otopathologic studies of patients with clinically well defined auditory and vestibular pathologies, so as to permit direct comparisons between clinical and experimental results. Through the application of new immunologic and molecular biological techniques to both clinical and experimental inner ear material, it will be possible to further clarify the mechanisms by which viruses damage the labyrinth.

Once a viral etiology for inner ear pathology has been established, the "best treatment" would be to prevent future infections through either the initiation of vaccination programs (Davis and Johnsson 1983) or elimination of the vector responsible for viral transmission. International programs of vaccination and disease control have been shown to dramatically reduce the incidence of mumps, rubella, and rubeola, with concomitant declines in the occurrence of associated deafness and vestibular pathologies. Unfortunately, once viral labyrinthitis has become established, the only effective form of intervention may be antiviral chemotherapy. However, new antiviral drugs currently under investigation have shown considerable promise for the treatment of viral infections, such as congenital cytomegalovirus and HIV infections.

Finally, a critical factor in the future control of viral induced auditory and vestibular deficits will be the availability of sensitive, specific, reliable, and inexpensive laboratory tests for rapid diagnosis of viral infection. New molecular biological techniques, such as polymerase chain reaction assay, offer the potential to simplify current testing procedures, reduce the time required to laboratory testing of samples, decrease the costs of tests through automation, and improve by orders of magnitude the sensitivity of existing clinical tests. Early diagnosis offers the best opportunity for antiviral intervention to prevent or minimize disease progression. Alternatively, if virus induced inner ear deficits are irreversible, then early diagnosis would permit the initiation of rehabilitation when optimal results can still be achieved (NIH Consensus Development Conference, 1993).

Acknowledgments. The technical assistance of Fred Koehrn, Thecla Bennett, and Abbyann Sisk in preparing the electron micrographs is gratefully acknowledged. This work was supported by grants from NIDCD DC00386, the NIH Office of Research on Women's Health, and the Research Service of the Veterans Administration.

References

Aleksic SN, Budzilovich GN, Lieberman AN (1973) Herpes zoster oticus and facial paralysis (Ramsay hunt syndrome). J Neuro Sci 20:149–159.
Alford BR (1968) Rubella: la bete noire de la medecine. Laryngoscope, 78:1623–1659.
Arnold W, Pfaltz R, Altermatt HJ (1985) Evidence of serum antibodies against inner ear tissues in the blood of patients with certain sensorineural hearing disorders. Acta Otolaryngol 99:437–444.
Atkinson MA, Kaufman DL, Newman D, Tobin AJ, et al. (1993) Islet cell cytoplasmic autoantibody reactivity to glutamate decarboxylase in insulin-dependent diabetes. J Clin Inves 91:350–356.
Azimi PH, Cramblett HG, Haynes RE (1969) Mumps meningoencephalitis in children. J Am Med Assoc 207:509–512.
Bang HO, Bang J (1943) Involvement of the central nervous system in mumps. Acta Med Scand 113:487–505.
Beal DD, Davey PR, Lindsay JR (1967) Inner ear pathology of congenital deafness. Arch Otolaryngol 85:134–142.
Beickert P (1961) Zur Frage der Empfindungsschwerhorigkeit und Autoallergie. Z Laryngol Rhinol Otol 40:837–842.
Belman AL, Lantos G, Horoupian D, Novick BE, et al. (1986) AIDS: Calcification of ganglia in infants and children. Neurol 36:1192–1199.
Berger P, Koja S, Rogowski M, Vollrath M (1989) Der lymphocytentransformationstest zum Nachweis einer immunologischen innenohrschwerhorigkeit. HNO 37:153–157.

Billings PB, Keithley EM, Harris JP (1995) Evidence linking the 68 kilodalton antigen identified in progressive sensorineural hearing loss patient sera with heat shock protein 70. Ann Otol Rhinol Laryngol 104:181–188.

Blackley B, Friedmann I, Wright I (1967) Herpes zoster auris associated with facial nerve palsy and auditory nerve symptoms. Acta Otolaryngol 63:533–550.

Bordley JE, Alford BR (1970) The pathology of rubella deafness. Int Audiol 9:58–67.

Bordley JE, Hardy JMB (1969) Laboratory and clinical observations on prenatal rubella. Ann Otol 78:917–928.

Bordley JE, Kapur MD (1972) Histopathological changes in the temporal bone in acute infections of smallpox and chicken pox. Laryngoscope 82:1477–1490.

Bordley JE, Brookhauser PE, Worthington EL (1972) Viral infections and hearing. Laryngoscope 82:557–572.

Borton TE, Stark EW (1970) Audiological findings in hearing loss secondary to maternal rubella. Pediatr 45:225–229.

Breda SD, Hammerschlag PE, Gigliotti F, Schinella R (1988) Pneumocystis carinii in the temporal bone as a primary manifestation of the acquired immunodeficiency syndrome. Ann Oto Rhino Laryngol 97:427–431.

Brivio L, Tornaghi R, Museti L, Marchisio P, et al. (1991) Improvement of auditory brainstem responses after treatment of AIDS with zidovudine. Pediatr Neurol 7:53–55.

Brookhouser PE, Bordley JE (1973) Congenital rubella deafness: Pathology and pathogenesis. Arch Otolaryngol 98:252–257.

Brunell PA (1989) Transmission of chickenpox in a school setting prior to the observed exanthem. Am J Dis Child 143:1451–1452.

Buddingh GJ, Schrum DI, Lanier JC, Guidry DJ (1953) Studies of the natural history of herpes simplex infections. Pediatr 11:595–609.

Byl FM (1977) Seventy-six cases of presumed sudden hearing loss occurring in 1973: prognosis and incidence. Laryngoscope 87:817–25.

Byl FM (1984) Sudden hearing loss: eight years experience and suggested prognostic table. Laryngoscope 94:647–661.

Byl FM, Adour KK (1977) Auditory symptoms associated with herpes zoster or idiopathic facial paralysis. Laryngoscope 87:372–379.

Carruthers DG (1945) Congenital deaf-mutism as a sequela of a rubella-like maternal infection during pregnancy. Med J Aust 1:315–320.

Catalano LW, Sever JL (1971) The role of viruses as causes of congenital defects. Annu Rev Microbiol 25:255–282.

Chandrasekhar SS, Siverls V, Sekhar HKC (1992) Histopathologic and ultrastructural changes in the temporal bones of HIV-infected human adults. Am J Otol 13:207–214.

Choi YC, Hsiung GD (1978) Cytomegalovirus infection in guinea pigs. II. Transplacental and horizontal transmission. J Infect Dis 138:197–202.

Cole RR, Jahrsdoerfer RA (1988) Sudden hearing loss: An update. Am J Otol 9:211–215.

Collier AC, Marra C, Coombs RW, Claypoole K, et al. (1992) Central nervous system manifestations in HIV infection without AIDS. J Acq Imm Def Synd 5:229–241.

Craft JL, Hilding DA, Bnatt PN, Jonas AM (1973) Spiral ganglion virus-like particles of guinea pigs. Adv Otorhinolaryngol 20:178–190.

Croen KD, Ostrove JM, Dragovic LJ, Strauss SE (1988) Patterns of gene expression and sites of latency in human nerve ganglia are different for varicella-zoster virus and herpes simplex viruses. Proc Natl Acad Sci USA 85:9733-9737.

Cruz OLM, Miniti A, Cossermelli W, Oliveira RM (1990) Autoimmune sensorineural hearing loss: A preliminary experimental study. Am J Otol 11:342-346.

Cummins D, McCormick JB, Bennett D, Samba JA, et al. (1990) Acute sensorineural deafness in lassa fever. J Am Med Assoc 264:2093-2096.

Dahle AJ, McCollister FP, Hamner BA, Reynolds DW, et al. (1979a) Subclinical congenital CMV infection and hearing impairment. J Speech Hear Sci 44:220-229.

Dahle AJ, McCollister FP, Stagno S, Reynolds DW, et al. (1979b) Progressive hearing impairment in children with congenital cytomegalovirus infection. J Speech Hear Res 44:220-229.

Davis A, Wood S (1992) The epidemiology of childhood hearing impairment: Factors relevant to planning of services. Br J Audiol 26:77-90.

Davis GL (1979) Congenital cytomegalovirus and hearing loss: Clinical and experimental observations. Laryngoscope 89:1681-1688.

Davis GL (1981) In vitro model of viral-induced congenital deafness. Am J Otol 3:156-160.

Davis GL, Hawrisiak MM (1977) Experimental cytomegalovirus infection and the developing mouse inner ear: In vivo and in vitro studies. Lab Invest 37:20-29.

Davis GL, Strauss M (1973) Viral disease of the labyrinth II. An experimental model using mouse cytomegalovirus. Ann Otol Rhinol Laryngol 82:584-594.

Davis LE (1981) Experimental viral infections of the facial nerve and geniculate ganglion. Ann Neurol 9:120-125.

Davis LE (1990) Comparative experimental viral labyrinthitis. Am J Otolaryngol 11:382-388.

Davis LE (1993) Viruses and vestibular neuritis: Review of human and animal studies. Acta Otolaryngol Suppl 503:70-73.

Davis LE, Johnson RT (1976) Experimental viral infections of the inner ear. I. Acute infections of the newborn hamster labyrinth. Lab Invest 34:349-356.

Davis LE, Johnsson LG (1983) Viral infections of the inner ear: Clinical, virologic, and pathological studies in humans and animals. Am J Otolaryngol 4:347-362.

Davis LE, Tweed GV, Steward JA, Stewart JA, et al. (1971) Cytomegalovirus mononucleosis in a first trimester pregnant female with transmission to the fetus. Pediatr 48:200-206.

Davis LE, James CG, Fiber F, McLaren LC (1979) Cytomegalovirus isolation from a human inner ear. Ann Otol Rhinol Laryngol 88:424-426.

Davis LE, Johnsson LG, Kornfeld M (1981) Cytomegalovirus labyrinthitis in an infant: Morphological, virological, and immunofluorescent studies. J Neuropathol Exp Neurol 40:9-19.

Davis LE, Rarey KE, Stewart JA, McLaren LC (1987) Recovery and probable persistence of cytomegalovirus in human inner ear fluid without cochlear damage. Ann Oto Rhino Laryngol 96:380-383.

Dawes JF (1963) Virus lesions of the cranial nerves with special reference to the eighth nerve. Proc Royal Soc Med 56:777-780.

Denny-Brown D, Adams RD, Fitzgerald PJ (1944) Pathologic features of herpes zoster. Arch Neurol Psychiatr 11:216-231.

Desmond MM, Montgomery JR, Melnick JL, Cochran GG, et al. (1969) Congenital rubella encephalitis. Am J Dis Child 118:30-31.

Desmond MM, Wilson JS, Verniaud WM, Melnick JL, et al. (1970) The early

growth and development of infants with congenital rubella. Adv Teratol 4:39–63.
Elder GA and Sever JL (1988) AIDS and neurological disorders. Ann Neurol 23:4–6.
Esterly JR, Oppenheimer EH (1973) The pathologic manifestations of intrauterine rubella infections. Arch Otolaryngol 98:246–248.
Evenberg G (1957) Deafness following mumps. Acta Otol 48:397–403.
Fitzgerald MD, Sitton AB, McConnell F (1970) Audiometric, developmental, and learning characteristics of a group of rubella deaf children. J Speech Hear Dis 35:218–228.
Freund J, Stern ER, Pisani TM (1947) Isoallergic encephalomyelitis and radiculitis in guinea pigs after one injection of brain and mycobacteria in water-in-oil emulsion. J Immunol 57:179–194.
Friedman I, Wright MI (1966) Histopathologic changes in the fetal and infantile inner ear caused by rubella. Br Med J 504:20–23.
Fujinami RS, Oldstone MBA (1985) Amino acid homology and immune responses between the encephalitogenic site of myelin basic protein and virus: A mechanism for autoimmunity. Science 230:1043–1045.
Fujinami RS, Oldstone MBA (1989) Molecular mimicry as a mechanism for virus-induced autoimmunity. Immunol Res 8:3–15.
Fukuda S (1986) Experimental paramyxovirus induced labyrinthitis in the guinea pig. Ear Res Japan 17:52–58.
Fukuda S, Keithley EM, Harris JP (1988) The development of endolymphatic hydrops following CMV inoculation of the endolymphatic sac. Laryngoscope 98:439–443.
Gilden DH, Rozenman Y, Murray R, Devlin M, et al. (1987) Detection of varicella-zoster virus nucleic acid in neurons of normal human thoracic ganglia. Ann Neurol 22:377–380.
Gregg NM (1941) Congenital cataract following German measles in the mother. Trans Ophthalmol Soc Aust 3:35–46.
Hanshaw JB, Dudgeon JA (1978) Congenital cytomegalovirus. Major Probl Clin Pediatr 17:97–152.
Hanshaw JB, Scheiner AP, Moxley AW, et al. (1976) School failure and deafness after "silent" congenital cytomegalovirus infection. N Engl J Med 295:468–470.
Hardy JB, McCracken Jr GH, Gilkeson MR, Sever JL (1969) Adverse fetal outcome following maternal rubella after the first trimester of pregnancy. J Am Med Assoc 207:2414–2420.
Harris JP (1987) Experimental autoimmune sensorineural hearing loss. Laryngoscope 97:63–76.
Harris JP (1993) Immunologic mechanisms in disorders of the inner ear. In: Cummings CW (ed) Otolaryngology—Head and Neck Surgery, 2nd ed. St. Louis: Mosby Yearbook, pp 2926–2942.
Harris JP, Sharp PA (1990) Inner ear autoantibodies in patients with rapidly progressive sensorineural hearing loss. Laryngoscope 100:516–524.
Harris JP, Woolf NK, Ryan AF, Butler DM, Richman DD (1984) Immunologic and electrophysiological response to cytomegaloviral inner ear infection in the guinea pig. J Infect Dis 150:523–530.
Harris JP, Woolf NK, Ryan AF (1986) A reexamination of experimental type II collagen autoimmunity: Middle and inner ear morphology and function. Ann Oto Rhino Laryngol 95:176–180.
Harris JP, Keithley EM, Al-Mansour M, Yeo SW, et al. (1993) Correlation of

antibody titers and hearing loss in an autoimmune model of sensorineural hearing loss. Abstracts of the Sixteenth Midwinter Meeting of the Association for Research in Otolaryngology, p. 149.

Hausler R, Vibert D, Koralnik IJ, Hirschel B (1991) Neuro-otological manifestations in different stages of HIV infection. Acta Otolaryngol 481:515-521.

Heathfield KWG, Mee AS (1978) Prognosis of the Ramsay Hunt syndrome. Br Med J 1:343-344.

Hirata T, Sekitani T, Okinaka Y, Matsuda Y (1989) Serovirological study of vestibular neuronitis. Acta Otolaryngol Suppl 468:371-373.

Hodgson WR (1969) Auditory characteristics of post-rubella impairment. Volta Rev 71:97-103.

Huang CC, Yi ZX, Abramson M (1986) Type II collagen-induced otospongiosis-like lesions in rats. Am J Otolaryngol 7:25-66.

Hughes GB, Kinney SE, Barna BP, Calabrese LH (1984) Practical versus theoretical management of autoimmune inner ear disease. Laryngoscope 94:758-766.

Hughes GB, Kinney SE, Hamid MA, Barna BP (1985) Autoimmune vestibular dysfunction: Preliminary report. Laryngoscope 95:893-897.

Hughes GB, Barna BP, Kinney SE, Calabrese LH (1986) Predictive value of laboratory tests in 'autoimmune' inner ear disease: Preliminary report. Laryngoscope 96:502-505.

Hughes GB, Kinney SE, Barna BP, Calabrese LH (1987) Autoimmune inner ear disease: Laboratory tests and audio-vestibular treatment responses. In: Veldman JE, McCabe BF (eds) Oto-Immunology. Amsterdam/Berkeley: Kugler Publications, pp. 149-155.

Hughes GB, Barna BP, Kinney SE, Calabrese LH, et al. (1988) Clinical diagnosis of immune inner-ear disease. Laryngoscope 98:251-253.

Hunt JF (1907) Herpetic inflammations of the geniculate ganglion, a new syndrome and its aural complications. Arch Otolaryngol 36:371-381.

Hyden D, Odkvist LM, Kylen P (1979) Vestibular symptoms in mumps deafness. Acta Otolaryngol Suppl 360:182-183.

Jaffe BF (1973) Clinical studies in sudden deafness. Adv Otorhinolaryngol 20:221-228.

Jaffe BF (1978) Viral causes of sudden inner ear deafness. Otolaryngol Clin N Am 11:63-69.

Johnson KP (1969) Mouse cytomegalovirus: Placental infection. J Infect Dis 120:445-450.

Johnson RT, Gibbs CJ (1974) Koch's postulates and slow infections of the nervous system. Arch Neurol 30:36-38.

Johnson RT, Griffin DE, Hirsch RL, Wolinsky JS, et al. (1984) Measles encephalomyelitis-Clinical and immunologic studies. N Engl J Med 310:137-141.

Karmody CS (1968) Subclinical maternal rubella and congenital deafness. N Engl J Med 278:809-814.

Katsuhiko T, Satoshi F, Tohru S, Yoshihiko T (1988) Experimental mumps virus-induced labyrinthitis. Immunohistochemical and ultrastructural studies. Acta Otolaryngol Suppl 456:98-105.

Keith CG (1991) Congenital rubella infection from reinfection of previously immunized mothers. Aust and N Zeal J of Ophthalmol 19:291-293.

Keithley EM, Sharp P, Woolf NK, Harris JP (1988) Temporal sequence of viral antigen expression in the cochlea induced by cytomegalovirus. Acta Otolaryngol 106:46-54.

Keithley EM, Woolf NK, Harris JP (1989) Development of morphological and physiological changes in the cochlea induced by cytomegalovirus. Laryngoscope 99:409–414.

Kelemen G, Gottlieb BN (1959) Pathohistology of fetal ears after maternal rubella. Laryngoscope 69:385–397.

Khetarpal U, Nadol JB, Glynn RJ (1990) Idiopathic sudden sensorineural hearing loss and postnatal viral labyrinthitis: A statistical comparison of temporal bone findings. Ann Oto Rhino Laryngol 99:969–976.

Kinney CE (1953) Hearing impairments in children. Laryngoscope 63:220–226.

Kohan D, Hammerschlag PE, Holliday RA (1990) Otologic disease in AIDS patients: CT correlation. Laryngoscope 100:1326–1330.

Kohan D, Rothstein SG, Cohen NL (1988) Otologic disease in patients with acquired immunodeficiency syndrome. Ann Oto Rhino Laryngol 97:636–640.

Koropchak CM, Graham G, Palmer J, Winsberg M, et al. (1991) Investigation of varicella-zoster virus infection by polymerase chain reaction in the immunocompetent host with acute varicella. J Infect Dis 163:1016–1022.

Koskiniemi M, Donner M, Pettay O (1983) Clinical appearance and outcome in mumps encephalitis in children. Acta Pediatr Scand 72:603–609.

Kumagami H (1972) Experimental facial nerve paralysis. Arch Otolaryngol 95:305–312.

Kwartler JA, Linthicum FH, Jahn AF, Hawke M (1991) Sudden hearing loss due to AIDS-related cryptococcal meningitis—A temporal bone study. Otolaryngol Head Neck Surg 104:265–269.

Lehnhardt VE (1958) Plotzliche Horstorungen, auf beiden Seiten gleichzeitig oder nacheinander aufgetreten. Laryngologie 37:1–28.

Levy RM, Bredesen DE, Rosenblum ML (1985) Neurological manifestations of the acquired immunodeficiency syndrome. J Neurosurg 62:475–495.

Lindsay JR (1973) Histopathology of deafness due to postnatal viral disease. Arch Otolaryngol 98:258–264.

Lindsay JR, Hemenway WG (1954) Inner ear pathology due to measles. Ann Otol Rhinol Laryngol 63:754–771.

Lindsay JR, Carruthers DG, Hemenway WG, Harrison MS (1953) Inner ear pathology following maternal rubella. Ann Oto Rhino Laryngol 62:1201–1218.

Lindsay JR, Davey PR, Ward PH (1960) Inner ear pathology in deafness due to mumps. Ann Otol Rhinol Laryngol 69:918–935.

Lohle E, Kistler GS, Riede UN, Merck W, et al. (1982) Virus particles in the cochlear spiral ganglion of guinea pigs. Acta Otolaryngol 94:233–239.

Lowry PW, Sabella C, Koropchak CM, Watson BN, et al. (1993) Investigation of the pathogenesis of varicella-zoster virus infection in guinea pigs by using polymerase chain reaction. J Infect Dis 167:78–83.

Maassab HF (1973) The role of virus in sudden deafness. Adv OtoRhino-Laryngol 20:229–235.

Marennikova SS, Maltseva NN, Shelukhina EM, Shenkman LS, et al. (1974) A generalized herpetic infection simulating smallpox in a gorilla. Intervirol 2:280–286.

Matsunaga Y, Yamanishi K, Takahashi M (1982) Experimental infection and immune response of guinea pigs with varicella-zoster virus. Infect Immun 27:407–412.

Mawson SR (1963) Diseases of the Ear. Baltimore: Williams & Wilkins Co.

McCabe BF (1979) Autoimmune sensorineural hearing loss. Ann Otol Rhinol

Laryngol 88:585-589.

McCabe BF (1987) Autoimmune inner ear disease: Clinical varieties of presentation. In: Veldman JE and McCabe BF (eds) Oto-Immunology. Amsterdam/Berkeley: Kugler Publications, pp. 143-148.

McCabe BF (1989) Autoimmune inner ear disease: Therapy. Am J Otol 10:196-197.

McCabe BF (1991) Autoimmune inner ear disease: Results of therapy. In: Pfaltz CR, Arnold W, Kleinsasser O (eds) Bearing of Basic Research on Clinical Otolaryngology. Advances in Otorhinolaryngology. Basel: Karger, 46:78-81.

Meyers EN, Stool S (1968) Cytomegalic inclusion disease of the inner ear. Laryngoscope 78:1904-1915.

Michaels L, Soucek S, Liang J (1994) The ear in the acquired immunodeficiency syndrome: I. temporal bone histopathologic study. Am J Otol 15:515-522.

Miller HG, Stanton JB, Gibbons JL (1956) Para-infectious encephalomyelitis and related syndromes. Quart J Med 25:427-505.

Moe R (1947) Transitory deafness. Acta Otol 35:20-41.

Monif GR, Avery GB, Korones SB, Sever JL (1965) Postmortem isolation of rubella virus from three children with rubella-syndrome defects. Lancet (1):723-724.

Moscicki RA, Ramadan H, Castro OJ, Nadol JB, et al. (1988) Corticosteroid response and immunologic studies in idiopathic progressive sensorineural hearing loss. J Allergy Clin Immunol 81:217. (abstract)

Moscicki RA, Martin JES, Quintero CH, Rauch SD, et al. (1994) Serum antibody to inner ear proteins in patients with progressive hearing loss. J Am Med Assoc 272:611-616.

Myers MG, Connelly BL (1992) Animal models of varicella. J Infect Dis Suppl 166:48-50.

Myers EN, Stool S (1968) Cytomegalic inclusion disease of the inner ear. Laryngoscope 78:1904-1915.

Myers MG, Stanberry LR, Edmond BJ (1985) Varicella-zoster virus infection of strain 2 guinea pigs. J Infect Dis 151:106-113.

Myers MG, Kramer LW, Stanberry LF (1987) Varicella in a gorilla. J Med Virol 23:317-322.

Myers MG, Connelly BL, Stanberry LR (1991) Varicella in hairless guinea pigs. J Infect Dis 163:746-751.

Nair TS, Parrett T, Wang Y, Raphael Y, et al. (1993) Evidence that KHRI-3 antibody causes decreased auditory brainstem responses. Abstracts of the Sixteenth Midwinter Meeting of the Association for Research in Otolaryngology, p. 148.

Najoo FL, Wertheim-van Dillen P, Devriese PP (1988) Serology in facial paralysis caused by clinically presumed herpes zoster infection. Arch Otorhinolaryngol 245:230-233.

Nankervis GA, Kumar M (1978) Diseases produced by CMV. Med Clin N Am 62:1021-1035.

Nelson JA, Reynolds-Kohler C, Oldstone MBA, Wiley CA (1988) HIV and HCMV coinfect brain cells in patients with AIDS. Virology 165:286-290.

NIH Consensus Development Conference (1993) Early identification of hearing impairment in infants and young children. Washington, D.C.: National Institutes of Health.

Nomura Y, Kurata T, Saito K (1985) Cochlear changes after herpes simplex virus infection. Acta Otolaryngol 99:419-427.

Nomura Y, Hara M, Kurata T (1988) Experimental herpes simplex virus and cytomegalovirus labyrinthitis. Acta Otolaryngol Suppl 457:57–66.

Ohashi T, Tomoda K, Yoshie N (1989) Electrocochleographic changes in endolymphatic hydrops induced by type II collagen immunization through the stylomastoid foramen. Ann Oto Rhino Laryngol 98:556–562.

Oldstone MBA (1972) Virus-induced autoimmune disease: Viruses in the production and prevention of autoimmune disease. In: Membranes and Viruses in Immunopathology. SB Dey and RA Good, eds. New York: Academic Press, pp. 469–475.

Oldstone MBA (1987) Molecular mimicry and autoimmune disease. Cell 50:819–820.

Oldstone MBA (1989) Virus-induced autoimmunity: Molecular mimicry as a route to autoimmune disease. J Autoimmun Suppl 2:187–194.

Oliveira CA, Schuknecht HF (1990) Pathology of profound sensorineural hearing loss in infancy and early childhood. Laryngoscope 100:902–909.

Orenstein WA, Greaves WL (1982) Congenital rubella syndrome: A continuing problem. J Am Med Assoc 247:1174–1175.

Orozco CR, Niparko JK, Richardson BC, Dolan DF, et al. (1990) Experimental model of immune-mediated hearing loss using cross-species immunization. Laryngoscope 100:941–947.

Palva T, Dammert K (1969) Human cochlear aqueduct. Acta Otolaryngol 246:1–58.

Pass RF, Stagno S, Myers GJ, Alford CA (1980) Outcome of symptomatic congenital cytomegalovirus infection: Results of long-term longitudinal follow-up. Pediatr 66:758–762.

Peng B, Yoshitoshi T, Shichi H (1992) Suppression of experimental autoimmune uveoretinitis by intraorchidic administration of S-antigen. Autoimmunity 14:149–153.

Petito CK (1988) Review of central nervous system pathology in human immunodeficiency virus infection. Ann Neurol Suppl 23:54–57.

Plotkin SA, Oski FA, Harnett EM, Hervada AR, et al. (1965) Some recently recognized manifestations of the rubella syndrome. J Pediatr 67:182–191.

Proctor L, Perlman H, Lindsay J, Matz G (1979) Acute vestibular paralysis in herpes zoster oticus. Ann Oto Rhino Laryngol 88:303–310.

Quinnan GV, Masur H, Rook AH, Armstrong G, et al. (1984) Herpes virus infections in acquired immune deficiency syndrome. J Am Med Assoc 252:72–77.

Rarey KE, Davis LE (1985) A case report of cytomegalovirus isolation associated with normal inner ear histology. Abstracts of the Eighth Midwinter Meeting of the Association for Research in Otolaryngology, p. 10.

Rarey KE, Davis LE (1993) Temporal bone histopathology 14 years after cytomegalic inclusion disease: a case study. Laryngoscope 103:904–909.

Rarey KE, Davis JA, Deshmukh DR (1984) Inner ear changes in the ferret model for Reye's syndrome. Am J Otolaryngol 5:191–202.

Reichert CM, O'Leary TJ, Levens DL, Simrell CR, et al. (1983) Autopsy pathology in the acquired immune deficiency syndrome. Am J Pathol 112:357–382.

Rivers TM (1937) Viruses and Koch's postulates. J Bacteriol 33:1–12.

Rose FC, Brett EM, Burston J (1964) Zoster encephalomyelitis. Arch Neurology (Chicago) 11:155–172.

Sadzot-Delvauz C, Merville-Louis MP, Delree P, Piette J (1990) An in vivo model of varicella-zoster virus latent infection of dorsal root ganglia. J Neurosci Res 26:83–89.

Sando I, Harada T, Loehr A, Sobel JH (1977) Sudden deafness. Histopathologic correlation in temporal bone. Ann Oto Rhino Laryngol 86:269-279.

Schuknecht HF (1974) Pathology of the Ear. Cambridge MA: Harvard University Press, pp 259-261.

Schuknecht HF, Donovan ED (1986) The pathology of idiopathic sudden sensorineural hearing loss. Arch Otolaryngol 243:1-15.

Schuknecht HF, Kitamura K (1981) Vestibular neuritis. Ann Otol Rhinol Laryngol 90:1-19.

Shaia FT, Sheehy JL (1976) Sudden sensorineural hearing impairment: A report of 1,220 cases. Laryngoscope 96:389-398.

Shambaugh GE, Hagens EW, Holderman JW (1928) Statistical studies of the children in the public schools for the deaf. Arch Otolaryngol 7:424-513.

Shikani AH, Richtsmeier WJ (1991) Effect of viruses and interferon on chick embryo otocyst cultures. Ann Oto Rhino Laryngol 100:1020-1023.

Shimokata K, Nishiyama Y, Ito Y, Kimura Y, et al. (1977) Affinity of sendai virus for the inner ear of mice. Infect Immun 16:706-708.

Silverstein A, Steinberg G, Nathanson M (1972) Nervous system involvement in infectious mononucleosis. Arch Neurol 26:353-358.

Smith GA, Gussen R (1976) Inner ear pathologic features following mumps infection. Arch Otolaryngol 102:108-111.

Soliman AM (1988) Optimizing immunofluorescence for the testing of autoantibodies in inner ear disorders. Arch Otorhinolaryngol 245:28-35.

Soliman AM (1989) Experimental autoimmune inner ear disease. Laryngoscope 99:188-193.

Soliman AM (1990) Type II collagen-induced inner ear disease: Critical evaluation of the guinea pig model. Am J Otol 11:27-32.

Solvesten-Sorensen M, Nielson LP, Bretlau P, Jorgensen MB (1988) The role of type II collagen autoimmunity in otosclerosis revisited. Acta Otolaryngol 105:242-247.

Stagno S, Reynolds DW, Amos CS, Dahle AJ, Thames SD et al. (1977a) Auditory and visual defects resulting from symptomatic and subclinical congenital cytomegaloviral and Toxoplasma infections. Pediatr 59:669-678.

Stagno S, Reynolds DW, Huang ES, et al. (1977b) Congenital cytomegalovirus infection: Occurrence in an immune population. N Engl J Med 296:1254-1258.

Stagno S, Pass RF, Dworsky ME, Henderson RE, et al. (1982) Congenital cytomegalovirus infection: The relative importance of primary and recurrent maternal infection. N Engl J Med 306:945-949.

Stagno S, Pass RF, Dworsky ME, Alford CA (1983) Congenital and perinatal cytomegalovirus infections. Sem Perinatol 7:31-42.

Stagno S, Pass RF, Cloud G, Britt WJ, et al. (1986) Primary cytomegalovirus infection in pregnancy. J Am Med Assoc 256:1904-1908.

Strauss M (1990) Human cytomegalovirus labyrinthitis. Am J Otolaryngol 11:292-298.

Strauss M, Davis GL (1973) Viral disease of the labyrinth: I. Review of the literature and discussion of the role of cytomegalovirus in congenital deafness. Ann Otol 82:577-583.

Stulting AA, Hendricks ML, Marais AF (1987) Ophthalmological findings in AIDS. A case report. South African Medical Journal 72:715-716.

Suboti R (1976) Histopathological findings in the inner ear caused by measles. J Laryngol Otol 90:173-178.

Sunil S, Eto K, Singh VK, Shinohara T (1993) Oligopeptides of three to five residues derived from uveitopathogenic sites of retinal S-antigen induce experimental autoimmune uveitis (EAU) in Lewis rats. Cell-Immunol 148(1):198-207.

Swan C, Tostevin AL, Moore B, et al. (1943) Congenital defects in infants following infectious diseases during pregnancy with special reference to relationship between German measles and cataracts, deaf-mutism, heart disease and microcephaly and to period of pregnancy in which occurrence of rubella is followed by congenital abnormalities. Med J Aust 2:201-210.

Tanaka K, Fukuda S, Terayama Y, Toriyama M, et al. (1988) Experimental mumps labyrinthitis in monkeys (*Macaca irus*). Immunohistochemical and ultrastructural studies. Auris Nasus Larynx 15:89-96.

Terayama Y, Sasaki Y (1964) Studies on experimental allergic (isoimmune) labyrinthitis in guinea pigs. Acta Otolaryngol 58:49-64.

Trybus RJ, Karchmer MA, Kerstetter PP, Hicks W (1980) The demographics of deafness resulting from maternal rubella. Am Ann Deaf 125:977-984.

Van Dishoeck HAE, Bierman TA (1957) Sudden perceptive deafness and viral infection. Ann Oto Rhino Laryngol 66:963-980.

Veldman JE (1988) The immune system in hearing disorders. Acta Otolaryngol Suppl 458:67-75.

Veldman JE, Hughes GB, Roord JJ (1987) Immune-mediated inner ear disorders: Immunological investigations in the neurotology clinic. In: House J, O'Connor AF (eds) Handbook of Neurotological Diagnosis. New York: Marcel Dekker, pp 339-372.

Veltri RW, Wilson WR, Sprinkle PM, Rodman SM, et al. (1981) The implication of viruses in idiopathic sudden hearing loss: A primary infection or reactivation of latent viruses? Otolaryngol Head Neck Surg 89:137-141.

Vuori M, Lahkainen EA, Peltonen T (1962) Perceptive deafness in connection with mumps. Acta Otolaryngol 55:231-236.

Waksman BH (1962) Auto-immunization and the lesions of autoimmunity. Medicine 41:93-141.

Ward PH, Honrubia V, Moore BS (1968) Inner ear pathology in deafness due to maternal rubella. Arch Otolaryngol 87:22-28.

Westmore GA, Pickard BH, Stern H (1979) Isolation of mumps virus from the inner ear after sudden deafness. Br Med J 1:14-15.

Wiley CA, Nelson JA (1988) Role of human immunodeficiency virus and cytomegalovirus in AIDS encephalitis. Am J Pathol 133:73-81.

Wilson WR, Veltri RW, Laird N, Sprinkle P (1982) Viral and epidemiological studies of idiopathic sudden hearing loss. Otolaryngol Head Neck Surg 90:146-157.

Woolf NK (1990) Experimental congenital cytomegalovirus labyrinthitis and sensorineural hearing loss. Am J Otolaryngol 11:299-303.

Woolf NK (1991) Guinea pig model of congenital CMV-induced hearing loss: A review. Transplant Proceed 23:32-34.

Woolf NK (1994) Inner ear fibronectin expression following viral infection and noise trauma. Abstracts of the Seventeenth Midwinter Meeting of the Association for Research in Otolaryngology, p. 4.

Woolf NK, Harris JP, Ryan AF, Butler DM, Richman DD (1985) Hearing loss in experimental cytomegalovirus infection of the guinea pig inner ear: Prevention by systemic immunity. Ann Oto Rhino Laryngol 94:350-356.

Woolf NK, Ochi JW, Silva EJ, Sharp PA, et al. (1988) Ganciclovir prophylaxis for cochlear pathophysiology during experimental guinea pig cytomegalovirus labyrinthitis. Antimicr Agents Chemother 32:6865-6872.

Woolf NK, Koehrn FJ, Harris JP, Richman DD (1989) Congenital cytomegalovirus labyrinthitis and sensorineural hearing loss in guinea pigs. J Infect Dis 160:929-937.

Yamanobe S, Harris JP (1992) Spontaneous remission in experimental autoimmune labyrinthitis. Ann Oto Rhino Laryngol 101:1007-1014.

Yamanobe S, Harris JP (1993) Extraction of inner ear antigens for studies in inner ear autoimmunity. Ann Oto Rhino Laryngol 102:22-27.

Yoo TJ, Stuart A, Kang A, Townes K, et al. (1982) Type II collagen autoimmunity in otosclerosis and Meniere's disease. Science 217:1153-1155.

Yoo TJ, Tomoda K, Stuart JM, Cremer MA, et al. (1983a) Type II collagen-induced autoimmune sensorineural hearing loss and vestibular dysfunction in rats. Ann Oto Rhino Laryngol 92:267-271.

Yoo TJ, Yazawa Y, Tomoda K, Floyd R (1983b) Type II collagen-induced autoimmune endolymphatic hydrops in the guinea pigs. Science 222:65-67.

Yoo TJ, Tomoda K, Hernandez AD (1984) Type II collagen-induced autoimmune inner ear lesions in guinea pig. Ann Oto Rhino Laryngol (Suppl 113) 93:3-5.

Yow MD (1989) Congenital cytomegalovirus disease: A now problem. J Infect Dis 159:163-167.

Zajtchuk JT, Matz GJ, Lindsay JR (1972) Temporal bone pathology in herpes oticus. Ann Otol Rhinol Laryngol 81:331-339.

7
Otoacoustic Emissions: Animal Models and Clinical Observations

MARTIN L. WHITEHEAD, BRENDA L. LONSBURY-MARTIN, GLEN K. MARTIN, AND MARCY J. MCCOY

1. Introduction

Within the organ of Corti, an active mechanical process responds to sound by utilizing metabolic energy to increase the sound induced motion of the basilar membrane, at low sound levels, near the characteristic frequency place. The action of this active cochlear process results in enhanced sensitivity and frequency selectivity of basilar membrane vibration (Davis 1983; Johnstone, Patuzzi, and Yates 1986; Ruggero and Rich 1991). The active process appears particularly vulnerable to a variety of traumas that affect the inner ear, resulting in reduced basilar membrane motion in response to low-level stimuli and, thus, elevated hearing thresholds (Johnstone, Patuzzi, and Yates 1986; Ruggero and Rich 1991). Thus, many forms of sensorineural hearing loss are thought to result from a reduction of the action of the active process. The active process is thought to be based in the outer hair cells which, in in vitro preparations, demonstrate cycle-by-cycle motile responses to acoustic frequency electrical stimulation (Brownell et al. 1985; Brownell 1990).

Under certain conditions, audio-frequency vibratory energy produced at locations along the cochlear partition propagates toward the base of the cochlea, and through the middle ear to the ear canal. Sounds produced in this manner are termed otoacoustic emissions (OAEs) and can be measured in the external ear canal by a sensitive microphone (Kemp 1978; Wilson 1980a). In general, OAEs are thought to be generated by the active cochlear process responsible for enhancing basilar membrane vibration. Consistent with this notion are the results of recent studies in both humans and laboratory mammals that indicate that sensorineural hearing losses are often associated with a reduction or absence of OAEs in the ear canal (e.g., Kemp 1978; Zwicker 1983a; Kemp and Brown 1984; Kemp et al. 1986; Lonsbury-Martin, Probst, Coats, and Martin 1987; Lonsbury-Martin and Martin 1990; Lonsbury-Martin, Whitehead, and Martin 1991; Probst, Lonsbury-Martin, and Martin 1991). Because the measurement of OAEs is fast, objective and noninvasive, and because reduced OAE amplitudes are associated with certain types of hearing loss, OAEs have great potential for use in the audiology clinic. Thus, clinical tests utilizing OAEs may permit rapid diagnosis of a specific sensory component of sensorineural hearing losses by detecting dysfunction of the mechanism, thought to be based in the outer hair cells, that is responsible for enhancing cochlear-partition vibration. In addition, OAEs may act as an indicator of the presence of hearing loss in subjects for whom hearing cannot be directly tested, e.g., in newborns.

Over the past few years, studies conducted in patient groups have demonstrated that OAEs can be used in clinical settings to detect sensorineural hearing losses resulting from such factors as ototoxic drug administration, noise overexposure, Meniere's disease, sudden idiopathic hearing loss, presbycusis, and hereditary hearing disorders (e.g., Kemp et al. 1986; Bray and Kemp 1987; Bonfils, Uziel and Pujol 1988; Harris 1990; Lonsbury-Martin and Martin 1990; Martin et al. 1990; Lonsbury-Martin, Whitehead and Martin 1991). In parallel, the effects on OAEs of induced cochlear traumas, including administration of ototoxic drugs, hypoxia and anoxia, noise overexposure, endolymphatic hydrops, and genetically determined hearing disorders, have been studied in animal models, allowing direct comparison of the observed changes in OAEs to changes in electrophysiological measures of cochlear function, and to the underlying histopathology (e.g., Zurek, Clark, and Kim 1982; Kemp and Brown 1984; Horner, Lenoir, and Bock 1985; Schmiedt 1986a; Brown, McDowell, and Forge 1989; Lonsbury-Martin et al 1989; Martin et al. 1989; Ohlms, Lonsbury-Martin, and Martin 1991; Schrott, Puel, and Rebillard 1991). In the present chapter, the basic properties of OAEs are described for humans, some monkey species, and those small mammals commonly found in hearing research laboratories, i.e., cats, rabbits, and various rodents (Section 2). The results of studies of the effects of a variety of pathologies on OAEs in both human patient groups and animal models are then reviewed in order to provide insight into the mechanisms responsible for the reduction of OAEs in pathological ears (Section 3). The findings of both human and animal experiments clearly indicate that OAEs allow the separation of cochlear from retrocochlear disorders, and further allow differential diagnosis of a specific sensory component of cochlear trauma, presumed to be due to dysfunction of the outer hair cell system. Additionally, the data demonstrate that OAEs have the potential to provide information regarding both the precise frequency pattern of cochlear damage, and changes in cochlear condition over time.

2. Basic Properties of Otoacoustic Emissions

Essentially all normal human ears demonstrate evoked OAEs, which are elicited by presenting sound stimuli to the ear via miniature loudspeakers, and measured by means of a sensitive microphone sealed to the ear canal. Evoked emissions are grouped into three separate classes on the basis of the type of stimuli used to elicit them: (1) transiently evoked OAEs (TEOAEs) are elicited by transient acoustic stimuli such as clicks or tonepips, (2) stimulus-frequency OAEs (SFOAEs) are elicited by a single, continuous pure tone, and (3) distortion-product OAEs (DPOAEs) are typically evoked by two simultaneously presented, continuous pure tones of different frequencies. In addition to these three types of evoked emission, many ears also demonstrate spontaneous OAEs (SOAEs), which occur in the absence of deliberate acoustic stimulation of the ear and can be detected simply by sealing a sensitive microphone into the external ear canal. These four OAE types will be discussed in turn in Sections 2.1-2.4, below.

The schematic in Figure 7.1 shows a typical equipment setup that can be used to measure each of the four OAE types in human ears. The microphone is typically housed in a probe that is sealed into the ear canal with a foam or rubber eartip. The loudspeakers may also be housed in the probe, or may deliver sound to the ear canal via narrow bore tubes that pass through the probe. Suitable microphone and speaker arrangements for measuring OAEs are commercially available or can be constructed in the laboratory. The speaker and microphone arrangement is sealed into the ear canal for two major reasons, both related to the small amplitude of OAEs. First, the seal forms a closed cavity of small volume, resulting in a substantial increase of emission sound pressure relative to the open field situation. Second, the seal acts to exclude much external noise, resulting in improved signal-to-noise ratios over much of the frequency range of interest. Nevertheless, ambient noise remains a major limiting factor in the detection of OAEs, particularly at low frequencies, where both environmental and subject generated noise levels are greatest.

2.1 Transiently Evoked OAEs (TEOAEs)

To extract TEOAEs from the ambient noise present in the external ear canal, the amplified output of a sensitive microphone is averaged over a period of milliseconds (typically 20 ms in humans, but shorter periods are suitable in most laboratory mammals) following the stimulus. Personal computer based equipment for the measurement of TEOAEs, incorporating sophisticated noise rejection and response verification procedures, is commercially available. A typical click-evoked TEOAE response from a normal human ear, collected using a commercially available device, is shown in Figure 7.2A. The large, lower panel shows the sound pressure in the ear canal for 20 ms following a 79 dB$_{peak}$ SPL click stimulus presented at 0 ms (the initial 2.5 ms has been blanked to remove the stimulus waveform). The TEOAE appears as a complex waveform present for many milliseconds following the stimulus. The top-left panel shows the stimulus waveform on a much compressed pressure (y axis) scale. The spectrum of the TEOAE

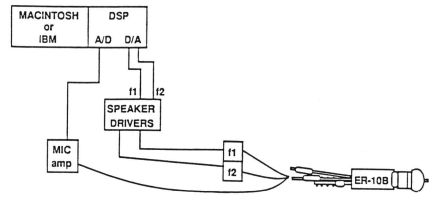

FIGURE 7.1 Schematic of a typical personal-computer-based equipment set-up allowing measurement of any of the four OAE types. The commercially available probe (Etymotic Research, ER-10B) contains a sensitive miniature microphone and is sealed into the ear canal with a rubber eartip. The microphone output is fed via a preamplifier and amplifier (MIC amp) and analog-to-digital (A/D) converter to a digital signal processor (DSP) mounted in the personal computer. This is all the hardware that is necessary to detect spontaneous OAEs. To allow measurement of evoked OAEs, two loudspeakers (f_1 and f_2) can be used to deliver sound to the ear canal via tubes passing through the probe. Voltage commands for each loudspeaker are generated by the DSP and passed via separate digital-to-analog (D/A) circuits, amplifiers, and impedance-matching devices (speaker drivers) to each loudspeaker. To evoke TEOAEs (click or tonepip stimuli) or SFOAEs (swept pure tone stimulus) only one speaker is required. For DPOAEs, two pure tones are presented, one over each speaker (to prevent generation of artifactual distortion products by a single speaker driven by two sinusoids). Analysis of the microphone output typically involves spectral averaging (i.e., averaging of spectra obtained by fast-Fourier transformation of samples of the microphone output) to detect spontaneous OAEs, conventional stimulus-locked averaging to detect TEOAEs, and conventional stimulus-locked averaging following by fast-Fourier transformation to reveal DPOAEs.

waveform is shown as the open area in the panel labeled "Response FFT." The shaded area in this panel is an indication of the background noise in the measurement. The inset below the "Response FFT" panel shows the spectrum of the stimulus click. The stimulus spectrum was flat from 0.5 to 4 kHz. Components of the TEOAE response were observed only at frequencies present in the stimulus spectrum, but the response spectrum was highly irregular even in the region where the stimulus spectrum was flat.

Virtually all normal human ears exhibit TEOAEs (Kemp 1978; Kemp et al. 1986; Probst et al. 1986; Bonfils, Bertrand, and Uziel 1988; Stevens 1988; Smurzynski and Kim 1992). The morphology of the response waveform and of the corresponding spectrum varies greatly between ears for reasons that are obscure. In any one normal human ear, however, in response to broadband click stimuli, TEOAE components are typically detectable over most of the 0.5 to 4 kHz frequency range, with the greatest amplitudes occurring in the 1 to 1.5 kHz region. At very low stimulus levels, TEOAE amplitude grows linearly, i.e., 1 dB per dB increase of stimulus level. However over most of the stimulus level range, TEOAE growth is compressively nonlinear, i.e., less than 1 dB per dB increase in stimulus level, and at high stimulus levels TEOAE amplitudes saturate at magnitudes that rarely exceed 20 dB SPL (Kemp 1978; Wilson 1980a; Zwicker, 1983b). The latency of TEOAEs, i.e., the time between presentation of the stimulus and detection of the response, varies from <3 ms to >15 ms in humans and decreases as the frequency of the OAE component increases (Kemp 1978; Wilson 1980a; Wit and Ritsma 1980). This frequency dependence of latency is seen as frequency dispersion in click evoked TEOAEs. The decrease in latency with increasing frequency is thought to reflect the reduction in travel times associated with the fact that progressively higher frequencies are processed by progressively more basal regions of the cochlea.

In macaque (*Macaca irus, M. mulatta, M. nemestrina*) and patas (*Erythrocebus patas*) monkeys, TEOAEs appear very similar to those in humans (Anderson and Kemp 1979; Wit and Kahmann 1982; Martin et al. 1988), although with slightly shorter latencies, which may be expected from the shorter cochleae of monkeys. Additionally, TEOAEs in these monkey species demonstrate shorter durations than observed in humans, perhaps indicating a greater damping of the emission generator. The top panels of Figure 7.3A show an example of a TEOAE obtained from a normal ear of a pigtail macaque (*Macaca nemestrina*) using equipment similar to that used to measure a TEOAE from a human ear in Figure 7.2A.

In the common, nonprimate laboratory mammals, TEOAEs have been measured in cats (Wilson 1980b) and guinea pigs (Zwicker and Manley 1981; Avan et al 1990; Ueda et al. 1992). An example of a TEOAE measured from a guinea pig using equipment similar to that used to acquire the human TEOAEs of Figure 7.2A is shown in Figure 7.4. Guinea pig TEOAEs have smaller amplitudes and much shorter latencies and durations than those in humans and monkeys. The very short latencies will result in a loss of some of the early, higher frequency components of the response during measurement because these components will be omitted by the measurement time window, which must be set to exclude the decaying "tail" of the acoustic stimulus. This factor, and the short durations, may contribute to the smaller measured amplitudes of TEOAEs in guinea pigs than in primates. These considerations may account for the detection of TEOAEs in only some guinea pig ears in the studies of Zwicker and Manley (1981), Avan et al. (1990), and Ueda et al. (1992), and for the failure of other studies to find TEOAEs in the ears of guinea pigs (Wit and Ritsma 1980) and gerbils (Schmiedt and Adams 1981), in contrast to the essentially 100% prevalence of TEOAEs in primate ears. With the equipment used to obtain the data shown in Fig. 7.4, TEOAEs are present in the great majority of guinea pig ears.

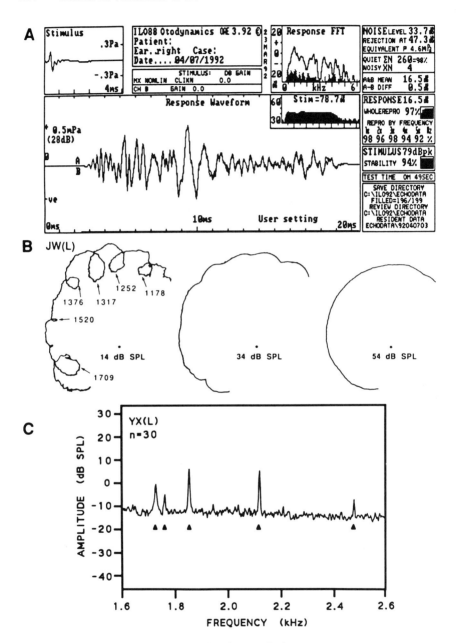

FIGURE 7.2. Caption on facing page.

2.2 Stimulus Frequency OAEs (SFOAEs)

Stimulus frequency OAEs are a continuous emission of low level sound from the cochlea at the frequency of a continuous pure tone stimulus. SFOAEs appear to be generated by the same mechanism as that which produces TEOAEs in response to transient stimuli. Thus, SFOAEs appear to be present in the ears of virtually all normally hearing humans, are robust at those frequencies where TEOAEs are strong, and demonstrate latency and compressive nonlinearity features that are similar to the properties of TEOAEs (Kemp and Chum 1980; Wilson 1980a; Kemp and Brown 1983;

←

FIGURE 7.2 Measurement of otoacoustic emissions. (A) Measurement of TEOAEs. The large panel labeled "Response Waveform" shows the temporal waveform of a click evoked TEOAE from a normal human ear collected with a commercially available device, the IL088 (Otodynamics, Ltd). Two independently averaged responses (A and B) are overplotted to show the extremely high repeatability of the emission. A 79 dB$_{peak}$ SPL click stimulus was presented at time zero. The initial 2.5 ms of the waveform has been blanked to remove the stimulus. However, the click stimulus waveform is shown on the same time scale, but a much compressed pressure scale, in the upper left panel. The frequency spectrum of the TEOAE response is indicated by the unshaded region in the upper right panel ("Response FFT"), and the background noise of the measurement is indicated by the shaded area in this panel. The response signal-to-noise ratio is 10–20 dB over most of the range in which it is present. The small panel below the "Response FFT" panel shows the frequency spectrum of the stimulus click, for comparison. The set of panels to the right display information about the stimulus parameters, the TEOAE response and background noise, and also details concerning test parameters [see Robinette (1992) for an explanation of these values]. (B) Measurement of SFOAEs. Vector diagrams (Nyquist plots) showing sound pressure in the sealed ear canal of a normally hearing human (JW) upon stimulation with a swept frequency pure tone at three stimulus levels, corresponding to 14 (left), 34 (middle), and 54 dB SPL (right) at 1 kHz. For each plot, a constant voltage was applied to the ear canal speaker. The centrally located dot represents the origin of each plot, and stimulus frequency increases anti-clockwise from approximately 1,100 to 1,800 Hz. The length and direction of a line (vector) joining the origin to any point on a plot represents the amplitude and phase of the sound pressure in the ear canal at that frequency. The pronounced loops present at the 14 dB SPL stimulus level represent interference of a SFOAE with the swept pure tone stimulus. The magnitude of these loops reflects the amplitude of the SFOAE relative to the stimulus. As stimulus level increases, the interference loops decrease and eventually disappear as the amplitude of the SFOAE relative to the stimulus decreases, indicating the compressively nonlinear growth of SFOAEs. The arrows indicate the frequencies of maximum cancellation of the stimulus by the SFOAE, i.e., the frequencies at which the SFOAE was 180° out of phase with the stimulus in the ear canal. (C) Measurement of SOAEs. Frequency spectrum of the sound field in the sealed ear canal of a normally hearing human (YX) in the absence of deliberate sound stimulation. The five sharp peaks above the noise floor, marked with arrowheads, are SOAEs.

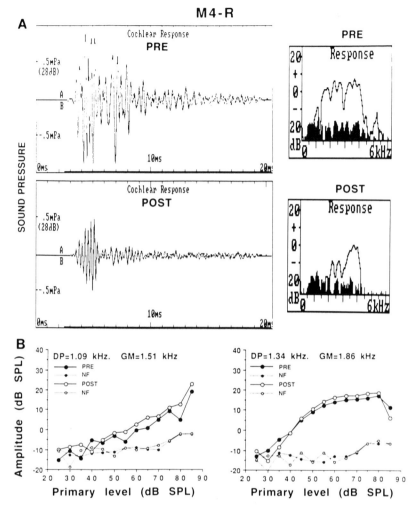

FIGURE 7.3 (A) Time waveforms (left), and associated frequency spectra (right), of click evoked TEOAEs obtained from the right ear of pigtail macaque M4 before (PRE), and approximately 6 hours after (POST), a subcutaneous injection of sodium salicylate (100 mg/kg). The TEOAEs were substantially reduced, especially at low frequencies, post-salicylate. The TEOAEs were obtained with equipment similar to that used for Figure 7.2A. (B) Growth functions of $2f_1-f_2$ DPOAEs obtained before (PRE, filled circles) and approximately 6 hours after (POST, open circles) the salicylate injection at frequencies within the region of maximum TEOAE reduction. The corresponding noise floors (NF) are also shown. Both the frequency of the DPOAE (DP), and the geometric mean (GM) of the stimulus tone frequencies, i.e., $(f_1 \times f_2)^{0.5}$, are given in each panel. The DPOAEs were not reduced despite a greater than 20 dB reduction of the TEOAEs at some of the frequencies of DPOAE measurement. The blood plasma salicylate level obtained shortly after the post injection OAE measurements was 22.1 mg%.

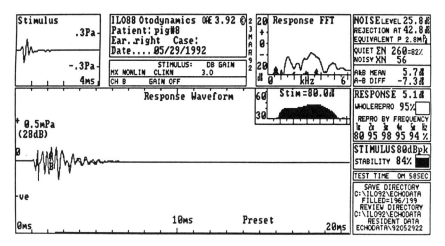

FIGURE 7.4 A click-evoked TEOAE from a normal guinea-pig ear, collected with the IL088 device also used to obtain the data of Figure 7.2A. The probe was sealed to the ear canal via a short length of tapered plastic tubing. The response window was 1–7 ms, with a cosine rise-time of 2.6 ms duration, and the stimulus level was 80 dB$_{peak}$ SPL. The "Response FFT" panel shows TEOAE components present over most of the range of the stimulus spectrum, i.e., 1–5 kHz, with substantial signal-to-noise ratios.

Rutten and Buisman 1983; Zwicker and Schloth 1984; Dallmayr 1987; Long and Tubis 1988; Kemp, Brass, and Souter 1990; Lonsbury-Martin et al. 1990b; Brass and Kemp 1991).

Because the stimulus is larger than the emission and simultaneously present at the same frequency in the ear canal, SFOAEs are more difficult to measure than the other emission types. The existence of SFOAEs is typically revealed by slowly sweeping the frequency of the low-level stimulus tone. The phase lag of the emission relative to the stimulus increases with increasing frequency, resulting in the physical interference of the stimulus and the emission in the sealed ear canal. This process gives rise to ripples in the otherwise smooth frequency response of the ear canal sound pressure as the stimulus and the SFOAE move alternately into and out of phase (Kemp and Chum 1980; Wilson 1980a). The three plots of Figure 7.2B are vector diagrams showing the sound pressure in the sealed ear canal of a swept-frequency pure tone as a function of frequency (increasing anticlockwise around the plot), for three different stimulus levels, corresponding to 14 (left), 34 (middle), and 54 (right) dB SPL at 1 kHz. In each case, a constant voltage was applied to the ear canal speaker. The regularly spaced interference loops and ripples at 14 dB SPL indicate the presence of SFOAEs. At the higher stimulus levels, the magnitude of the interference decreases as the SFOAE amplitude decreases relative to the stimulus, due to the compressive nonlinearity of the emission generator.

In macaque monkeys, SFOAEs appear to be essentially identical to those in humans, although the frequency spacing of the ripples caused by the interference of the SFOAE and the stimulus tone is larger in the macaque (Martin et al. 1988; Lonsbury-Martin et al. 1991). The greater frequency spacing of the interference ripples is to be expected as this spacing is inversely proportional to the latency of the emission, and monkeys demonstrate shorter latency TEOAEs than humans (Anderson and Kemp 1979; Wit and Kahmann 1982; Martin et al. 1988).

In gerbils, SFOAEs with properties similar to those observed in humans have been reported in response to stimulus levels <40 dB SPL (Kemp and Brown 1983). These SFOAEs demonstrated compressive nonlinearity, with maximum amplitudes of about −5 dB SPL, i.e., about 30 dB below maximum amplitudes in humans, and latencies comparable to those of SFOAEs and TEOAEs in guinea pigs (Zwicker and Manley 1981; Avan et al. 1990; Ueda et al. 1992), but much shorter than the latencies of TEOAEs and SFOAEs in humans. A similar SFOAE component, again smaller than that in humans, appears to be present in guinea pigs (Avan et al. 1990). The finding of smaller SFOAEs in guinea pigs than in humans is consistent with the smaller amplitudes of TEOAEs in these animals than in humans. In guinea pigs (Avan et al. 1990), there are much larger frequency spacings between the interference ripples that indicate the presence of SFOAEs than in humans consistent with the much shorter latencies of the emissions in guinea pigs than humans. A different SFOAE component, demonstrating expansive nonlinearity and almost zero latency unlike that reported in humans, appears to dominate the gerbil ear canal signal above a stimulus level of 40 dB SPL (Kemp and Brown 1983). SFOAEs have also been measured in cats (Guinan 1986, 1990).

2.3 Spontaneous OAEs (SOAEs)

Spontaneous OAEs are narrowband acoustic signals generated within the cochlea in the absence of deliberate sound stimulation. In Figure 7.2C, the SOAEs in a normal human ear can be observed as sharp peaks above the noise floor in the spectral analysis of the output of the ear canal microphone. SOAEs are detected at one or more discrete frequencies in approximately 65% of normal human ears (although they are more prevalent in females than males). These frequencies are unique to each ear and typically vary by only a few Hertz even over several years (Kemp 1981; Zurek 1981; Schloth 1983; Dallmayr 1985; Whitehead 1989; Lonsbury-Martin et al. 1990b; Burns, Arehart, and Campbell 1992; Whitehead et al. 1993a). Most SOAEs occur in the 0.8 to 4 kHz region, with the majority between 1 and 2 kHz.

Spontaneous OAEs are thought to be produced by the same process as TEOAEs and SFOAEs, apparently as a result of continual feedback of the output of the emission generator into its input. At frequencies where this feedback is positive, if the loop gain is sufficient, self-sustaining oscillation will result, which is observed in the ear canal as an SOAE (Kemp 1981; Wilson and Sutton 1981). Thus, SOAEs can be thought of as continuously self-eliciting evoked OAEs. Consistent with this view, SOAEs are found in regions of strong evoked response (Wilson 1980a; Kemp 1981; Zwicker and Schloth 1984), and rarely exceed 20 dB SPL, presumably due to the compressive nonlinearity of the OAE generator.

TEOAEs, SFOAEs and SOAEs can be suppressed, i.e., reduced in amplitude, by external tones. Plotting the level of the suppressor tone required to reduce the OAE amplitude by a criterion amount yields an isosuppression contour. These contours demonstrate similar shapes to psychophysical tuning curves and to the frequency tuning curves of auditory nerve fibers, indicating that these OAEs are generated at a late stage of cochlear filtering (Kemp and Chum 1980; Wilson 1980a; Zurek 1981).

There appears to be a rare subtype of SOAE (Wilson and Sutton 1983; Yamamoto et al. 1987; Mathis et al. 1991) that differs in several respects from the SOAEs routinely observed in normal human ears. Specifically, these rare "atypical" SOAEs are much larger (up to 60 dB SPL) than normal SOAEs (usually less than 20 dB SPL), occur in a higher frequency range than the majority of SOAEs, and tend to be associated with regions of audiometric abnormality. These atypical SOAEs also differ from most typical SOAEs in that they often demonstrate considerable short-term instability of frequency and level, and may also have broad multi-lobed, rather than sharp, single-lobed isosuppression contours (Wilson and Sutton 1983).

In macaque monkeys, the prevalence of SOAEs is somewhat lower than in humans (Martin et al. 1985; Lonsbury-Martin and Martin 1988). Otherwise, the properties of monkey SOAEs appear to be essentially identical to the typical SOAEs in normally hearing humans (Martin et al. 1988).

In the nonprimate laboratory mammals, SOAEs are less common than in humans. In systematic surveys of normal animals, Zurek and Clark (1981) found no SOAEs in 26 ears of 17 chinchillas, and Ohyama et al. (1991) found SOAEs in nearly 21% of 248 ears of 124 Hartley strain albino guinea pigs. These guinea pig SOAEs demonstrated similar properties to the typical, low-level SOAEs observed in about 65% of normal human ears. The lower prevalence of SOAEs in guinea pigs is consistent with the hypothesis that the typical low-level SOAEs are caused by feedback of the TEOAE and SFOAE generator's output into its input, and the observation that, in the nonprimate laboratory mammals, this generator's output is weaker than it is in the primates, which would result in reduced ability to sustain a feedback oscillation. There have been other reports of SOAEs in nonprimate mammals, not arising from systematic surveys of normal ears. Whereas some of these SOAEs displayed properties similar to the typical SOAEs in human ears (Evans, Wilson, and Borerwe 1981; Brown, Woodward, and Gaskill 1990), the other SOAEs reported in these animals showed properties resembling those of the high-level, atypical SOAEs observed in some pathological human ears, rather than the low-level SOAEs common in normal human ears (Zurek and Clark 1981; Decker and Fritsch 1982; Clark et al. 1984; Ruggero, Kramek, and Rich 1984). In particular, these SOAEs had large amplitudes (up to 59 dB SPL), and most of them were unstable in frequency and level, and/or demonstrated relatively broad or multilobed isosuppression contours. It is noteworthy that in the nonprimate laboratory mammals, some of the particularly large SOAEs have been associated with cochlear lesions (Zurek and Clark 1981; Clark et al. 1984), or apparent audiometric abnormalities (Ruggero, Kramek, and Rich 1984). Thus, both humans and nonprimate mammals demonstrate both the typical low-level, and atypical high-level SOAEs. However, it appears that in the nonprimates the typical type are less prevalent than in humans, whereas the atypical type, although rare, may be more common in nonprimates than in humans.

The different properties of the atypical SOAEs and their apparent association with pathology suggest that the mechanism of generation of these SOAEs in both humans and nonprimate laboratory mammals may be different to that of the more typical SOAEs.

2.4 Distortion Product OAEs (DPOAEs)

Distortion product OAEs are tones emitted by the cochlea at frequencies not present in a single tone or multiple tone stimulus. Although ears produce harmonic distortion product OAEs in response to a single tone stimulus, DPOAEs are typically measured at intermodulation-distortion frequencies upon stimulation of the ear with two simultaneously presented pure tones of frequencies f_1 and f_2, of appropriate frequency separation (f_2/f_1) and levels (L_1, L_2). The stimulus tones are often referred to as "primary" tones. DPOAEs can be present at many intermodulation-distortion frequencies, including f_2-f_1, $2f_2-f_1$, and $3f_1-2f_2$, but the largest DPOAE in all mammalian species tested occurs at the frequency $2f_1-f_2$. DPOAEs are typically measured by spectral analysis of the microphone output. Figure 7.5A provides an example of a $2f_1-f_2$ DPOAE from a human ear. Whereas $2f_1-f_2$ DPOAEs can be detected in essentially all normal human ears (Kemp et al. 1986; Gaskill and Brown 1990; Harris 1990; Lonsbury-Martin et al. 1990a; Smurzynski and Kim 1992), they are typically small, i.e., usually ~60 dB below equilevel ($L_1 = L_2$) stimulus tones. The amplitude of the $2f_1-f_2$ DPOAE depends systematically upon the parameters of the stimulus tones (i.e., frequencies, levels, frequency separation, and level difference). In humans, the $2f_1-f_2$ DPOAE increases in amplitude with increasing $L_1 = L_2$ at a growth rate of around 1 dB/dB, and saturates above stimulus levels of ~75 dB SPL (Lonsbury-Martin et al. 1990a). Mean DPOAE growth functions from human ears at three frequencies are illustrated in Figure 7.5B (squares). The amplitude of the $2f_1-f_2$

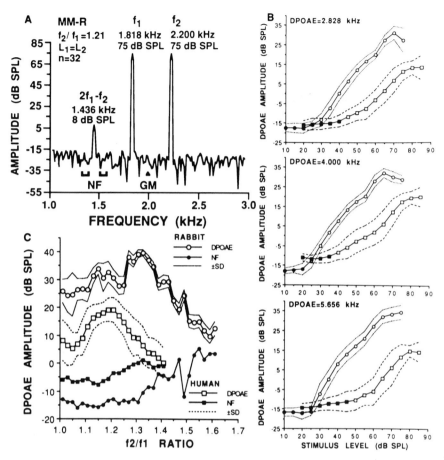

FIGURE 7.5 (A) Frequency spectrum of the sound field in the sealed ear canal of a normal human (MM) ear upon stimulation by two pure tones of frequencies $f_1 = 1.818$ and $f_2 = 2.2$ kHz (i.e., $f_2/f_1 = 1.21$), and levels $L_1 = L_2 = 75$ dB SPL. The sharp peak at 1.436 kHz is the $2f_1$-f_2 DPOAE. The geometric mean (GM) frequency of the stimulus tones was 2 kHz. The spectrum was obtained from the stimulus-locked average of 32 samples. The noise floor (NF) is estimated by taking the mean level of several spectral bins above and below the DPOAE frequency bin (indicated by the brackets surrounding the 1.436 kHz DPOAE region). In this example, the 8 dB SPL response at the $2f_1$-f_2 frequency was well above the associated noise floor estimate of -24 dB SPL. (B) Mean (± 1 s.d.) amplitude of the $2f_1$-f_2 DPOAEs at 2.8, 4, and 5.6 kHz, measured from 24 human (squares and dashed lines) and 24 rabbit ears (circles and solid lines), as a function of $L_1 = L_2$, with $f_2/f_1 = 1.25$. Filled symbols indicate that the mean amplitudes at the $2f_1$-f_2 frequency were <3 dB above the corresponding mean noise floor. (C) Mean (± 1 s.d.) amplitude of the $2f_1$-f_2 DPOAE at 4 kHz measured from 10 human (squares) and 12 rabbit (circles) ears, as a function of f_2/f_1, with $L_1 = L_2 = 75$ dB SPL. Filled symbols indicate noise floors (NF). The data from human ears in this panel were taken from Harris et al. (1989).

DPOAE is also systematically dependent on the relative levels of the stimulus tones, and can be increased slightly by decreasing L_2 below L_1 (Gaskill and Brown 1990; Hauser and Probst 1990; Whitehead et al. 1995a,b). Finally, as illustrated in Figure 7.5C, DPOAE amplitude (squares) is strongly dependent on the frequency separation of the stimulus tones, being largest when f_2/f_1 is around 1.22 (Harris et al. 1989; Gaskill and Brown 1990).

The properties of $2f_1$-f_2 DPOAEs in rhesus (*Macaca mulatta*) and pigtail (*M. nemestrina*) macaques are very similar to those in humans (Martin et al. 1988; Lonsbury-Martin et al. 1991). Some DPOAE growth functions obtained with equilevel primary tones from a pigtail macaque ear are shown in Figure 7.3B. In the nonprimate laboratory mammals, including cats (Kim 1980; Schmiedt 1986a; Wiederhold, Mahoney, and Kellogg 1986), rabbits (Lonsbury-Martin et al. 1987; Whitehead, Lonsbury-Martin, and Martin 1992a), and several rodent species (Kim 1980; Schmiedt and Adams 1981; Zurek, Clark, and Kim 1982; Kemp and Brown 1983; Horner, Lenoir, and Bock 1985; Brown and Gaskill 1990; Norton and Rubel 1990), $2f_1$-f_2 DPOAEs are considerably larger than in humans and monkeys, often being less than 30 dB below equilevel stimulus tones.

Despite the large difference in absolute amplitudes of DPOAEs between humans and the nonprimate laboratory mammals, the DPOAEs in these species demonstrate considerable qualitative similarity in their amplitude variations with changes in stimulus parameters such as the frequency separation (f_2/f_1), level (L_1, L_2), and level difference (L_1-L_2) of the stimulus tones (Brown and Gaskill 1990; Whitehead et al. 1995b). This similarity suggests commonality of the underlying generation mechanisms. Mean data from rabbit (circles) and human (squares) ears are compared in Figures 7.5B and C. Whereas the general similarity of DPOAE amplitude variation in rabbits and humans are apparent, there are differences. For example, in the rabbit, as in other nonprimate laboratory mammals, DPOAE growth with increasing stimulus levels tends to saturate at lower stimulus levels (Fig. 7.5B) than in humans, and the maximum DPOAE amplitude occurs at a greater frequency separation of the stimulus tones (Fig. 7.5C) than in humans (Brown and Gaskill 1990; Whitehead, Lonsbury-Martin, and Martin 1992a). These observations, combined with the large difference in the absolute amplitudes of DPOAEs between humans and the nonprimate laboratory mammals, indicate that there are some differences in the processes responsible for DPOAE generation in humans and the nonprimate laboratory mammals.

Evidence from rabbits and various rodent species indicates that the $2f_1$-f_2 DPOAE is produced by at least partially discrete mechanisms below and above stimulus levels of 60 to 70 dB SPL. Thus, this DPOAE shows differential parametric properties below and above these stimulus levels (Whitehead, Lonsbury-Martin, and Martin 1990, 1992a). Additionally, sharp notches are present in some DPOAE growth functions at stimulus levels of between 55 and 70 dB SPL. Such DPOAE amplitude notches are associated with rapid DPOAE phase reversals, suggesting a cancellation of two distinct components of approximately equal amplitude, ~180° out of phase with each other, at the stimulus level of the notch (Brown 1987; Whitehead, Lonsbury-Martin, and Martin 1990; 1992a,b). Moreover, as discussed in Sections 3.1 and 3.2, below, DPOAEs evoked by primary tones below 60 to 70 dB SPL are considerably more vulnerable to a variety of physiological insults than are DPOAEs elicited by stimuli above these levels (Schmiedt and Adams 1981; Kemp and Brown 1984; Schmiedt 1986a; Brown, McDowell, and Forge 1989; Norton and Rubel 1990; Whitehead, Lonsbury-Martin, and Martin 1992b), again suggesting that the $2f_1-f_2$ DPOAE is dominated by different generator mechanisms below and above 60–70 dB SPL in rabbits and rodents.

As discussed above, there are significant differences in the properties of DPOAEs and the other OAE types between the primate and nonprimate mammal species tested. The most pronounced difference is that DPOAEs are very much smaller in primates than in nonprimate mammals, whereas TEOAEs and SFOAEs tend to be larger, and SOAEs more common, in primates. These differences are cause for caution in generalizing from nonprimate animal models to the human situation. Thus, although there appear to be discrete mechanisms producing DPOAEs above and below 60 to 70 dB SPL in rabbits and rodents, it should not be assumed that there are corresponding discrete low- and high-level DPOAE generators in humans, or, if there are such discrete mechanisms in humans, that they contribute the ear canal DPOAE over the same stimulus level ranges as in nonprimate species.

2.5 Summary

The four classes of OAE can be measured in humans, monkeys, and the common, nonprimate laboratory mammals (cats, rabbits, and various rodent species). Thus, laboratory animals may serve as practical experimental models of OAE generation in human ears, and it is probable that future breakthroughs in our understanding of the mechanisms underlying OAE generation will derive from experimental observations in animal models. However, whereas the basic properties of the four OAE types appear very similar in humans and monkeys, there are some significant quantitative and qualitative differences between primates and the common nonprimate laboratory mammals that should be borne in mind when generalizing from results obtained in animal models to the human situation.

3. Otoacoustic Emissions and Hearing Pathology

In the following, we describe the deviations of OAEs measured in patients with hearing losses of various etiologies from those measured in normal human ears. Additionally, we describe the effects on OAEs in animal models of manipulations related to the pathologies observed in human ears in order to provide an experimental basis upon which the observations in patients can be interpreted.

3.1 Ototoxic Drugs

Ototoxic drugs administered to humans include cisplatin, loop diuretics, aminoglycoside antibiotics, and aspirin. The effects of each of these agents on OAEs have been studied in animal models. In addition to providing practical information concerning the adverse effects of these chemical agents on cochlear function, these studies have also yielded some insights into the fundamental basis of emission generation.

3.1.1 Cisplatin

Cisplatin is a known ototoxin utilized in antitumor therapy. In animal experiments, cisplatin caused loss of outer hair cells, and also damage to inner hair cells, supporting cells of the organ of Corti, and the stria vascularis, associated with high-frequency threshold elevation (Fleischmann et al. 1975; Estrem et al. 1981; Anniko and Sobin, 1986; Laurell and Engström 1989a). Some reports (Komune et al. 1981; Laurell and Engström, 1989b), but not others (Konishi, Gupta, and Prazma 1983; Taudyl et al. 1989), indicated that cisplatin reduced the endolymphatic potential. Cisplatin therapy causes a permanent high-frequency hearing loss in some patients (e.g., Schaefer et al. 1985; Pasic and Dobie 1991). The literature contains examples of reductions of TEOAEs and DPOAEs (Lonsbury-Martin et al. 1993; Zorowka, Schmitt and Gutjahr 1993) in patients undergoing cisplatin treatment. In these patients, the OAE reductions tended to occur at high frequencies, corresponding to the cisplatin-induced hearing losses. The findings depicted in Figure 7.6 illustrate the effects of several courses of cisplatin administration on both DPOAEs and TEOAEs in an 8-year-old female patient with an inoperable brain tumor involving the thalamus. Testing carried out before the start of therapy (PRE, filled symbols) revealed normal behavioral audiograms bilaterally (top). The DPOAE amplitudes (middle) were also normal in both ears, in that they were within one standard deviation (s.d.) of the mean DPOAE amplitudes from a group of normal ears (± 1 s.d. indicated by bold dashed lines in Fig. 7.6). The TEOAEs (bottom) were also normal bilaterally, in that robust emissions were detected at most frequencies between 0.5 and 5.5 kHz in both ears. Post-treatment testing (POST, open symbols) demonstrated hearing-threshold elevation at frequencies above 2 kHz (top), and a reduction of both DPOAE (middle) and TEOAE (bottom) components above 2 kHz. Thus, both evoked emission types effectively reflected the cisplatin-induced frequency-specific decrease of hearing sensitivity.

These clinical data are consistent with the results of McAlpine and Johnstone (1990), who found that DPOAEs in guinea pigs were substantially decreased by acute injection of cisplatin (12 mg/kg), or of cisplatin (6 mg/kg) coadministered with furosemide (160 mg/kg) or aminooxyacetic acid (15 mg/kg), and that the elevation of the DPOAE detection threshold correlated well with the loss in neural sensitivity assessed by the threshold of the compound action potential. In rats, high-frequency DPOAEs were reduced by cisplatin (McPhee et al. 1994).

3.1.2 Loop Diuretics

It is well-known that acute doses of the loop diuretics furosemide and ethacrynic acid, used in the treatment of kidney failure in human patients, produce a temporary hearing loss. From basic research in animals, these drugs are known to act primarily upon the cochlea's stria vascularis in such a way as to produce a rapid, reversible reduction of the endolymphatic potential. Such reductions in the effective energy supply of the organ of Corti result in corresponding decreases in the sound evoked responses of the cochlea (e.g., Matz 1976; Rybak 1986).

Although there have been no published studies of the effects of these drugs on OAEs in human ears, Anderson and Kemp (1979) demonstrated that acute administrations of both furosemide and ethacrynic acid produced temporary reductions of TEOAEs in patas monkeys. In the common nonprimate laboratory mammals, acute administration of furosemide temporarily reduced TEOAEs in cats (Wilson and Evans 1983) and guinea pigs (Ueda et al. 1992), and DPOAEs evoked by low level, but not high level, stimulus tones in gerbils (Kemp and Brown 1984; Mills, Norton, and Rubel 1993). Figure 7.7A illustrates the effects of an injection of ethacrynic acid (40 mg/kg, iv) on DPOAEs in a rabbit. The DPOAEs evoked by low level stimuli (45 to 65 dB SPL) were temporarily greatly reduced by ethacrynic acid, whereas those evoked by high level stimuli (75 dB SPL) were relatively unaffected (Whitehead, Lonsbury-Martin, and Martin 1992b). This level dependence of the vulnerability of DPOAEs is similar to that reported upon furosemide administration in gerbils, and the time course of the initial diuretic induced decrease of low level DPOAEs is similar to the associated decrease of the cochlear microphonic and endolymphatic potential in the same animals (Kemp and Brown 1984; Mills 1993). These data suggest that the mechanism that generates DPOAEs in response to low level stimuli is dependent upon the endolymphatic potential, whereas the mechanism underlying DPOAEs evoked by high-level stimuli is relatively independent of the endolymphatic potential. Interestingly, in gerbils, the recovery of the low-level DPOAEs occurred while the endolymphatic potential was still greatly reduced, indicating that the generator of these DPOAEs can adapt to the reduced endolymphatic potential (Mills 1993).

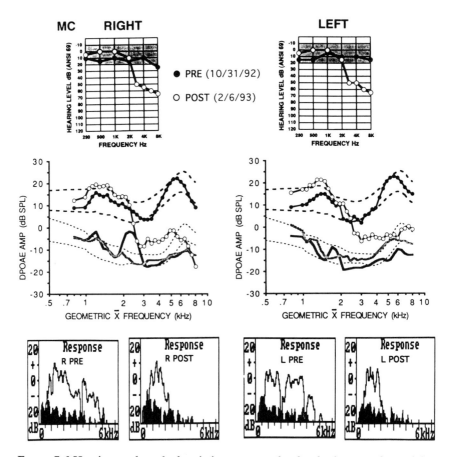

FIGURE 7.6 Hearing and evoked-emission test results for both ears of an eight-year-old girl (MC) with a brain tumor. Data were obtained before (PRE, 10/31/92) and following (POST, 2/6/93) drug treatment with the anti-neoplastic agent cisplatin. The top plots show conventional behavioral audiograms obtained pre- and post-treatment from the right and left ears. The stippling indicates the region considered clinically normal. Before treatment (filled circles), hearing was normal in both ears. Three months following the initial treatment (open circles), hearing thresholds above 2 kHz were elevated to 50–65 dB HL bilaterally. The middle two panels show DPOAE amplitudes from the two ears measured as a function of the geometric mean frequency of the primary tones, with $f_2/f_1 = 1.21$, and $L_1 = L_2 = 75$ dB SPL. The pair of bold dashed lines around 10–15 dB SPL show ±1 s.d. of the mean amplitude of DPOAEs measured with these stimulus parameters from 33 ears of normal hearing adults to indicate expected DPOAE amplitudes (see, Lonsbury-Martin et al. 1990). The lower pair of dashed lines show ±1 s.d. of the corresponding noise levels. Before treatment (filled circles), the DPOAE amplitudes of this patient were within the normal range bilaterally. After treatment (open circles), the DPOAE amplitudes were within the normal range up to 2 kHz, but were substantially below the normal range above 2 kHz. This pattern of DPOAE-amplitude reduction corresponds well to the hearing-threshold elevation observed

3.1.3 Aminoglycoside Antibiotics

Chronic administration of some aminoglycoside antibiotics can produce permanent hearing loss in humans, particularly at high frequencies (see Rybak 1986; Brummett and Fox 1989; Govaerts et al. 1990). Consistent with this finding, histological studies in animal models have demonstrated damage at almost every level of the cochlea upon chronic aminoglycoside treatment, with basal outer hair cells apparently the most vulnerable cochlear element (Rybak 1986; Govaerts et al. 1990). The only published study to date of the effects of aminoglycosides on OAEs in humans (Hotz, Harris, and Probst 1994) reported a small but significant reduction of TEOAEs in leukemia patients undergoing amikacin-sulfate therapy for more than twelve days. In guinea pigs chronically treated with the aminoglycoside gentamicin, Brown, McDowell, and Forge (1989) demonstrated that in frequency regions where there was substantial outer hair cell pathology, DPOAEs elicited by stimulus tones below 60 to 70 dB SPL were greatly reduced, but those produced by higher level stimuli were relatively unaffected. This finding, in combination with the loop diuretic results discussed above, suggests that the mechanism that generates DPOAEs in response to low-level stimuli requires normal outer hair cells, and is in some way dependent upon the endolymphatic potential, whereas the mechanism underlying high-level DPOAEs is relatively independent of both of these factors. It is notable that the active cochlear process responsible for enhancement of basilar membrane motion and, thus, for normal hearing sensitivity, also appears to be dependent upon both normal outer hair cell function and the endolymphatic potential (Ruggero and Rich 1991).

Aminoglycoside antibiotics and loop diuretics administered in combination are known to have a powerful ototoxic effect in humans, which can result in profound, permanent hearing loss (West, Brummett, and Himes 1973). Animal studies have demonstrated that acute dosages of these drugs in combination have a very disruptive effect on the function and anatomy of the organ of Corti, resulting in severe decrements of sound evoked cochlear potentials, and widespread hair cell loss, occurring a few hours after administration (West, Brummett, and Himes 1973; Russell, Fox, and

←

FIGURE 7.6 (*Continued*) in the behavioral audiograms. The lower, pale and dark hatched lines indicate the noise floors associated with the open and filled symbols, respectively. The bottom panels show spectra of TEOAE responses evoked by click stimuli of approximately 80 dB$_{peak}$ SPL, measured using the IL088 device (see Fig. 7.2A). Before treatment (PRE), robust TEOAE components (open area) were present above the background noise (filled area) between approximately 0.5 to 5.5 kHz in both ears, as expected from a normal eight-year-old ear. However, after treatment (POST), no TEOAEs were detectable above 2.2 kHz. Note that in the other examples of patient test results presented in figures in this chapter, the stimuli used to evoke DPOAEs and TEOAEs are the same as in this figure.

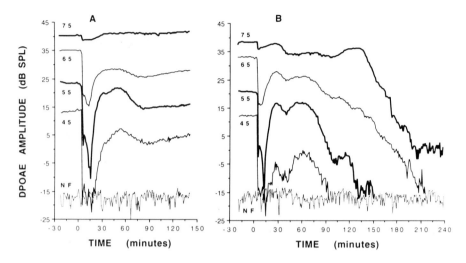

FIGURE 7.7 The effects of the loop diuretic ethacrynic acid on $2f_1-f_2$ DPOAEs (A) Rabbit #746 The amplitude of the DPOAE at 5.66 kHz is plotted as a function of time for stimulus tone levels of $L_1 = L_2 = 75, 65, 55$, and 45 dB SPL, as indicated by the number associated with each trace. The bottom line, around -17 dB SPL, indicates the noise floor (NF). Ethacrynic acid (40 mg/kg) was injected intravenously at 0 min and caused a temporary, stimulus level dependent reduction of DPOAE amplitude. (B) As in A, for rabbit #173, except that a subcutaneous dose of the aminoglycoside antibiotic gentamicin (100 mg/kg) was administered two hours prior to the ethacrynic-acid injection (40 mg/kg). Over the first hour postinjection of diuretic, the effects were very similar to those produced by ethacrynic acid alone, i.e., a temporary, stimulus level dependent reduction of DPOAE amplitude. However, at later times there was a decrease of DPOAE amplitudes at all stimulus levels, which was permanent.

Brummett 1979; Hayashida et al. 1989). Figure 7.7B illustrates the effects of ethacrynic acid (40 mg/kg, iv) administered two hours after an acute administration of the aminoglycoside antibiotic gentamicin (100 mg/kg, subcutaneous), on rabbit DPOAEs. This dosage of gentamicin alone has no detectable effect on DPOAEs in rabbits. In the first 30 mins after the ethacrynic-acid injection (at 0 min), the effects were similar to those produced by ethacrynic acid alone (Fig. 7.7A). However, at later times postinjection, the DPOAEs at all stimulus levels declined and were permanently abolished (except for a small, residual component in response to 75 dB SPL stimuli) by ~200 mins after the ethacrynic acid injection. The time course of this effect was similar to that seen for the decline of cochlear electrophysiological responses in guinea pigs after administration of an aminoglycoside antibiotic and a loop diuretic (West, Brummett, and Himes 1973; Russell et al. 1979; Hayashida et al. 1989). These data demonstrate that, whereas high-level DPOAEs are much less susceptible to trauma than are low-level DPOAEs, they are nevertheless vulnerable to some physiological traumas. It is possible that generation of DPOAEs evoked by high-level stimuli requires the structural integrity of some component(s) of the organ of Corti other than the outer hair cells (Whitehead, Lonsbury-Martin, and Martin 1992b).

3.1.4 Salicylate

Oral administration of aspirin (acetylsalicylic acid) can cause a reversible, moderate (≤ 40 dB) threshold elevation in humans, often associated with tinnitus (McFadden, Plattsmier, and Pasanen 1984). Oral administration of aspirin has been shown to reduce TEOAE, SFOAE, and SOAE amplitudes in humans (Johnsen and Elberling 1982; McFadden and Plattsmier 1984; Long and Tubis 1988). However, Wier, Pasanen, and McFadden (1988) found that aspirin had less effect on $2f_1$-f_2 DPOAEs evoked by low-level primary tones than on SOAEs at nearby frequencies in the same ears, suggesting a dissociation between the mechanisms responsible for the generation of DPOAEs and those underlying the other emission types. Similarly, in rhesus and pigtail macaques, acute injections of sodium salicylate (100 mg/kg, sc) that greatly reduced TEOAEs, SFOAEs, and SOAEs, had relatively little effect on DPOAEs over a wide range of stimulus levels (Martin et al. 1988; Lonsbury-Martin et al 1991). The differential effects of aspirin on TEOAEs and DPOAEs are illustrated for a pigtail macaque ear in Figure 7.3, where it can be seen that at frequencies where TEOAEs (Fig. 7.3A) were greatly reduced by a salicylate injection, $2f_1$-f_2 DPOAE amplitudes (Fig. 7.3B) were little influenced over a wide range of stimulus levels. In other species, salicylate has been shown to reduce DPOAEs. Thus, in guinea pigs, Kujawa, Fallon, and Bobbin (1992) found that DPOAEs in response to low level stimuli, but not high level stimuli, were reduced by perilymphatic perfusion of sodium salicylate. Moreover, in the bat, moderate injections of sodium salicylate (100 mg/kg, sc) had a greater effect on DPOAEs evoked by low level than high level stimuli, whereas very large dosages (400 mg/kg) greatly reduced DPOAEs at all stimulus levels (Kössl 1992). In the cat, DPOAEs evoked by 70 dB SPL stimulus tones were reduced approximately 5 dB by intravenous injections of 200 mg/kg sodium salicylate (Stypulkowski 1990). The finding that salicylate had little or no effect on DPOAEs in monkeys, but reduced DPOAEs in other species, may be due to different dosages administered and, possibly, to species differences in the metabolism of salicylate and its transport into and out of the inner ear fluids.

Whereas the mechanism of salicylate ototoxicity has not been established, salicylate is known to influence outer hair cell ultrastructure both in vivo and in vitro (Douek, Dodson, and Bannister 1983; Dieler, Shehata-Deiler, and Brownell 1991) and to reduce outer hair cell motility in response to electrical stimuli in vitro (Shehata, Brownell, and Dieler 1991). The finding that OAEs can be reduced by salicylate, therefore, is consistent with the hypothesis that the generation of OAEs is dependent upon the normal function of outer hair cells.

3.1.5 Other Drugs Reported to Reduce OAE Amplitudes

Quinine causes reversible, mild to moderate hearing loss, through an unknown mechanism. In two studies using small numbers of normal volunteer subjects, quinine was found to reduce TEOAEs, SOAEs, and DPOAEs in humans (Karlsson, Berninger, and Alvan 1991; McFadden and Pasanen, 1994).

The solvent toluene causes outer hair cell loss, and elevation of auditory brainstem response and behavioral-hearing thresholds, in rats (Pryor et al. 1984; Sullivan, Rarey, and Conolly 1989; Johnson, 1993; Crofton, Lassiter, and Rebert, 1994). The hearing loss caused by toluene inhalation in rats is unusual in that it is most pronounced at mid-frequencies, rather than high frequencies. Several days of inhalation of toluene elevated auditory brainstem response thresholds and reduced DPOAE amplitudes in rats (Johnson 1993; Johnson and Canlon, 1994). The frequencies of DPOAE amplitude reductions appear to coincide with the frequencies of auditory brainstem response threshold elevation, and outer hair cell loss.

3.1.6 Implications of Ototoxic Drug Effects on DPOAEs

The data discussed above demonstrate that OAEs can be used to detect drug induced ototoxicity and can provide information regarding the frequency, extent, and time course of the ototoxicity. However, in rabbits and rodents, loop diuretics, aminoglycoside antibiotics, and salicylate each have a much greater influence on DPOAEs elicited by stimuli below 60 to 70 dB SPL than on those evoked by higher level stimuli. These findings, and other data discussed in Sections 2.4, above, and 3.2, below, suggest that different processes underlie the generation of DPOAEs evoked by low-level and high-level stimuli. The DPOAEs evoked by low level stimuli appear to be influenced similarly to electrophysiological and behavioral measures of cochlear function, whereas DPOAEs elicited by higher level stimuli are much less affected. Thus, these animal studies indicate that relatively low level stimuli should be employed if DPOAEs are to be utilized to monitor cochlear condition.

Brown, McDowell, and Forge (1989) found that $2f_1$-f_2 DPOAEs evoked by closely spaced stimulus tones ($f_2/f_1 = 1.03$) were less sensitive to the trauma induced by chronic gentamicin treatment than those evoked by more widely spaced stimulus tones ($f_2/f_1 = 1.3$). This latter finding is consistent with the conclusion of Whitehead, Lonsbury-Martin, and Martin (1992a), based on studies of the parametric properties of DPOAEs, that the high-level DPOAE generator, which is much less affected by cochlear trauma than the low-level mechanism, dominates the ear canal $2f_1$-f_2 signal to lower stimulus levels at the smallest f_2/f_1 ratios than at moderate or large f_2/f_1 ratios. Thus, it is possible that attention will have to be paid to the frequency spacings, as well as the levels, of the primary tones used to evoke DPOAEs in order to optimize the sensitivity of these tests to cochlear trauma.

It is not yet known whether human ears possess distinct low-level and high-level DPOAE generators similar to those in rabbit and rodent ears. Moreover, if these two distinct mechanisms are present in human ears, they may dominate the ear canal $2f_1-f_2$ signal over different stimulus level ranges than in rabbits and rodents. In this regard, it is noted that in clinical tests utilizing DPOAEs, our laboratory successfully uses primary levels of 75 dB SPL (e.g., Fig. 7.6, above, and later figures) to detect cochlear trauma, whereas this stimulus level is in the range dominated by the high-level, less vulnerable DPOAE generator in rabbits and rodents (e.g., Fig. 7.7). This issue is clearly important for the design and interpretation of clinical tests utilizing DPOAEs. However, studies of this issue in humans are complicated by the much smaller amplitudes of DPOAEs in humans relative to rabbits and rodents, by the inability to use in human subjects many of the ototoxic drugs employed in animal experiments, and by the interactions of DPOAEs with SFOAEs and SOAEs that occur in human ears.

In nonprimate mammals, the relative invulnerability of DPOAEs evoked by high-level stimuli to a variety of cochlear traumas that greatly reduce both DPOAEs evoked by low-level stimuli and the other OAE types suggest that the high-level DPOAEs are generated by a mechanism distinct from the mechanism(s) responsible for the low-level DPOAEs, and for TEOAEs, SFOAEs, and SOAEs. The observation in primates that salicylate can substantially reduce TEOAEs, SFOAEs, and SOAEs while having little effect on DPOAEs elicited by relatively low level stimuli (see Fig. 7.3; Wier, Pasanen, and McFadden 1988; Lonsbury-Martin et al. 1991) further suggests that there may be some dissociation of the mechanism(s) responsible for the generation of DPOAEs from those underlying TEOAEs, SFOAEs, and SOAEs. A differentiation of the DPOAE generator from the mechanism(s) underlying TEOAEs and SFOAEs is also suggested by the observation that DPOAEs are much smaller, whereas TEOAEs and SFOAEs are larger, in primates than in the nonprimate laboratory mammals (Section 2.4, above). In summary, it appears that the mechanisms underlying the generation of TEOAEs, SFOAEs, and SOAEs are at least partially separate from those responsible for low-level DPOAEs, and that each of these mechanisms are at least partially separate from those underlying generation of the high-level DPOAEs. This conclusion suggests the possibility that the pattern of trauma-induced amplitude reductions among the different OAE types could provide information regarding the dysfunctions underlying certain cochlear hearing losses.

3.2 Hypoxia and Anoxia

In animal experiments, limiting the animal's oxygen intake effectively reduces the energy supply to the ear. In nonprimate laboratory mammals, temporary hypoxia causes a rapid, reversible reduction of TEOAEs (Zwicker and Manley 1981), SFOAEs (Avan et al. 1990), and SOAEs (Evans, Wilson, and Borerwe 1981), and lethal anoxia rapidly abolishes these emissions (Wilson 1980b; Evans, Wilson, and Borerwe 1981). The $2f_1$-f_2 DPOAEs evoked by stimulus tones below approximately 60 to 70 dB SPL demonstrate the expected rapid, reversible reduction upon temporary hypoxia in cats (Kim 1980), guinea pigs (Rebillard and Lavigne-Rebillard 1992), and gerbils (Kemp and Brown 1984). Lethal anoxia resulted in a rapid, permanent reduction of DPOAEs evoked by stimulus tones below 60 to 70 dB SPL, whereas DPOAEs elicited by stimulus tones above these levels declined much more slowly postmortem (Schmiedt and Adams 1981; Kemp and Brown 1984; Schmiedt 1986b; Lonsbury-Martin et al. 1987; Whitehead, Lonsbury-Martin, and Martin 1992b). These data are consistent with the observation that in rabbits and rodents DPOAEs evoked by stimuli below 60 to 70 dB SPL are much more susceptible to the effects of various ototoxic drugs than are DPOAEs evoked by higher level stimuli (Section 3.1, above).

It is to be expected that OAEs in humans would also be compromised by reduction of the oxygen supply to the ear. This may occur as a result of disruption of the blood supply to cochlea, resulting, e.g., from compression of the cochlear artery by tumors of the auditory nerve or by damage to this artery during surgery to remove such tumors (Widick et al. 1994).

3.3 Acoustic Overexposure

Exposure to intense sound can result in temporary or permanent behavioral threshold elevations in humans and other mammals, associated with damage to the organ of Corti (e.g., Salvi et al. 1986). In humans, sound exposures that produced a temporary behavioral threshold elevation reduced the amplitudes of SOAEs (Kemp 1981; Norton, Mott, and Champlin 1989) and TEOAEs (Kemp 1981, 1982; Zwicker 1983a). These OAE amplitude reductions were frequency specific and demonstrated a correlate of the psychophysical half-octave shift, i.e., the maximum amplitude reduction tended to occur for OAE components at a frequency approximately one-half octave above the exposure frequency. Additionally, the postexposure recovery of OAE amplitude followed, in general, a time course similar to that of the behavioral threshold.

In human patient groups, Probst et al. (1987) determined that noise induced high-frequency hearing loss was associated with a reduction in the number of prominent peaks in the spectra of TEOAEs. Frequency-specific reductions of $2f_1$-f_2 DPOAE amplitudes in frequency regions of behaviorally measured noise induced permanent hearing loss have been demonstrated by Harris (1990), Lonsbury-Martin and Martin (1990), and Martin et al. (1990). Figure 7.8 shows the behavioral hearing and evoked emission test results for a forty-nine-year-old male laborer with a symmetrical, high-frequency hearing loss (top left) apparently caused by noise overexposure associated with working in the building-construction industry for more than 20 years. In the presence of normal tympanometric (top right) and acoustic reflex (top middle) function, DPOAEs (bottom left) in both ears were symmetrically reduced below normative population averages at frequencies above 2 kHz, i.e., in the region of hearing loss. The TEOAEs (bottom right) were also reduced above 2 kHz in that no components were detected in this region, whereas TEOAEs are typically present out to about 3 kHz in normal adult ears. As OAEs are thought to be produced by the operation of the outer hair cell system, these findings suggest that reduced function of the outer hair cell system is a primary contributor to this patient's hearing loss. This suggestion is consistent with histological studies of the noise damaged cochlea, which typically document outer hair cell pathology.

Subramaniam et al. (1994) and Subramaniam, Henderson, and Spongr (1994) exposed chinchillas to a 95-dB SPL octave-band noise centered at 0.5 kHz for 6 hours a day for 10 to 15 days. The noise exposure reduced DPOAE amplitudes. After recovery, however, DPOAE amplitudes, and the thresholds of evoked potentials measured in the inferior colliculus, were in the normal range despite moderate loss and stereociliar disarray of third row outer hair cells, and small losses of second- and first-row outer hair cells. This finding indicates that DPOAEs are resistant to small to moderate amounts of outer hair cell pathology. The finding that the evoked potential thresholds were normal suggests that the active cochlear process was also functioning normally, despite the presence of some outer hair cell pathology. Previous reports (Bohne and Clark 1978; Clark and Bohne 1978; Hamernik et al. 1989) have also indicated that electrophysiological-response and behavioral-hearing thresholds can be normal in the presence of noise-induced moderate ($<30\%$) loss of outer hair cells in the apical turns of the chinchilla cochlea.

An advantage of animal studies utilizing systematic acoustic overexposures is that by careful selection of the exposure stimulus, functional deficiencies can be restricted to a relatively narrow frequency region, reflecting anatomical damage to a limited area on the cochlear partition. The methodical use of sound overexposure to produce localized cochlear trauma has allowed determination of the ability of OAEs to provide information about the frequency pattern of a primarily sensory cochlear dysfunction. Experiments on DPOAEs in the cat demonstrated that, in general, acoustic overexposures producing a frequency-specific elevation of the compound action potential (CAP) threshold also produced a corresponding frequency-specific reduction of DPOAE amplitude (e.g., Schmiedt 1986a). However, there was not always a clear correlation between the amount of CAP threshold elevation and the degree of DPOAE amplitude reduction (Schmiedt 1986a; Wiederhold, Mahoney, and Kellogg 1986).

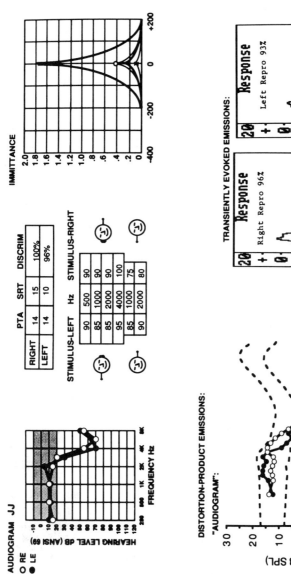
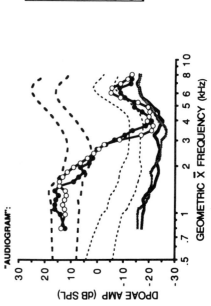

Experiments in the rabbit (Martin et al. 1987) and chinchilla (Zurek, Clark, and Kim 1982) demonstrated that the $2f_1$-f_2 DPOAE is maximally reduced by noise exposures that damage the organ of Corti in the region corresponding to the stimulus tone frequencies, rather than that corresponding to the distortion-product frequency. This finding is consistent with the results of studies of the suppression-tuning properties of this DPOAE (Brown and Kemp 1984; Martin et al. 1987; Harris and Glattke 1992), and with studies of electrophysiologically measured distortion products (Kim, Molnar, and Matthews 1980), which also suggest that the $2f_1$-f_2 DPOAE is produced at a place on the cochlear partition corresponding to the stimulus tone frequencies. These experiments indicate that clinical tests utilizing DPOAEs provide information about the place along the cochlea corresponding to the primary tone frequencies, not to the DPOAE frequency. For this reason, DPOAE amplitude data obtained from patients are plotted as a function of the primary tone frequencies, i.e., either f_2, or, as in several of the figures in this chapter, the geometric mean of f_1 and f_2.

Because acoustic overexposures can be selected to produce temporary reductions of function, information can also be obtained regarding the ability of OAEs to track the time course of cochlear dysfunction. Temporary reductions of OAEs by sound overexposures have been reported for TEOAEs in patas monkeys (Anderson and Kemp 1979) and cats (Wilson and Evans 1983), SFOAEs in guinea pigs (Avan et al. 1990), and $2f_1$-f_2 DPOAEs in gerbils (Schmiedt 1986a), cats (Kim, Molnar, and Matthews 1980; Dolan and Abbas 1985; Schmiedt 1986a; Wiederhold, Mahoney, and Kellogg 1986), rabbits (Lonsbury-Martin et al. 1987; Martin et al. 1987) and humans (Sutton et al. 1994). Short term (i.e., seconds and minutes) recovery of DPOAEs from noise overexposure followed a similar time course to the

←

FIGURE 7.8 Hearing and evoked emission test results for the left (LE) and right (RE) ears of a forty-nine-year-old male (JJ), who had a history of acoustic overexposure associated with having worked in the building construction industry for about 20 years. The top part of the figure gives the results of standard clinical tests, including: LEFT, the behavioral audiogram; MIDDLE, pure tone average (PTA), speech-reception threshold (SRT), and speech discrimination (DISCRIM) scores and acoustic-reflex test results, indicating normal contralateral and ipsilateral acoustic-reflex thresholds; RIGHT, tympanometry, indicating normal middle-ear pressure. Note that the symmetrical pattern of high-frequency hearing loss in the behavioral audiogram in the left (filled symbols) and right (open symbols) ears at and above 3 kHz, is accurately reflected by the high-frequency reduction of the DPOAE responses (bottom left) below the expected range. Also note that the TEOAE responses (bottom right) were detected only below about 2 kHz bilaterally. The percentage values ('Repro') given in each TEOAE panel are measures of the reproducibility of the TEOAE response, determined by cross-correlation of two independently averaged TEOAE waveforms. In general, a reproducibility of greater than 50% indicates the presence of a genuine TEOAE response.

recovery of sound induced auditory nerve activity under some conditions, but not others (Kim, Molnar, and Matthews 1980; Dolan and Abbas 1985). In chinchillas exposed to intermittent, low-frequency octave-band noise over 10 days, Subramaniam, Henderson, and Spongr (1994) found that DPOAE amplitudes tracked the loss and recovery of thresholds of evoked potentials measured in the inferior colliculus.

Permanent DPOAE amplitude reductions resulting from chronic noise overexposures have been described in chinchilla (Zurek, Clark, and Kim 1982) and rabbit (Stainback et al. 1987; Franklin et al. 1991). In general, $2f_1$-f_2 DPOAEs were reduced at the frequencies of corresponding behaviorally measured threshold increases (Zurek, Clark, and Kim 1982; Franklin et al. 1991) or in frequency regions associated with areas of histologically observed damage to the organ of Corti (Zurek, Clark, and Kim 1982; Clark et al. 1984; Stainback et al. 1987). In rabbits, the time course of the DPOAE amplitude reduction during, and the recovery after, chronic noise overexposure reflected the temporal behavior of the behavioral threshold elevations observed in the same ears over periods of days and weeks (Franklin et al. 1991).

The results of sound overexposure experiments, therefore, indicate that DPOAEs have the capability to track, over both frequency and time, the temporary and permanent impairments of cochlear function resulting from acoustic trauma. This finding is consistent with the results of studies utilizing ototoxic drugs (Section 3.1, above).

The results of several of the acoustic trauma studies demonstrate a tendency for noise overexposures to have less effect on DPOAEs evoked by higher level stimuli than on those evoked by lower level stimuli (Zurek, Clark, and Kim 1982; Schmiedt 1986a; Wiederhold, Mahoney, and Kellogg 1986; Stainback et al. 1987). This observation is illustrated in Figure 7.9, in which the amplitude of a $2f_1$-f_2 DPOAE from a rabbit ear is plotted as a function of stimulus level, before (open circles) and after (filled squares) a series of nine 10 s exposures (separated by 55 s intervals) to a 100 dB SPL pure tone at a frequency half an octave below the f_2 primary. The intermittent pure tone overstimulation produced a temporary increase of about 7 dB in the DPOAE detection threshold immediately after cessation of the exposure episode. At stimulus levels above the DPOAE detection threshold, DPOAE amplitude was reduced more at low than at high stimulus levels, resulting in a recruitment-like effect, i.e., steepening of the slope of the DPOAE growth function. Although the stimulus level dependent sensitivity to trauma observed with noise overexposure is often not as clear-cut as that seen upon administration of certain ototoxic drugs, or during anoxia (Sections 3.1. and 3.2, above), this observation again underscores the need to exercise care in the choice of stimulus levels to be employed when using DPOAEs to monitor cochlear condition (Sutton et al. 1994; Whitehead et al. 1995a,b).

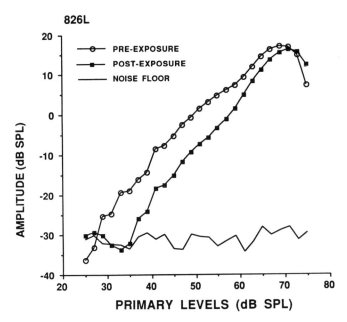

FIGURE 7.9 Growth function of the $2f_1-f_2$ DPOAE at 1.5 kHz ($f_1 = 2.0$, $f_2 = 2.5$ kHz) in rabbit #826 (from Mensh 1991). DPOAE amplitude was measured in 2 dB steps of stimulus level ($L_1 = L_2$) before (open circles), and immediately after (filled squares) a noise overexposure episode consisting of nine 10 s exposures (i.e., a total exposure of 90 s) to a 100 dB SPL pure tone. The exposures were separated by 55 s intervals. The exposure frequency was 1.768 kHz, i.e., 0.5 octave below the f_2 primary. The line without symbols indicates the noise floor. The degree of DPOAE amplitude reduction postexposure increased with decreasing stimulus level.

3.4 Meniere's Disease and Endolymphatic Hydrops

The exact mechanisms responsible for the fluctuating hearing loss observed in ears with Meniere's disease are unclear, as changes are known to occur at many levels of the cochlea, including the hair cells, the cochlear innervation, and the volume and chemical composition of the cochlear fluids (see Nadol 1989). Measurements of TEOAEs in human patient groups with sensorineural hearing losses of various etiologies have demonstrated that these emissions are typically absent in frequency regions in which the hearing level is greater than 25 to 30 dB (e.g., Kemp et al. 1986). This observation is also true for TEOAEs in some Meniere's ears. However, TEOAEs have been measured in Meniere's ears with corresponding hearing thresholds, at the time of measurement, exceeding 35 to 40 dB HL (Bonfils, Uziel and Pujol 1988; Harris and Probst 1992). Similarly, whereas in the majority of Meniere's ears DPOAE amplitudes are reduced below normative population averages at frequencies corresponding to those of behavioral-threshold increase, consistent with data obtained from ears with sensorineural hearing loss of other etiologies, in some patients DPOAE amplitudes are normal in the presence of equally large behavioral threshold increases (Lonsbury-Martin and Martin 1990; Martin et al. 1990). As OAEs are thought to reflect the operation of the outer hair cell system, one interpretation of these findings is that only in some cases of Meniere's disease, i.e., those demonstrating reduced OAEs, is the operation of outer hair cells compromised, and that in ears with normal emissions the hearing loss does not result from dysfunction of the outer hair cells, but from dysfunction of other sensory or neural elements.

Administration of the hyperosmotic agents glycerol or urea, which often produces a temporary improvement of hearing sensitivity in Meniere's ears, has also been shown to increase both TEOAE and DPOAE amplitudes in these ears (Bonfils, Uziel, and Pujol 1988; Lonsbury-Martin and Martin 1990; Martin et al. 1990). To illustrate these effects, Figure 7.10 shows test results obtained from a fifty-four-year-old female pre- (filled symbols) and post- (open symbols) urea administration. Fifteen years earlier, this patient had been diagnosed, according to classic symptomatology, with Meniere's disease in the left ear. Recently, she had begun to experience similar problems with the right ear. In the left ear, with long-standing disease and hearing thresholds greater than 55 dB, both DPOAEs (middle right) and TEOAEs (bottom right) were greatly reduced or absent across the whole frequency range, and there were no consistent urea induced improvements in either hearing thresholds (top center) or DPOAE amplitudes. In the more recently involved right ear, the urea administration produced a uniform improvement in behavioral hearing (top left) at all audiometric test frequencies, which was associated with enhanced amplitudes of DPOAEs (middle left), although there was little change of the TEOAEs (bottom left). These data again indicate that DPOAEs may be used to dynamically track temporal changes in cochlear condition. Interestingly, the robust TEOAE observed in the right ear of this patient prior to urea administration, where hearing thresholds were elevated to approximately 30 dB HL from 1 to 4 kHz, is larger than that expected in ears with similar hearing threshold elevations resulting from other sensorineural hearing losses. Thus, this ear exemplifies a Meniere's case representing the category noted above, in which ears with documented sensorineural hearing losses expected to be associated with substantially reduced or absent TEOAEs, may demonstrate normal evoked emissions (normal TEOAEs in this example).

Experimental endolymphatic hydrops produced by surgical destruction of the endolymphatic sac or duct is commonly used as an animal model of Meniere's disease (Nadol 1989). Horner and Cazals (1989) found that $2f_1-f_2$ DPOAEs, evoked by high level stimuli in hydropic guinea pigs, were reduced when the stimulus tones fell in frequency regions demonstrating CAP threshold elevation. In hydropic rabbits, behavioral threshold measures showed an appreciable low-frequency hearing loss, which often varied

7. Experimental and Clinical Studies of Emissions 229

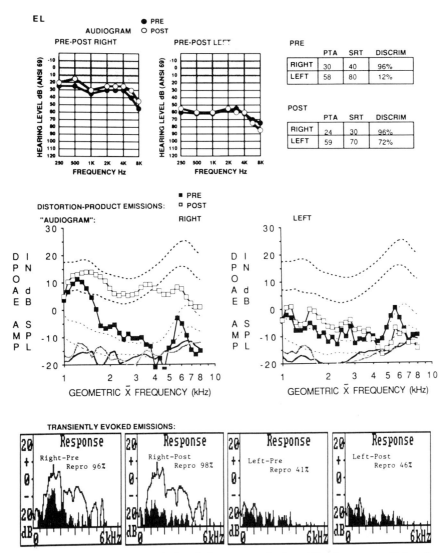

FIGURE 7.10 Hearing and evoked-emission test results for a fifty-four-year-old female (EL) acquired before (PRE, filled symbols) and after (POST, open symbols) ingestion of urea. This subject had been diagnosed as having Meniere's disease in the left ear 15 y previously and had recently developed similar symptoms in the right ear. In the right, recently involved ear, with mild hearing loss (top left), the slight improvement of hearing sensitivity observed post-urea was associated with a substantial improvement of DPOAE amplitudes (middle left), but little change of TEOAE amplitudes (bottom left two panels). In the left ear, with long-standing disease, in which the hearing loss was greater, neither hearing thresholds (top center) or DPOAE amplitudes (middle right) appeared systematically changed by urea administration, and TEOAEs were not clearly present above the background noise (bottom right two panels), with 'repro' scores of less than 50% both pre- and post-urea administration.

over several days postsurgery (Martin et al. 1989), and responses of single cochlear afferent neurons also demonstrated threshold elevation relative to control animals (Lonsbury-Martin et al. 1989). In general, measurements of DPOAEs in these rabbits demonstrated changes qualitatively parallel to those seen behaviorally and electrophysiologically, indicating that DPOAEs are capable of providing useful information concerning cochlear dysfunction in endolymphatic hydrops. However, postsurgery DPOAE amplitudes sometimes varied independently of behavioral threshold (Martin et al. 1989), suggesting that the two measures may be monitoring partially independent aspects of cochlear function. Additionally, both behavioral (Martin et al. 1989) and electrophysiological (Lonsbury-Martin et al. 1989) measures appeared to demonstrate greater sensitivity losses than did DPOAEs in the same ears. These findings are consistent with the interpretation, similar to that discussed above for Meniere's disease in humans, that hydrops induced hearing loss involves components due to both outer hair cell dysfunction, reflected as reduced DPOAEs, and abnormal operation of other cochlear components (e.g., the inner hair cells or afferent neurons) that are not reflected in reduced DPOAE amplitudes. This interpretation is consistent with the multiple cochlear changes known to occur in both Meniere's disease in humans and in surgically induced endolymphatic hydrops in animal models (Nadol 1989).

Glycerol appears to have contrasting effects on cochlear sensitivity in humans with Meniere's disease and animals with surgically induced hydrops. As noted above, glycerol often temporarily improves behavioral-hearing thresholds in Meniere's ears, and this has been reported to have a correlate in increases of OAE amplitude (Bonfils, Uziel and Pujol 1988; Lonsbury-Martin and Martin 1990; Martin et al. 1990). However, in hydropic guinea pigs, glycerol did not improve the CAP (Horner and Cazals 1987). In rabbits, DPOAEs already reduced in hydropic animals were further decreased by glycerol administration, whereas glycerol had no effect on DPOAEs in normal animals (Martin et al. 1989).

3.5 *Hereditary Hearing Loss*

In patients with both autosomal-dominant and autosomal-recessive based hereditary hearing loss, DPOAEs have been shown to be absent or reduced below normative population averages in the frequency regions of hearing loss (Lonsbury-Martin and Martin 1990; Martin et al. 1990; Ohlms, Lonsbury-Martin, and Martin 1990). Horner, Lenoir, and Bock (1985) reported that, whereas normal mice demonstrated strong $2f_1$-f_2 DPOAEs, no DPOAEs were detected in strains of mice with hereditary dysplasia of the stria vascularis or of the whole cochlea. In contrast, mutants with central deafness, but anatomically normal cochleae, demonstrated DPOAEs similar to those in normal mice. Both naturally occurring mutants

(Horner, Lenoir, and Bock 1985; Schrott, Puel, and Rebillard 1991) and transgenic mice (Katz et al. 1989) with abnormal organ of Corti anatomy demonstrate DPOAEs that are reduced in amplitude relative to those in normal mice.

In the Bronx waltzer mutant mouse, about 70% of the inner hair cells are missing, but there is a normal number of outer hair cells. However, the outer hair cells show some anatomical and morphological disruption. In this mutant, $2f_1-f_2$ DPOAEs are reduced by 10 to 20 dB, and DPOAE detection thresholds are elevated, relative to normal mice (Horner, Lenoir, and Bock 1985; Schrott, Puel, and Rebillard 1992). It is not clear whether the DPOAE reduction results from some disruption of the outer hair cells, or from the absence of 70% of the inner hair cells. It is possible that, even if the outer hair cells function normally, the generation and/or propagation to the cochlear base of OAEs may be impaired by changes of the mechanical properties of the organ of Corti, for example, due to absence of inner hair cells.

The waltzing strain of guinea pig has a genetically induced progressive deafness, in which hearing sensitivity rapidly deteriorates over the first month after birth, associated with a systematic degeneration of outer hair cells and, to a lesser extent, inner hair cells. In waltzing guinea pigs, in the first two weeks after birth, the auditory brainstem response deteriorated much more rapidly than did DPOAEs (Canlon, Marklund and Borg 1993). One interpretation of this finding is that neural or cochlear elements other than outer hair cells were affected. At 15 days after birth, DPOAE amplitudes were close to normal in response to both high and low level stimuli, but at 30 days after birth, DPOAEs were not detectable to low level stimuli, and were greatly reduced for high level stimuli (Canlon, Marklund and Borg 1993). The DPOAE-amplitude reduction from 15 to 30 days after birth corresponded to the progressive loss of hair cells. Thus, at 15 days after birth, in the cochlear region corresponding to the primary-tone frequencies, there was moderate loss of third row outer hair cells, but only small losses of second and first row outer hair cells, and no obvious loss of inner hair cells. However, at 30 days after birth, there was substantial but highly variable loss of outer hair cells from all three rows, and also moderate loss of inner hair cells (Canlon, Marklund and Borg 1993).

In the waltzing guinea pigs, at 15 days after birth, DPOAE amplitudes were normal despite moderate loss of third row outer hair cells, and small losses of second and first row outer hair cells. This finding suggests that DPOAE amplitudes are robust in the presence of moderate loss of third-row outer hair cells, and small losses of second and first row outer hair cells. As noted in Section 3.3, Subramaniam et al. (1994) found that, in noise-exposed chinchillas, DPOAEs were in the normal range despite moderate loss of third-row outer hair cells, and small losses of second and first row outer hair cells.

3.6 Other Cochlear Hearing Losses

In human patient groups, both TEOAEs (Robinette 1992) and DPOAEs (Gaskill and Brown 1990; Lonsbury-Martin and Martin 1990) have been shown to be reduced in cases of sudden idiopathic sensorineural hearing loss, and, to some extent, to be able to track the recovery of behavioral thresholds in these cases. Other studies have indicated that TEOAEs in some ears with sudden idiopathic hearing loss are larger than expected on the basis of the corresponding threshold elevation (Sakashita et al. 1991). This observation is similar to that discussed for Meniere's disease (Section 3.4, above) and suggests that in some cases of idiopathic sensorineural hearing loss the outer hair cell system is compromised (resulting in hearing threshold elevation and reduced OAE amplitudes), whereas in other cases the hearing loss may result from dysfunction of components other than the outer hair cell system (resulting in hearing threshold elevation but little reduction of OAE amplitudes).

It has been demonstrated that TEOAEs, SOAEs and DPOAEs decrease with age over about 40 to 50 years (Bonfils, Bertrand, and Uziel 1988; Bonfils 1989; Collet et al. 1990a; Lonsbury-Martin, Cutler, and Martin 1991; Stover and Norton, 1993), suggesting that OAEs may provide a new means of studying presbycusis.

3.7 Retrocochlear Hearing Loss

In humans, both SFOAEs and TEOAEs can be detected at stimulus levels well below the threshold of hearing (Wilson 1980c; Zwicker 1983b), and the TEOAE waveform shows precise inversion with stimulus polarity (Anderson 1980). These properties indicate that these emissions are produced without direct afferent neural involvement. In animal studies, after perfusion of the cochlea with potassium cyanide, Kim (1980) measured normal $2f_1-f_2$ DPOAEs in the absence of the CAP. Furthermore, visually confirmed surgical section of the auditory portion of the eighth cranial nerve resulted in no change in DPOAEs in rabbits over periods of up to ten weeks, by which time significant retrograde neural degeneration would be expected to have occurred (Ohlms et al. 1991). Thus, animal studies have demonstrated convincingly that DPOAE generation does not require primary afferent activity, i.e., DPOAEs are generated in the cochlea with no direct neural or retrocochlear involvement.

Because of the preneural origin of OAEs, it is expected that retrocochlear disorders, such as neuromas of the eighth cranial nerve, would not influence OAEs even if the tumor involved the auditory portion of the nerve. Consistent with this expectation, in some patients TEOAEs have been shown to be present in ears with sensorineural hearing losses apparently due to retrocochlear disorders (Bonfils and Uziel 1988; Lutman et al. 1989). An example of this anticipated outcome is illustrated by Figure 7.11, which shows the hearing and evoked-emission test results obtained for a thirty-nine-year-old male diagnosed by magnetic resonance imaging to have an acoustic neuroma on the left side. For this patient, the asymmetry of the hearing losses in the right and left ears is not reflected in the evoked-emission measures, which do not indicate any functional condition greater than the mild loss present in the right ear despite measured hearing thresholds of 65 to 70 dB HL at and below 2 kHz in the left ear.

In contrast to the findings of normal or near-normal OAEs in some patients with hearing loss associated with acoustic neuroma, in the majority of such patients, both TEOAEs (Kemp et al. 1986; Bonfils and Uziel 1988) and DPOAEs (Lonsbury-Martin and Martin 1990; Martin et al. 1990; Ohlms et al. 1991) are reduced. This more usual outcome is illustrated by the hearing and evoked-emission findings displayed in Figure 7.12 for a fifty-year-old female who was also diagnosed by magnetic resonance imaging to have an acoustic neuroma on the left side. In this case, the affected ear showed substantial hearing threshold elevation (top left) and essentially no detectable DPOAEs or TEOAEs, whereas all responses from the unaffected right ear were normal. The absence of emissions in the affected ear in cases like this indicates that the retrocochlear pathology interfered with normal outer hair cell function, probably as a result of cochlear injury due to pressure exerted by the tumor on the vascular supply to the cochlea.

Studies of $2f_1$-f_2 DPOAEs and SFOAEs in nonprimate laboratory mammals (Siegel and Kim 1982; Guinan 1986; Puel and Rebillard 1990) and of SOAEs, TEOAEs, and DPOAEs in humans (Mott et al. 1989; Collet et al. 1990b; Moulin, Collet, and Morgon 1992) indicate that the medial olivocochlear-efferent system, which primarily innervates the outer hair cells, may influence OAE amplitudes. However, such changes are small, being substantially less than the range of variability of OAE amplitudes in the normal human population. Thus, OAEs may be expected to demonstrate normal amplitudes in cases of pure retrocochlear dysfunction. This conclusion is supported by the observation that a strain of mutant mice with central deafness, but apparently normal cochlear function, demonstrated DPOAEs similar to those observed in normal mice (Horner, Lenoir, and Bock 1985).

3.8 Middle Ear Disorders

Because OAEs generated within the cochlea must propagate out through the middle ear to their measurement site in the external ear canal, they are reduced when sound conduction through the middle ear system is compromised. Indeed, OAEs tend to be more susceptible to middle-ear dysfunction than measures that involve transmission of sound only once through the middle ear. Some of the earliest studies of OAEs in humans demonstrated that increasing middle ear stiffness by changes of static air pressure in the

234 Martin L. Whitehead et al.

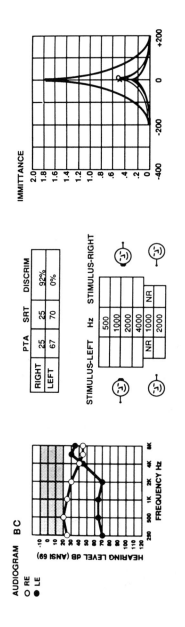
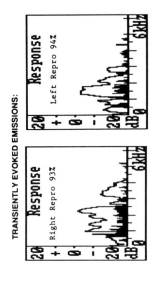
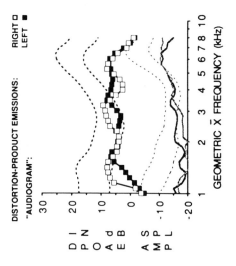

ear canal (Kemp 1981; Wilson and Sutton 1981; Schloth and Zwicker 1983) or by eliciting the contralateral acoustic reflex (Schloth and Zwicker 1983) decreased the amplitude of SOAEs. In ears with normal middle ear pressure, altering the static pressure in the ear canal by ±200 daPa caused a mean reduction of approximately 7 dB for TEOAE frequencies <1.5 kHz, with little effect at frequencies >3.5 kHz (Naeve et al. 1992). Moreover, in ears with abnormal negative middle ear pressure, equalization of the middle ear pressure increased TEOAE amplitudes, especially at low frequencies (Trine, Hirsch, and Margolis 1993).

Dysfunction of the middle ear conduction system has also been shown to be associated with reduced TEOAEs in adult patients with otosclerosis (Rossi et al. 1989) and reduced TEOAEs and DPOAEs in children with otitis media (Owens et al. 1992; Prieve 1992). Figure 7.13 shows data from a five-year-old female patient exhibiting an acute episode of bilateral tubal dysfunction, associated with significant negative pressure in both middle ears (right ear = -250 daPa, left ear = -320 daPa, see immittance data in top-right panel). Although the behavioral audiograms in both ears were in the normal range across all frequencies, TEOAEs were not detected, and there was a large reduction below the normal range of low- and mid-frequency DPOAEs, but relatively little effect at the highest frequencies, above 6 kHz. Clearly, deviations of middle ear pressure from normal can result in substantial reductions of OAE amplitude.

Animal experiments have also demonstrated that OAEs are extremely susceptible to changes in middle ear status. Evans, Wilson, and Borerwe (1981) found that changes in middle ear pressure of ±50 mm H_2O decreased the ear canal amplitude of a guinea pig SOAE by 20 dB, while having little effect on the amplitude of the round window recorded, cochlear microphonic electrical correlate of this emission. This finding suggests that, whereas the SOAE was reduced in the ear canal due to a stiffening of the middle ear transmission system, the amplitude of the SOAE within the cochlea remained unchanged. Both mass loading of the tympanic membrane in the cat (Wiederhold 1990) and increased stiffness of the middle ear in the rabbit, induced by eliciting the contralateral acoustic reflex (Whitehead, Martin, and Lonsbury-Martin 1991), reduced $2f_1$-f_2 DPOAE amplitudes. Figure 7.14A illustrates the effects of the acoustic reflex, elicited by a 4 kHz tone in the contralateral ear, on the amplitude of a 1.97 kHz DPOAE in a rabbit. The action of the acoustic reflex is apparent in the changes of the levels of the primary tones, f_1 and f_2 (Figure 7.14B and

←

FIGURE 7.11 Hearing and evoked-emission test results for a thirty-nine-year-old male (BC) with a 2 cc acoustic tumor involving the left ear. Although the hearing loss (top left) is much greater in the left ear (filled symbols) than in the right ear (open symbols), both DPOAE (bottom left) and TEOAE (bottom right) levels are similar in the two ears. For a further description, see legend to Fig. 7.8.

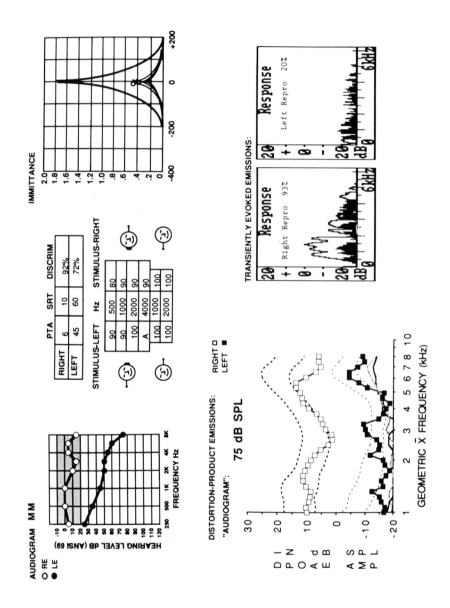

C), indicating altered middle ear admittance upon presentation of the contralateral tone at levels above 70 dB SPL. The middle ear reflex reduced DPOAE amplitudes, and the reduction increased with the level of the contralateral tone such that when the reflex eliciting tone was at 90 dB SPL, the DPOAE was reduced by about 23 dB. There is evidence suggesting that some manipulations of the middle ear may cause a greater attenuation of outward transmission of OAEs than of inward transmission of the stimuli (Wiederhold 1990; Whitehead, Martin, and Lonsbury-Martin 1991).

In combination, the animal and human findings clearly indicate that great care must be taken to ensure normal middle ear status if OAEs are to be utilized as an index of cochlear condition. Acute middle ear disorders present at the time of testing can substantially reduce DPOAE amplitudes, confounding interpretation of OAE test results in terms of cochlear status. Thus, the use of OAEs in the clinic must be accompanied by tests of middle ear function, especially tympanometry. Moreover, although the influences of chronic middle ear disease on OAE measurements remain to be methodically determined, based on clinical experience to date, it appears that the practical utility of OAEs will be limited in ears with middle ear abnormalities. It is possible that changes of OAE amplitudes associated with medical and/or surgical treatments of otitis media with effusion could be used to monitor the effectiveness of these treatments (Owens et al. 1992).

4. Clinical Applications of Otoacoustic Emissions

Because SOAEs are not ubiquitous in humans, and as they occur only at a few, idiosyncratic frequencies in those ears in which they are found, their utility as tools for the detection of hearing impairment is severely limited. In contrast, evoked OAEs are present in essentially all normal human ears, can be reliably measured over wide frequency ranges, and are systematically reduced in ears with certain hearing losses. Thus, because evoked OAEs can be measured objectively, noninvasively, rapidly, and inexpensively, they demonstrate great potential for several clinically useful applications. Because SFOAEs are more difficult and time-consuming to measure than TEOAEs and appear to provide little additional information over that provided by TEOAEs, there is less interest in developing SFOAEs for

---

FIGURE 7.12 Hearing and evoked-emission test results for a fifty-year-old female (MM) with a 2 cc acoustic tumor involving the left ear. In the right ear (open symbols), hearing is normal, and both DPOAEs (bottom left) and TEOAEs (bottom right) are robust. In the affected left ear (filled symbols), the elevated hearing thresholds are associated with an absence of DPOAEs and TEOAEs. For a further description, see legend to Fig. 7.8.

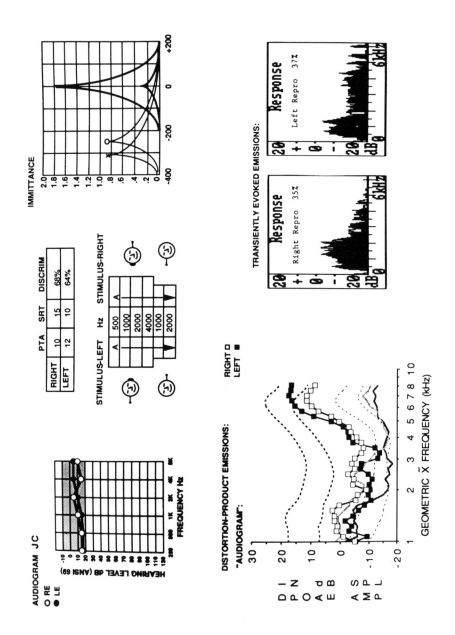

clinical use than is the case for TEOAEs and DPOAEs. Both TEOAEs and DPOAEs appear to be capable of providing information regarding the frequency extent and time course of sensorineural hearing losses of various etiologies in humans and laboratory animals. However, in normal human ears, reliable TEOAE measurement is limited to frequencies below about 4 kHz, whereas DPOAEs are routinely measured to stimulus frequencies above 8 kHz. This gives DPOAEs an advantage over TEOAEs in some applications because most sensorineural hearing losses preferentially involve the high frequencies.

4.1 Differential Diagnosis

Because OAEs are produced within the cochlea, apparently without direct central or retrocochlear involvement, they can contribute to the differential diagnostic work-up of patients complaining of hearing loss of unknown etiology. The clinical hearing and evoked-emission results depicted in Figure 7.15 illustrate the usefulness of this "site-of-lesion" information in the case of a fifteen-year-old female who complained of hearing problems of recent origin in the left ear. In this case, the decrement in DPOAE amplitudes from the left ear for frequencies above 3 kHz, i.e., in the region of hearing loss, allow the clinician to confirm with confidence that the observed hearing loss is primarily due to cochlear, rather than retrocochlear or centrally based, pathology. As emphasized in Section 3.8, clinical findings of reduced OAE amplitudes can be interpreted reliably only if accompanying tests indicate normal middle ear function (top right in Fig. 7.15).

Because OAEs are expected to be normal in cases of pure retrocochlear hearing loss (Section 3.7, above), the presence of robust OAEs in ears with profound hearing loss of unknown etiology has been used to support a diagnosis of retrocochlear dysfunction (Lutman et al. 1989; Robinette and Facer 1991).

←
___

FIGURE 7.13 Hearing and evoked-emission test results for both ears of a five-year-old girl (JC) with significant negative pressure (top right) in the middle ear [X = left ear -320 daPa; 0 = right ear -250 daPa], and absent stapedial reflexes (top center). Although hearing thresholds (top left) were normal, DPOAEs (lower left) were absent, i.e., at the same level as the subject's noise floor, between 1 and 2 kHz. Above 2 and 3 kHz, DPOAEs were detected, but remained reduced below normal levels up to about 7 kHz. Similarly, the spectral plots (bottom right) show no evidence of TEOAEs above the (relatively high) background noise level, with reproducibility scores less than 50% bilaterally, whereas robust TEOAEs would be expected from 0.5 to 4.5 kHz in a normal five-year-old ear. For a further description, see legend to Fig. 7.8.

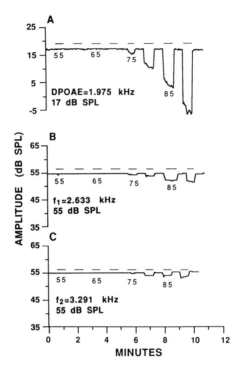

FIGURE 7.14 (A) Effects of a 4 kHz contralateral tone on the ear-canal sound pressure of a 1.975 kHz DPOAE in a rabbit. The primary tones were f_1 = 2.633 kHz, f_2 = 3.291 kHz, $L_1 = L_2$ = 55 dB SPL. Bars above the trace show the times of presentation of the 4 kHz contralateral tone, which was increased by 5 dB between each presentation, from 55 to 90 dB SPL, as indicated by the numbers just below the trace. In the absence of the contralateral stimulus, the DPOAE amplitude was 17 dB SPL, but presentation of the contralateral tone at levels above 70 dB SPL reduced DPOAE amplitude. (B) and (C) The levels of both f_1 and f_2 were also influenced by presentation of the contralateral tone at levels above 70 dB SPL, indicating changes of the admittance of the middle ear due to activation of the acoustic reflex.

4.2 Newborn Hearing Screening

Because OAE measurement is objective, i.e., requires no active participation by the patient, OAE testing allows evaluation of patients who are difficult to test because of communication problems, e.g., newborns and young children, and patients with certain mental or physical disabilities. Potentially the most important clinical application of OAEs is the screening of neonates for hearing impairment. Currently, auditory brainstem evoked response testing, which is also noninvasive and objective, is used to screen only those neonates considered at-risk for hearing impairment. However, approximately 50% of children later found to have severe or profound hearing loss are not considered at-risk at birth. Often, the presence of hearing loss is not determined until two to three years of age, by which time substantial harm may have been done to language acquisition and cognitive and academic development. Because OAE measurement is faster, and more amenable to computerized response collection and interpretation, than is auditory brainstem evoked response testing, OAEs may allow economically feasible screening of all newborns. Both TEOAEs and DPOAEs are present in normal neonate ears (Stevens et al. 1987, 1991; Lafreniere et al. 1991; Spektor et al. 1991; Lasky, Perlman, and Hecox 1992). Indeed, TEOAEs are typically substantially larger, and DPOAEs slightly larger, in neonates than in adults. Ongoing clinical research is assessing the capability of TEOAEs and DPOAEs to identify hearing impairment in newborns in order to determine the feasibility of using OAE based tests as a universal screening procedure for all newborns (White and Behrens 1993).

As in all clinical applications, accurate interpretation of the absence of OAEs in hearing screening applications requires assessment of middle ear function. It is noted that OAE testing will not identify those infants with hearing losses of retrocochlear or central origin.

4.3 Longitudinal Monitoring of Cochlear Status

In normal ears, both TEOAE and DPOAE properties are quite stable over periods of time extending from minutes to years. Thus, because OAE amplitudes can be rapidly measured in fine resolution, it is possible that small decreases of amplitude could be detected at an early stage of trauma. It is not known whether OAE amplitudes may be systematically reduced by minor cochlear damage, occurring in the early stages of certain traumas, that does not cause hearing threshold elevation. If this is the case, OAE testing may allow detection of the onset of cochlear hearing loss prior to any hearing threshold elevation. However, even if this is not the case, the greater resolution and long-term repeatability of OAE amplitude measurements than of routine clinical measures of hearing threshold, suggest the possibility that OAE based tests may be able to detect the onset of hearing loss of cochlear origin prior to routine behavioral testing. If so, OAEs could be used to monitor the cochlear condition of subjects at risk for hearing impairment, e.g., workers in noisy environments or patients receiving life-saving but ototoxic drugs (Section 3.1, above), to provide an early indication of trauma. Systematic longitudinal studies of OAEs and hearing thresholds in these subject groups are required to investigate this issue.

It is also possible that OAEs may be of use for intraoperative monitoring during surgery to remove tumors of the auditory nerve while attempting to preserve hearing in the affected ear. In this role, OAEs may serve as a rapid, objective means to assess cochlear status, allowing action to be taken to reduce possible cochlear trauma resulting from surgical manipulations (Balkany et al. 1994). In such applications, OAE components at low

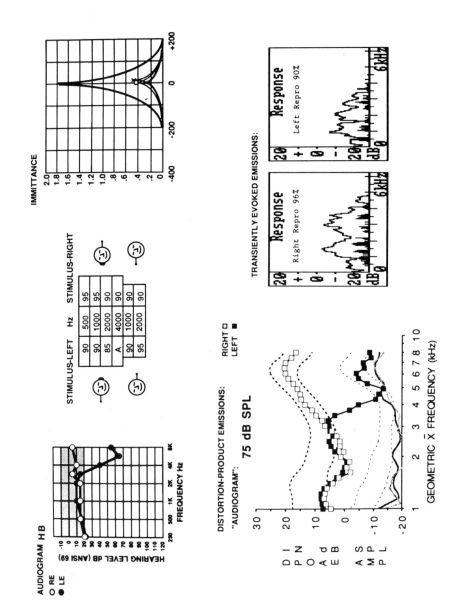

frequencies will be obscured by the high levels of low-frequency noise present in operating rooms. Unfortunately, a large proportion of candidates for these surgeries have greatly reduced high-frequency OAEs because of the effects of the tumor (Section 3.7, above), or for other reasons, e.g., presbyacusis. However, a significant number of patients may have OAEs large enough to be useful for intraoperative monitoring purposes. It remains to be demonstrated whether the reliable detection of an OAE amplitude reduction upon trauma in human ears is rapid enough to provide useful information to the surgeon.

If OAEs are to be used in intraoperative monitoring, it will be necessary to know the effects of anesthesia on OAEs. Hauser et al. (1992) reported variations of TEOAE amplitudes during general anesthesia in humans that appeared to arise primarily indirectly, from changes of middle ear status. Whereas more data are available concerning the effects of general anesthesia on OAEs in animal models, the data are confusing. There appear to be no differences in the properties of $2f_1-f_2$ DPOAEs between awake and ketamine anesthetized rabbits (Lonsbury-Martin et al. 1987; Whitehead et al. 1992b). However, Brown (1988) reported large variations in the amplitude of the f_2-f_1 DPOAE and smaller variations in the amplitude of $2f_1-f_2$ DPOAEs with depth of anesthesia induced by various drugs in guinea pigs and gerbils. Although both guinea pigs and gerbils are known to develop build-ups of middle ear negative pressure during anesthesia, it seems unlikely that the large differences in the variations of the two DPOAEs observed by Brown (1988) can be explained entirely by a middle ear mechanism, suggesting some cochlear effect of the anesthesia, either directly, or indirectly via a central influence mediated by the cochlear efferent innervation. In the bat *Pteronotus parnellii*, Kössl and Vater (1985) found that both halothane and Nembutal anesthesia decreased both the amplitude and frequency of SFOAEs. However, the bat cochlea is generally thought to be more susceptible to anesthesia than is the case for most other mammals.

5. Conclusions

In general, the results of studies in human patient groups and in animal models are quite consistent. Factors known to cause both temporary and

---

←

FIGURE 7.15 Hearing and evoked-emission test results for a fifteen-year-old female (HB), who complained of difficulties in hearing her teacher lecture in the classroom. Note that the elevated hearing thresholds measured in the left ear (top left, filled circles) at frequencies above 3 kHz, were reflected in a reduction to noise-floor levels of the DPOAEs (bottom left, filled squares) above 3 kHz, and that no TEOAEs were detected above 3.4 kHz in the left ear. For a further description, see legend to Fig. 7.8.

permanent sensorineural hearing loss have been found to reduce or abolish all OAE types in humans, monkeys, and nonprimate laboratory mammals. Those traumas that result primarily in damage to the organ of Corti (aminoglycosides, noise overexposure) or in reduced energy supply to the organ of Corti (loop diuretics, hypoxia, anoxia) tend to demonstrate a greater correlation of changes of OAEs with behavioral or electrophysiological measures of hearing impairment, whereas this correlation is less clear in the more complicated cases of Meniere's disease and surgically induced endolymphatic hydrops. The results of several studies discussed above (Sections 3.1 and 3.2) suggest that DPOAEs evoked by low-level stimuli, and also TEOAEs, SFOAEs, and SOAEs, require the integrity of the outer hair cells and are dependent upon the energy supply to the organ of Corti. These properties are also shared by the mechanism, thought to be based in outer hair cell electromotility, that is responsible for the enhancement of basilar membrane motion at low sound levels and, thus, for normal hearing sensitivity (Ruggero and Rich 1991). The finding that OAEs are influenced by activation of the medial olivocochlear-efferent system (Siegel and Kim 1982; Guinan 1986; Puel and Rebillard 1990), which preferentially innervates the outer hair cells, is also consistent with the proposed origin of OAEs in the electromotility of outer hair cells. The findings indicating that there is no direct afferent neural involvement in the generation of TEOAEs, SFOAEs, and DPOAEs, suggest that emissions may be particularly valuable as specific indicators of the status of this sensory component of the hearing process.

Thus, experimental studies have given rise to a model on which the interpretation of clinical OAE test results can be based. In this model, the outer hair cell system within the cochlea acts to enhance the sound induced motion of the cochlear partition at low stimulus levels. Reduced enhancement of cochlear partition vibration, due to dysfunction of the outer hair cell system, results in reduced stimulation of the inner hair cells and, thus, of the auditory afferents, resulting in a sensory hearing loss. Whereas the enhancement of basilar membrane motion is thought to be carried out by the outer hair cells, and presumably involves the outer hair cell electromotility demonstrated in vitro, the normal operation of this mechanism in vivo also appears to be dependent upon the function of the stria vascularis, and, possibly, structural integrity of other elements of the organ of Corti. Thus, each of these cochlear elements can be regarded as part of the functional outer hair cell system. TEOAEs, SFOAEs, SOAEs, and the DPOAEs evoked by low-level stimuli, appear to be produced as a by-product of the action of this mechanism. Thus, reduced operation of this mechanism is reflected as both elevated hearing thresholds and reduced amplitudes of these OAEs. The DPOAEs evoked by stimuli above 65 to 70 dB SPL in rabbits and rodents appear to be largely independent of the outer hair cells. It is not clear at what stimulus level the DPOAEs in human ears become independent of outer hair cell status.

This model of the relationship between OAEs, cochlear function, and hearing threshold has the following implications. First, reduced OAEs are an indication of only one specific sensory component of sensorineural hearing loss, i.e., dysfunction of the outer hair cell system. Second, whereas reduced OAE amplitudes indicate dysfunction of the outer hair cell system, the underlying pathology may not necessarily involve the outer hair cells themselves. These implications are consistent with the empirical findings that OAEs are typically normal in cases of purely central or retrocochlear hearing loss and are typically reduced in frequency regions of hearing loss due to factors known to involve the outer hair cells and/or stria vascularis, e.g., acoustic trauma and some ototoxic drugs. Third, dysfunction of elements within the cochlea that are not required for the normal operation of the outer hair cell system, e.g., the inner hair cells, may cause sensorineural hearing loss without reduced OAE amplitudes. This implication is consistent with the findings that OAEs are sometimes normal in sensorineural hearing losses known to have complex effects on various structures in the cochlea, e.g., Meniere's disease, endolymphatic hydrops (Section 3.4), and sudden idiopathic hearing loss. Many hearing losses involve dysfunction of both the outer hair cell system and other elements within the cochlea.

The model also suggests that, given normal middle ear function, OAEs should not be systematically reduced in the presence of normal hearing, unless (1) minor functional impairment of the outer hair cell system can reduce OAE amplitudes without elevating hearing threshold; or (2) some forms of cochlear pathology can influence the propagation of the OAE from the site of generation along the cochlear partition to the base of the cochlea. At present, there does not appear to be any systematic evidence in the literature indicating that any cochlear trauma or pathology consistently reduces OAE amplitudes in the presence of normal hearing.

Note that the model provides no indication of the quantitative relationship between the amount of OAE amplitude reduction and of hearing threshold elevation. Empirically, this relationship appears to be less than 1 dB/dB. Thus, Whitehead et al. (1993b) found that, in humans, the amplitudes of DPOAEs evoked by 75 dB SPL primaries and of TEOAEs evoked by an 80 dB SPL click stimulus were reduced by approximately 0.4 dB per 1 dB elevation of hearing threshold in sensorineural losses. In both humans and animal models, DPOAEs evoked by high-level stimuli are reduced less by trauma than those evoked by low-level stimuli, implying that the dB/dB relation between DPOAE amplitude reduction and hearing threshold elevation increases with decreasing stimulus level. In considering this dB/dB relationship, it should be remembered that OAE-amplitude reductions are thought to reflect primarily dysfunction of the outer hair cell system, whereas hearing-threshold elevations of sensorineural origin may also reflect dysfunction of other elements in addition to the outer hair cell system, e.g., the inner hair cells or the auditory nerve.

The model also provides no indication of the quantitative relationship between outer hair cell loss, OAE-amplitude reduction, and reduction of enhancement of basilar-membrane motion. Findings from 15-day old waltzing guinea pigs (Section 3.5; Canlon, Marklund and Borg 1993) and noise-exposed chinchillas (Section 3.3; Subramaniam et al. 1994; Subramaniam, Henderson and Spongr 1994) indicate that DPOAEs can be normal in the presence of moderate outer hair cell pathology. This finding suggests that the degree of outer hair cell pathology must exceed some 'threshold' before DPOAE amplitudes are reduced. In the 15-day old waltzing guinea pigs, auditory brainstem responses were much reduced, which may have reflected dysfunction of elements other than the outer hair cell system. In the noise-exposed chinchillas, evoked responses from the inferior colliculus were normal, suggesting that the active cochlear process responsible for enhancing cochlear-partition vibration was functioning normally, despite the outer hair cell pathology. Thus, the finding that DPOAEs were normal in the presence of moderate outer hair cell pathology is not necessarily inconsistent with the hypothesis that DPOAEs reflect the action of an outer-hair-cell based mechanism responsible for enhancement of cochlear-partition vibration.

Evidence from studies of human and monkey ears suggests at least a partial dissociation of the mechanism underlying the generation of DPOAEs from that responsible for TEOAEs, SFOAEs and SOAEs (Section 3.1). Moreover, studies in rabbits and rodents suggest that at least partially separate mechanisms underlie the generation of $2f_1$-f_2 DPOAEs at low- and high-stimulus levels (Sections 2.4, 3.1, and 3.2). Whereas TEOAEs, SFOAEs and SOAEs, and also DPOAEs evoked by stimuli below 60 to 70 dB SPL, appear to require the integrity of the outer hair cells and are dependent upon the energy supply to the organ of Corti, the DPOAEs evoked by primary tones above about 70 dB SPL appear relatively independent of these factors, although they can also be reduced by certain traumas (e.g., Fig. 7.7B). Because of the differences between DPOAEs in humans and the nonprimate laboratory mammals, in particular the large difference in the absolute amplitude of $2f_1$-f_2 DPOAEs and the even greater difference in the amplitude of DPOAEs relative to that of TEOAEs and SFOAEs, it is not clear that the finding of two distinct generators of $2f_1$-f_2 DPOAEs in rabbits and rodents can be generalized to humans. This issue needs to be resolved. If humans also possess distinct low-level/more-vulnerable, and high-level/less-vulnerable DPOAE generators, then the interpretation of clinical tests in terms of cochlear dysfunction may depend upon the stimulus parameters used. If humans do not possess separate DPOAE generators, this may limit the relevance of studies of DPOAEs in nonprimate mammals to the human situation. The great similarity of OAEs in humans and monkeys implies that monkeys represent a valid model of OAEs in human ears (Anderson and Kemp 1979; Wit and Kahmann 1982; Lonsbury-Martin and Martin 1988; Martin et al. 1988). However, for several practical reasons, nonprimate mammals are a more convenient preparation for the study of hearing than are monkeys.

Whereas the differences in the basic properties of OAEs in humans and the nonprimate laboratory mammals suggest that caution should be applied when inferring properties of human OAEs from those of OAEs in these animal models, the effects of a variety of cochlear traumas on OAEs are generally similar across species. These similarities indicate that continued study of the effects of cochlear traumas on OAEs in nonprimate laboratory mammals will be useful in the development of clinical applications of OAEs.

Acknowledgments. This work was partially supported by grants from the Public Health Service (DC00313, DC00613, DC01668, ES03500).

References

Anderson SD (1980) Some ECMR properties in relation to other signals from the auditory periphery. Hear Res 2:273–296.

Anderson SD, Kemp DT (1979) The evoked cochlear mechanical response in laboratory primates. Arch Otorhinolaryngol 224:47–54.

Anniko M, Sobin A (1986) Cisplatin: Evaluation of its ototoxic potential. Am J Otorhinolaryngol 7:276–293.

Avan P, Loth D, Menguy C, Teyssou M (1990) Evoked otoacoustic emissions in guinea pig: Basic characteristics. Hear Res 44:151–160.

Balkany TJ, Telischi FF, Lonsbury-Martin BL, Martin GK (1994): Otoacoustic emissions in clinical practice. Am J Otol 15, Suppl 1:29–38.

Bohne B, Clark WW (1982) Growth of hearing loss and cochlear lesion with increasing duration of noise exposure. In: Hamernik RP, Henderson D, Salvi R (eds), New Perspectives on Noise Induced Hearing Loss. New York: Raven Press, pp. 283–301.

Bonfils P (1989) Spontaneous otoacoustic emissions: Clinical interest. Laryngoscope 99:752–756.

Bonfils P, Uziel A (1988) Evoked otoacoustic emissions in patients with acoustic neuromas. Am J Otol 9:412–417.

Bonfils P, Bertrand Y, Uziel A (1988) Evoked otoacoustic emissions: Normative data and presbycusis. Audiology 27:27–35.

Bonfils P, Uziel A, Pujol R (1988) Evoked otoacoustic emissions from adults and infants: Clinical applications. Acta Otolaryngol 105:445–449.

Brass D, Kemp DT (1991) Time-domain observation of otoacoustic emissions during constant tone stimulation. J Acoust Soc Am 90:2415–2427.

Bray P, Kemp DT (1987) An advanced cochlear echo technique suitable for infant screening. Br J Audiol 21:191–204.

Brown AM (1987) Acoustic distortion from rodent ears: A comparison of responses from rats, guinea pigs and gerbils. Hear Res 31:25–38.

Brown AM (1988) Continuous low level sound alters cochlear mechanics: An efferent effect? Hear Res 34:27-38.

Brown AM, Gaskill SA (1990) Measurement of acoustic distortion reveals underlying similarities between human and rodent mechanical responses. J Acoust Soc Am 88:840-849.

Brown AM, Kemp DT (1984) Suppressibility of the $2f_1$-f_2 stimulated acoustic emissions in gerbil and man. Hear Res 13:29-37.

Brown AM, McDowell B, Forge A (1989) Acoustic distortion products can be used to monitor the effects of chronic gentamicin treatment. Hear Res 42:143-156.

Brown AM, Woodward S, Gaskill SA (1990) Frequency variations in spontaneous sound emissions from guinea pig and human ears. Eur Arch Otorhinolaryngol 247:24-28.

Brownell WE (1990) Outer hair cell electromotility and otoacoustic emissions. Ear Hear 11:82-92.

Brownell WE, Bader CR, Bertrand D, de Ribaupierre Y (1985) Evoked mechanical responses of isolated cochlear outer hair cells. Science 227:194-196.

Brummett RE, Fox KE (1989) Aminoglycoside-induced hearing loss in humans. Antimic Agents Chemother 33:797-800.

Burns EM, Arehart KH, Campbell SL (1992) Prevalence of spontaneous otoacoustic emissions in neonates. J Acoust Soc Am 91:1571-1575.

Canlon B, Marklund K, and Borg E (1993) Measures of auditory brain-stem responses, distortion-product emissions, hair cell loss, and forward masked tuning curves in the waltzing guinea pig. J Acoust Soc Am 94:3232-3243.

Clark WW, Bohne BA (1978) Animal model for the 4-kHz tonal dip. Ann Otol Rhinol Laryngol, Suppl 51, 87:1-16.

Clark WW, Kim DO, Zurek PM, Bohne BA (1984) Spontaneous otoacoustic emissions in chinchilla ear canals: Correlation with histopathology and suppression by external tones. Hear Res 16:299-314.

Collet L, Moulin M, Gartner M, Morgon A (1990a) Age-related changes in evoked otoacoustic emissions. Ann Otol Rhinol Laryngol 99:993-997.

Collet L, Kemp DT, Veuillet E, Duclaux R, Moulin A, Morgon A (1990b) Effect of contralateral auditory stimuli on active cochlear micro-mechanical properties in human subjects. Hear Res 43:251-262.

Crofton KM, Lassiter TL, Rebert CS (1994) Solvent-induced ototoxicity in rats: An atypical selective mid-frequency hearing deficit. Hear Res 80:25-30.

Dallmayr C (1985) Spontane oto-akustische emissionen. Statistik und Reaktion auf akustische Störtöne. Acustica 59:67-75.

Dallmayr C (1987) Stationary and dynamical properties of simultaneous evoked otoacoustic emissions (SEOAE) Acustica 63:243-255.

Davis H (1983) An active process in cochlear mechanics. Hear Res 9:79-90.

Decker TN, Fritsch JH (1982) Objective tinnitus in the dog. J Am Vet Med Assoc 180:74.

Dieler R, Shehata-Dieler WE, Brownell WE (1991) Concomitant salicylate-induced alterations of outer hair cell subsurface cisternae and electromotility. J Neurocytol 20:637-653.

Dolan TG, Abbas PJ (1985) Changes in the $2f_1$-f_2 acoustic emission and whole-nerve response following sound exposure: Long-term effects. J Acoust Soc Am 77:1475-1483.

Douek EE, Dodson HC, Bannister LH (1983) The effects of sodium salicylates on the cochlea of the guinea pig. J Laryngol Otol 93:793-799.

Estrem SA, Babin RW, Ryu JH, Moore KC (1981) *Cis*-diamminedichloroplatinum

(II) ototoxicity in the guinea pig. Arch Otolaryngol Head Neck Surg 89:638–645.

Evans EF, Wilson JP, Borerwe TA (1981) Animal models of tinnitus. In: Evered D, Lawrenson G (eds) Tinnitus. London: Pitman, pp. 108–138.

Fleischmann RW, Stadnicki SW, Ethier ME, Schaeppi U (1975) Ototoxicity of cis-dichlorodiammine platinum (II) in the guinea pig. Toxicol Appl Pharmacol 33:320–332.

Franklin DJ, Lonsbury-Martin BL, Stagner BB, Martin GK (1991) Altered susceptibility of $2f_1$-f_2 acoustic distortion products to the effects of repeated noise exposure in rabbits. Hear Res 53:185–208.

Gaskill SA, Brown AM (1990) The behavior of the acoustic distortion product, $2f_1$-f_2, from the human ear and its relation to auditory sensitivity. J Acoust Soc Am 88:821–839.

Govaerts PJ, Claes J, Van De Heyning PH, Jorens PhG, Marquet J, De Broe ME (1990) Aminoglycoside-induced ototoxicity. Toxicol Lett 52:227–251.

Guinan Jr JJ (1986) Effect of efferent neural activity on cochlear mechanics. In: Cianfrone G, Grandori F (eds) Cochlear Mechanics and Otoacoustic Emissions. Scand Audiol Suppl 25, pp. 53–62.

Guinan Jr JJ (1990) Changes in stimulus-frequency otoacoustic emissions produced by two-tone suppression and efferent stimulation in cats. In: Dallos P, Geisler CD, Matthews JW, Ruggero MA, Steele CR (eds) Mechanics and Biophysics of Hearing. New York: Springer, pp. 170–177.

Hamernik RP, Patterson JH, Turrentine GA, Ahroon WA (1989) The quantitative relation between sensory cell loss and hearing thresholds. Hear Res 38:199–212.

Harris FP (1990) Distortion-product otoacoustic emissions in humans with high frequency sensorineural hearing loss. J Speech Hear Res 33:594–600.

Harris FP, Glattke TJ (1992) The use of suppression to determine the characteristics of otoacoustic emissions. Sem Hear 13:67–80.

Harris FP, Probst R (1992) Transiently evoked otoacoustic emissions in patients with Meniere's disease. Acta Otolaryngol 112:36–44.

Harris FP, Lonsbury-Martin BL, Stagner BB, Coats AC, Martin GK (1989) Acoustic distortion products in humans: Systematic changes in amplitude as a function of f_2/f_1 ratio. J Acoust Soc Am 85:220–229.

Hauser R, Probst R (1990) The influence of systematic primary-tone level variation L_2-L_1 on the acoustic distortion product emission $2f_1$-f_2 in normal human ears. J Acoust Soc Am 89:280–286.

Hauser R, Harris FP, Probst R, Frei F (1992) Influence of general anesthesia on transiently evoked otoacoustic emissions in humans. Ann Otol Rhinol Laryngol 101:994–999.

Hayashida T, Hiel H, Dulon D, Erre J-P, Guilhaume A, Aran J-M (1989) Dynamic changes following combined treatment with gentamicin and ethacrynic acid with and without acoustic stimulation. Acta Otolaryngol 108:404–413.

Horner KC, Cazals Y (1987) Glycerol-induced changes in the cochlear responses of the guinea pig hydropic ear. Arch Otorhinolaryngol 244:49–54.

Horner KC, Cazals Y (1989) Distortion products in early stage experimental hydrops in the guinea pig. Hear Res 43:71–80.

Horner KC, Lenoir M, Bock GR (1985) Distortion product otoacoustic emissions in hearing-impaired mutant mice. J Acoust Soc Am 78:1603–1611.

Hotz MA, Harris FP, Probst R (1994) Otoacoustic emissions: An approach for monitoring aminoglycoside-induced ototoxicity. Laryngoscope 14:1130–1134.

Johnsen NJ, Elberling C (1982) Evoked acoustic emissions from the human ear. I.

Equipment and response parameters. Scand Audiol 11:3-12.

Johnson A-C (1993) The ototoxic effect of toluene and the influence of noise, acetyl salicylic acid, or genotype. Scand Audiol, Suppl 39:1-40.

Johnson A-C, Canlon B (1994) Toluene exposure affects the functional activity of the outer hair cells. Hear Res 72:189-196.

Johnstone BM, Patuzzi R, Yates GK (1986) Basilar membrane measurements and the traveling wave. Hear Res 22:147-153.

Karlsson KK, Berninger E, Alvan G (1991) The effect of quinine on psychoacoustic tuning curves, stapedius reflexes and evoked otoacoustic emissions in healthy volunteers. Scand Audiol 20:83-90.

Katz CD, Madors SJ, Henley CM, Overbeek PA, Kovak M, Martin GK, Lonsbury-Martin BL (1989) Cochlear dysfunction in a transgenic mouse family. Abstracts of the Annual Meeting of the Society for Neuroscience 15:212.

Kemp DT (1978) Stimulated acoustic emissions from within the human auditory system. J Acoust Soc Am 64:1386-1391.

Kemp DT (1981) Physiologically active cochlear micromechanics-one source of tinnitus. In: Evered D, Lawrenson G (eds) Tinnitus. London: Pitman, pp. 54-81.

Kemp DT (1982) Cochlear echoes: Implications for noise-induced hearing loss. In: Hamernik RP, Henderson D, Salvi R (eds) New Perspectives on Noise-Induced Hearing Loss. New York: Raven, pp. 189-207.

Kemp DT, Brown AM (1983) A comparison of mechanical nonlinearities in the cochleae of man and gerbil from ear canal measurements. In: Klinke R, Hartmann R (eds) Hearing-Physiological Bases and Psychophysics. Berlin: Springer Verlag, pp. 82-88.

Kemp DT, Brown AM (1984) Ear canal acoustic and round window correlates of $2f_1$-f_2 distortion generated in the cochlea. Hear Res 13:39-46.

Kemp DT, Chum R (1980) Observations on the generator mechanism of stimulus frequency acoustic emissions, two-tone suppression. In: van den Brink G, Bilsen FA (eds) Psychophysical, Physiological and Behavioural Studies in Hearing. Delft: Delft University Press, pp. 34-41.

Kemp DT, Bray P, Alexander L, Brown AM (1986) Acoustic emission cochleography—practical aspects. In: Cianfrone G, Grandori F (eds) Cochlear Mechanics and Otoacoustic Emissions. Scand Audiol Suppl 25, pp. 71-96.

Kemp DT, Brass DN, Souter M (1990) Observations on simultaneous SFOAE and DPOAE generation and suppression. In: Dallos P, Geisler CD, Matthews JW, Ruggero MA, and Steele CR (eds) Mechanics and Biophysics of Hearing. New York: Springer Verlag, pp. 202-209.

Kim DO (1980) Cochlear mechanics: Implications of electrophysiological and acoustical observations. Hear Res 2:297-317.

Kim DO, Molnar CE, Matthews JW (1980) Cochlear mechanics: Nonlinear behavior in two-tone responses as reflected in cochlear-nerve-fiber responses and in ear-canal sound pressure. J Acoust Soc Am 67:1704-1721.

Komune S, Asakuma S, Snow JB (1981) Pathophysiology of the ototoxicity of cis-diamminedichloroplatinum. Otolaryngol Head Neck Surg 89:275-282.

Konishi T, Gupta BN, Prazma J (1983) Ototoxicity of cis-dichlorodiammine platinum (II) in guinea pigs. Am J Otolaryngol 4:18-26.

Kössl M (1992) High frequency distortion products from the ears of two bat species, *Megaderma lyra* and *Carollia perspicillata*. Hear Res 60:156-164.

Kössl M and Vater M (1985) Evoked acoustic emissions and cochlear microphonics in the mustache bat, *Pteronotus parnellii*. Hear Res 19:157-170.

Kujawa SG, Fallon M, Bobbin RP (1992) Intracochlear salicylate reduces low-intensity acoustic and cochlear microphonic distortion products. Hear Res 64:73–80.

Lafreniere DC, Jung MD, Smurzynski J, Leonard G, Kim DO, Sasek J (1991) Distortion product and click-evoked otoacoustic emissions in healthy newborns. Arch Otolaryngol 117:1382–1389.

Lasky R, Perlman J, Hecox K (1992) Distortion-product otoacoustic emissions in human newborns and adults. Ear Hear 13:430–441.

Laurell G, Engström B (1989a) The ototoxic effect of cisplatin on guinea pigs in relation to dosage. Hear Res 38:27–38.

Laurell G, Engström B (1989b) The combined effect of cisplatin and furosemide on hearing function in guinea pigs. Hear Res 38:19–26.

Long GR, Tubis A (1988) Modification of spontaneous and evoked otoacoustic emissions and associated psychoacoustic microstructure by aspirin consumption. J Acoust Soc Am 84:1343–1353.

Lonsbury-Martin BL, Martin GK (1988) Incidence of spontaneous otoacoustic emissions in macaque monkeys: A replication. Hear Res 34:313–317.

Lonsbury-Martin BL, Martin GK (1990) The clinical utility of distortion-product otoacoustic emissions. Ear Hear 11:90–99.

Lonsbury-Martin BL, Probst R, Coats AC, Martin GK (1987) Acoustic distortion products in rabbits. I. Basic features and physiological vulnerability. Hear Res 28:173–189.

Lonsbury-Martin BL, Martin GK, Coats AC, Johnson KC (1989) Alterations in acoustic-distortion products and single nerve-fiber activity in hydropic rabbits. In: Nadol JB (ed) Meniere's disease: Pathogenesis, Pathophysiology, Diagnosis, and Treatment. Amsterdam: Kugler and Ghedini Publications, pp. 337–341.

Lonsbury-Martin BL, Harris FP, Hawkins MD, Stagner BB, Martin GK (1990a) Distortion-product emissions in humans: I. Basic properties in normally hearing subjects. Ann Otol Rhinol Laryngol Suppl 236:3–13.

Lonsbury-Martin BL, Harris FP, Stagner BB, Hawkins MD, Martin GK (1990b) Distortion-product emissions in humans: II. Relations to stimulated and spontaneous emissions and acoustic immittance in normally hearing subjects. Ann Otol Rhinol Laryngol Suppl 236:14–28.

Lonsbury-Martin BL, Cutler WM, Martin GK (1991) Evidence for the influence of aging on distortion-product otoacoustic emissions in humans. J Acoust Soc Am 89:1749–1759.

Lonsbury-Martin BL, Whitehead ML, Martin GK (1991) Clinical applications of otoacoustic emissions. J Speech Hear Res 34:964–981.

Lonsbury-Martin BL, Whitehead ML, Henley CM, Martin GK (1991) Differential effects of sodium salicylate on the distinct classes of otoacoustic emissions in rabbit and in monkey. Abstracts of the Fourteenth Midwinter Meeting of the Association for Research in Otolaryngology, p. 67.

Lonsbury-Martin BL, McCoy MJ, Whitehead ML, Martin GK (1993) Clinical testing of distortion-product otoacoustic emissions. Ear Hear 14:11–22.

Lutman ME, Mason SM, Sheppard S, Gibbin KP (1989) Differential diagnostic potential of otoacoustic emissions: A case study. Audiology 28:205–210.

Martin GK, Lonsbury-Martin BL, Probst R, Coats AC (1985) Spontaneous otoacoustic emissions in the nonhuman primate: A survey. Hear Res 20:91–95.

Martin GK, Probst R, Scheinin SA, Coats AC, Lonsbury-Martin BL (1987) Acoustic distortion products in rabbits. II. Sites of origin revealed by suppression and pure-tone exposures. Hear Res 28:191-208.

Martin GK, Lonsbury-Martin BL, Probst R, Coats AC (1988) Spontaneous otoacoustic emissions in a nonhuman primate. I. Basic features and relations to other emissions. Hear Res 33:49-68.

Martin GK, Stagner BB, Coats AC, Lonsbury-Martin BL (1989) Endolymphatic hydrops in rabbits: Behavioral thresholds, acoustic distortion products, and cochlear pathology. In: Nadol JB (ed) Meniere's disease: Pathogenesis, Pathophysiology, Diagnosis and Treatment. Amsterdam: Kugler and Ghedini Publications, pp. 205-219.

Martin GK, Franklin DJ, Harris FP, Ohlms LA, Lonsbury-Martin BL (1990) Distortion-product emissions in humans: III. Influence of hearing pathology. Ann Otol Rhinol Laryngol Suppl 236:29-44.

Mathis A, Probst R, De Min N, Hauser R (1991) A child with an unusually high-level spontaneous otoacoustic emission. Arch Otolaryngol Head Neck Surg 117:674-676.

Matz GJ (1976) The ototoxic effects of ethacrynic acid in man and animals. Laryngoscope 86:1065-1086.

McAlpine D, Johnstone BM (1990) The ototoxic mechanism of cisplatin. Hear Res 47:191-204.

McFadden D, Pasanen EG (1994) Otoacoustic emissions and quinine sulfate. J Acoust Soc Am 95:3460-3474.

McFadden D, Plattsmier HS (1984) Aspirin abolishes spontaneous oto-acoustic emissions. J Acoust Soc Am 76:443-448.

McFadden D, Plattsmier HS, Pasanen EG (1984) Aspirin-induced hearing loss as a model of sensorineural hearing loss. Hear Res 16:251-260.

McPhee JR, Sperling NM, Madell JR, Lucente FE (1994) Early detection of ototoxicity using high frequency distortion product emissions. Assoc Res Otolaryngol Abstr 14:49.

Mensh BD (1991) Changes in Cochlear Mechanical Non-Linearities in Response to Brief, Tonal Overstimulation in Rabbits: Dependence on Prior Acoustic Exposure and Relation to Chronic-Noise Effects [Doctoral Dissertation] Houston (TX): Baylor College of Medicine.

Mills DM, Norton SJ, Rubel EW (1993) Vulnerability and adaptation of distortion-product otoacoustic emissions to endocochlear potential variation. J Acoust Soc Am 94:2108-2122.

Mott JB, Norton SJ, Neely ST, Warr WB (1989) Changes in spontaneous otoacoustic emissions produced by acoustic stimulation of the contralateral ear. Hear Res 38:229-242.

Moulin A, Collet L, Morgon A (1992) Influence of spontaneous otoacoustic emissions (SOAE) on acoustic distortion product input/output functions: Does the medial efferent system act differently in the vicinity of an SOAE? Acta Otolaryngol 112:210-214.

Nadol JB (1989) Meniere's Disease: Pathogenesis, Pathophysiology, Diagnosis and Treatment. Amsterdam: Kugler and Ghedini Publications.

Naeve SL, Margolis RH, Levine SC, Fournier EM (1992) Effect of ear-canal air pressure on evoked otoacoustic emissions. J Acoust Soc Am 91:2091-2095.

Norton SJ, Rubel EW (1990) Active and passive ADP components in mammalian and avian ears. In: Dallos P, Geisler CD, Matthews JW, Ruggero MA, Steele CR (eds) Mechanics and Biophysics of Hearing. New York: Springer Verlag, pp. 219-226.

Norton SJ, Mott JB, Champlin CA (1989) Behavior of spontaneous otoacoustic emissions following intense ipsilateral acoustic stimulation. Hear Res 38:243-258.

Ohlms LA, Lonsbury-Martin BL, Martin GK (1990) The clinical application of acoustic distortion products. Arch Otolaryngol 102:315-322.

Ohlms LA, Lonsbury-Martin BL, Martin GK (1991) Acoustic-distortion products: Separation of sensory from neural dysfunction in sensorineural hearing loss in humans and rabbits. Arch Otolaryngol 104:159-174.

Ohyama K, Wada H, Kobayashi T, Takasaka T (1991) Spontaneous otoacoustic emissions in guinea pig. Hear Res 56:111-121.

Owens JJ, McCoy MJ, Lonsbury-Martin BL, Martin GK (1992) Influence of otitis media on evoked otoacoustic emissions in children. Sem Hear 13:53-66.

Pasic TR, Dobie RA (1991) Cis-platinum ototoxicity in children. Laryngoscope 101:985-991.

Prieve BA (1992) Otoacoustic emissions in infants and children: Basic characteristics and clinical application. Sem Hear 13:37-52.

Probst R, Coats AC, Martin GK, Lonsbury-Martin BL (1986) Spontaneous, click- and toneburst-evoked otoacoustic emissions from normal ears. Hear Res 21:261-275.

Probst R, Lonsbury-Martin BL, Martin GK, Coats AC (1987) Otoacoustic emissions in ears with hearing loss. Am J Otolaryngol 8:73-81.

Probst R, Lonsbury-Martin BL, Martin GK (1991) A review of otoacoustic emissions. J Acoust Soc Am 89:2027-2067.

Pryor GT, Dickinson J, Feeney E, Rebert CS (1984) Hearing loss in rats first exposed to toluene as weanlings or as young adults. Neurobehav Toxicol Teratol 6:111-119.

Puel J-L and Rebillard G (1990) Effect of contralateral sound stimulation on the distortion product $2f_1-f_2$: Evidence that the medial efferent system is involved. J Acoust Soc Am 87:1630-1635.

Rebillard G and Lavigne-Rebillard M (1992) Effect of reversible hypoxia on the compared time courses of endocochlear potential and $2f_1-f_2$ distortion products. Hear Res 62:142-148.

Robinette MS (1992) Clinical observations with transient evoked otoacoustic emissions with adults. Sem Hear 13:23-36.

Robinette MS and Facer GW (1992) Evoked otoacoustic emissions in differential diagnosis: A case report. Otolaryngol Head Neck Surg 105:120-123.

Rossi G, Solero P, Rolando M, Olina M (1989) Delayed oto-acoustic emissions evoked by bone-conduction stimulation: Experimental data on their origin, characteristics and transfer to the external ear in man. Scand Audiol Suppl 29:1-24.

Ruggero MA, Rich NC (1991) Furosemide alters organ of Corti mechanics: Evidence for feedback of outer hair cells upon the basilar membrane. J Neurosci 11:1057-1067.

Ruggero MA, Kramek B, Rich NC (1984) Spontaneous otoacoustic emissions in a dog. Hear Res 13:293-296.

Russell NJ, Fox KE, Brummett RE (1979) Ototoxic effects of the interaction between kanamycin and ethacrynic acid. Acta Otolaryngol 88:369-381.

Rutten WLC, Buisman HP (1983) Critical behaviour of auditory oscillators near feedback phase transitions. In: de Boer E, Viergever MA (eds) Mechanics of Hearing. Delft: Delft University Press, pp. 91-99.

Rybak LP (1986) Drug ototoxicity. Ann Rev Pharmacol Toxicol 26:79-99.

Sakashita T, Minowa Y, Hachikawa K, Kubo T, Nakai Y (1991) Evoked otoacoustic emissions from ears with idiopathic sudden deafness. Acta Otolaryngol 486:66–72.

Salvi RJ, Henderson D, Hamernik RP, Colletti V (1986) Basic and Applied Aspects of Noise-Induced Hearing Loss. New York: Plenum Press.

Schaefer SD, Post JD, Close LG, Wright CG (1985) Ototoxicity of low and moderate-dose cisplatin. Cancer 56:1934–1939.

Schloth E (1983) Relation between spectral composition of spontaneous otoacoustic emissions and fine-structure of threshold in quiet. Acustica 53:250–256.

Schloth E and Zwicker E (1983) Mechanical and acoustical influences on spontaneous oto-acoustic emissions. Hear Res 11:285–293.

Schmiedt RA (1986a) Acoustic distortion in the ear canal: I. Cubic difference tones: Effects of acute noise injury. J Acoust Soc Am 79:1481–1490.

Schmiedt RA (1986b) Effects of asphyxia on levels of ear canal emissions in gerbils. Abstracts of the Ninth Midwinter Meeting of the Association for Research in Otolaryngology, p. 112.

Schmiedt RA, Adams JC (1981) Stimulated acoustic emissions in the ear canal of the gerbil. Hear Res 5:295–305.

Schrott A, Puel JL, Rebillard G (1991) Cochlear origin of $2f_1-f_2$ distortion products assessed by using two types of mutant mice. Hear Res 52:245–253.

Shehata WE, Brownell WE, Dieler R (1991) Effects of salicylate on shape, electromotility and membrane characteristics of isolated outer hair cells from guinea pig cochlea. Acta Otolaryngol 11:707–718.

Siegel JH, Kim DO (1982) Efferent neural control of cochlear mechanics? Olivocochlear bundle stimulation affects cochlear biomechanical nonlinearity. Hear Res 6:171–182.

Smurzynski J, Kim DO (1992) Distortion-product and click-evoked otoacoustic emissions of normally-hearing adults. Hear Res 58:227–240.

Spektor Z, Leonard G, Kim DO, Jung MD, Smurzynski J (1991) Otoacoustic emissions in normal and hearing-impaired children and normal adults. Laryngoscope 101:965–976.

Stainback RF, Lonsbury-Martin BL, Stagner BB, Coats AC (1987) Noise damage and its relation to acoustic distortion-product generation in the rabbit. Abstracts of the Tenth Midwinter Meeting of the Association for Research in Otolaryngology, p. 68.

Staecker H, Lefebvre P, Malgrange B, Moonen G, Van de Water T (1995a) Technical Notes. Science 267:709–711.

Staecker H, Lefebvre P, Malgrange B, Moonen G, Van de Water T (1995b) The effects of neurotrophins on adult auditory neurons in vitro and in vivo. Abstracts of the Eighteenth Midwinter Meeting of the Association for Research in Otolaryngology.

Stevens JC (1988) Click-evoked oto-acoustic emissions in normal and hearing impaired adults. Br J Audiol 22:45–49.

Stevens JC, Webb HD, Smith MF, Buffin JT, Ruddy H (1987) A comparison of oto-acoustic emissions and brain stem electric response audiometry in the normal newborn and babies admitted to a special care baby unit. Clin Phys Physiol Meas 8:95–104.

Stevens JC, Webb HB, Hutchinson J, Connell J, Smith MF, Buffin JT (1991) Evaluation of click-evoked oto-acoustic emissions in the newborn. Br J Audiol 25:11–14.

Stover L, Norton SJ (1993) The effects of aging on otoacoustic emissions. J Acoust Soc Am 94:2670-2681.

Stypulkowski PH (1990) Mechanisms of salicylate ototoxicity. Hear Res 46:113-146.

Subramaniam M, Henderson D, Spongr VP (1994) The relationship among distortion-product otoacoustic emissions, evoked potential thresholds, and outer hair cells following interrupted noise exposures. Ear and Hearing 15:299-309.

Subramaniam M, Salvi RJ, Spongr VP, Henderson D, and Powers NL (1994) Changes in distortion-product otoacoustic emissions and outer hair cells following interrupted noise exposures. Hear Res 74:204-216.

Sullivan MJ, Rarey KE, Conolly RB (1989) Ototoxicity of toluene in rats. Neurotoxicol Teratol 10:525-520.

Sutton LA, Lonsbury-Martin BL, Martin GK, Whitehead ML (1994) Sensitivity of distortion-product otoacoustic emissions in humans to tonal over-exposure: Time course of recovery and effects of lowering L_2. Hear Res 75:161-174.

Taudyl M, Syka J, Popelar J, Ulehlova L (1989) Comparison of carboplatin and cisplatin ototoxicity in guinea pig. Proc. 26th Workshop on Inner Ear Biology, P66.

Trine MB, Hirsch JE, Margolis RH (1993) The effect of middle ear pressure on transient evoked otoacoustic emissions. Ear Hear 14:401-407.

Ueda H, Hattori T, Sawaki M, Niwa H, Yanagita N (1992) The effect of furosemide on evoked otoacoustic emissions in guinea pigs. Hear Res 62:199-205.

West BA, Brummett RE, Himes DL (1973) Interaction of kanamycin and ethacrynic acid. Arch Otolaryngol 98:32-37.

White KR, Behrens TR (1993) The Rhode Island Hearing Assessment Project: Implications for Universal Newborn Hearing Screening. Sem Hear 14.

Whitehead ML (1989) Some properties of otoacoustic emissions in vertebrate ears, and their relationship to other vertebrate hearing mechanisms. [Doctoral Dissertation] Staffordshire (United Kingdom): University of Keele.

Whitehead ML, Lonsbury-Martin BL, Martin GK (1990) Actively and passively generated acoustic distortion at $2f_1-f_2$ in rabbits. In: Dallos P, Geisler CD, Matthews JW, Ruggero MA, Steele CR (eds) Mechanics and Biophysics of Hearing. New York: Springer Verlag, pp. 243-250.

Whitehead ML, Martin GK, Lonsbury-Martin BL (1991) The effects of the crossed acoustic reflex on distortion-product otoacoustic emissions in rabbits. Hear Res 51:55-72.

Whitehead ML, Lonsbury-Martin BL, Martin GK (1992a) Evidence for two discrete sources of $2f_1-f_2$ distortion-product otoacoustic emission in rabbit: I. Differential dependence on stimulus parameters. J Acoust Soc Am 91:1587-1607.

Whitehead ML, Lonsbury-Martin BL, Martin GK (1992b) Evidence for two discrete sources of $2f_1-f_2$ distortion-product otoacoustic emission in rabbit: II. Differential physiological vulnerability. J Acoust Soc Am 92:2662-2682.

Whitehead ML, Kamal N, Lonsbury-Martin BL, Martin GK (1993a) Spontaneous otoacoustic emissions in different racial groups. Scand Audiol 22:3-10.

Whitehead ML, McCoy MJ, Lonsbury-Martin BL, Martin GK (1993b) Click-evoked and distortion-product otoacoustic emissions in adults: Detection of high-frequency sensorineural hearing loss. Abstracts of the Sixteenth Midwinter Meeting of the Association for Research in Otolaryngology, p. 100.

Whitehead ML, McCoy MJ, Lonsbury-Martin BL, Martin GK (1995a) Dependence of distortion-product otoacoustic emissions on primary levels in normal and impaired ears: I. Effects of decreasing L_2 below L_1-L_2. J Acoust Soc Am

97:2346-2358.

Whitehead ML, Stagner BB, Lonsbury-Martin BL, and Martin GK (1995b) Dependence of distortion-product otoacoustic emissions on primary levels in normal and impaired ears: II. Asymmetry in L_1,L_2 space. J Acoust Soc Am 97:2359-2377.

Widick MP, Telischi FF, Lonsbury-Martin BL, Stagner BB (1994) Early effects of cerebellopontine angle compression on rabbit distortion-product otoacoustic emissions: A model for monitoring cochlear function during acoustic neuroma surgery. Otolaryngol Head and Neck Surg 111:407-416.

Wiederhold ML (1990) Effects of tympanic membrane modification on distortion product otoacoustic emissions in the cat ear canal. In: Dallos P, Geisler CD, Matthews JW, Ruggero MA, Steele CR (eds) Mechanics and Biophysics of Hearing. New York: Springer Verlag, pp. 251-258.

Wiederhold ML, Mahoney JW, Kellogg DL (1986) Acoustic overstimulation reduces $2f_1-f_2$ cochlear emissions at all levels in the cat. In: Allen JB, Hall JL, Hubbard A, Neely ST, Tubis A (eds) Peripheral Auditory Mechanisms. New York: Springer Verlag, pp. 322-329.

Wier CC, Pasanen EG, McFadden D (1988) Partial dissociation of spontaneous otoacoustic emissions and distortion products during aspirin use in humans. J Acoust Soc Am 84:230-237.

Wilson JP (1980a) Evidence for a cochlear origin for acoustic re-emissions, threshold fine-structure and tonal tinnitus. Hear Res 2:233-252.

Wilson JP (1980b) Model for cochlear echoes and tinnitus based on an observed electrical correlate. Hear Res 2:527-532.

Wilson JP (1980c) Subthreshold mechanical activity within the cochlea. J Physiol 298:32P-33P.

Wilson JP, Evans EF (1983) Effects of furosemide, flaxedil, noise and tone over-stimulation on the evoked otoacoustic emission in cat. Proc Congr Intrntl Union Physiol Science 15:100.

Wilson JP, Sutton GJ (1981) Acoustic correlates of tonal tinnitus. In: Evered D, Lawrenson G (eds) Tinnitus. London: Pitman, pp. 82-107.

Wilson JP, Sutton GJ (1983) A family with high-tonal objective tinnitus – An update. In: Klinke R, Hartmann R (eds) Hearing-Physiological Bases and Psychophysics. Berlin: Springer Verlag, pp. 97-103.

Wit HP, Kahmann HF (1982) Frequency analysis of stimulated cochlear acoustic emissions in monkey ears. Hear Res 8:1-11.

Wit HP, Ritsma RJ (1980) Evoked acoustical responses from the human ear: Some experimental results. Hear Res 2:253-261.

Yamamoto E, Takagi A, Hirono Y, Yagi N (1987) A case of "spontaneous otoacoustic emission." Arch Otolaryngol 113:1316-1318.

Zorowka PG, Schmitt HJ, Gutjahr P (1993) Evoked otoacoustic emissions and pure tone threshold audiometry in patients receiving cisplatinum therapy. International Journal of Pediatric Otorhinolaryngology 25:73-80.

Zurek PM (1981) Spontaneous narrowband acoustic signals emitted by human ears. J Acoust Soc Am 62:514-523.

Zurek PM, Clark WW (1981) Narrow-band acoustic signals emitted by chinchilla ears after noise exposure. J Acoust Soc Am 70:446-450.

Zurek PM, Clark WW, Kim DO (1982) The behavior of acoustic distortion products in the ear canals of chinchillas with normal or damaged ears. J Acoust Soc Am 72:774-780.

Zwicker E (1983a) On peripheral processing in normal hearing. In: Klinke R, Hartmann R (eds) Hearing—Physiological Bases and Psychophysics. Berlin: Springer Verlag, pp. 104–110.

Zwicker E (1983b) Delayed evoked oto-acoustic emissions and their suppression by Gaussian-shaped pressure impulses. Hear Res 11:359–371.

Zwicker E, Manley G (1981) Acoustical responses and suppression-period patterns in guinea pigs. Hear Res 4:43–52.

Zwicker E, Schloth E (1984) Interrelation of different otoacoustic emissions. J Acoust Soc Am 75:1148–1154.

8
Tinnitus: Psychophysical Observations in Humans and an Animal Model

MERRILYNN J. PENNER AND PAWEL J. JASTREBOFF

1. Introduction

This chapter focuses on measurement and models of a clinical symptom called tinnitus. Tinnitus is defined here as the perception of sound originating in the head. The justification motivating this definition is presented later.

Tinnitus is often a concomitant of hearing loss but may occur without it. Fowler (1944) stated that 86% of 2000 consecutive patients with hearing loss reported experiencing tinnitus, while Heller and Bergman (1953) found that 73% of their patients did. Not only does tinnitus often accompany hearing loss, but it is often the only reason the patient consults the physician. Indeed, Reed (1960) surveyed a sample of 200 patients and found that 53% of them were actually unaware that they had a hearing loss, but were complaining, instead, about their tinnitus.

According to a National Health Survey conducted in the United States in 1968, 6.4% of the American population have troublesome tinnitus. This number is close to the estimate in Phase III of the National Study of Hearing in England (Smith and Coles 1987): 7.2% of the British population suffer from tinnitus (i.e., have consulted a doctor with a primary complaint about "head noise"). Tinnitus causes distress, is reported to provoke thoughts of suicide (Tyler and Baker 1983), and is a widespread medical problem that at present is treatable in a variety of ways but is not curable.

Suggested treatments for tinnitus include psychological counseling (Scott et al. 1985), relaxation therapy (Jakes et al. 1985), biofeedback (Carmen and Svihovec 1984), hypnotherapy (Marlowe 1973), electrical stimulation (Chouard, Meyer, and Maridat 1981; Dobie, Hoberg, and Rees 1986), lidocaine iontophoresis (Brusis and Loennecken 1985), masking (Vernon 1977; Hazell, Meerton, and Conway 1989), and various drugs (e.g., Emmett and Shea 1984). As the number and nature of the treatments suggest, the efficacy of the treatment often cannot be anticipated.

One reason for the unpredictability of a treatment is that, despite extensive scrutiny, the perceptual characteristics of tinnitus are baffling and its physiological basis is not understood. Even a precise and valid definition of the phenomenon called tinnitus is elusive. Because the physiological basis of tinnitus is not understood, the efficacy of palliatives for tinnitus cannot be predicted.

The first part of this chapter focuses on data from human subjects. After defining tinnitus, we (1) examine psychophysical data for subjects in whom an acoustical correlate of tinnitus has not been found, (2) review psychophysical data for subjects in whom an acoustical correlate of tinnitus has been found, and (3) consider the implications of the psychophysical data on the physiological basis of tinnitus. The second part of this chapter focuses on data from animals. In that part, we (1) present data on production and measurement of salicylate- or quinine-induced tinnitus in animals and (2) discuss the implications of animal models of tinnitus for understanding the physiological basis of salicylate- or quinine-induced tinnitus.

2. Tinnitus in Humans

2.1 Definitions of Tinnitus

Early definitions of tinnitus classified it as belonging to mutually exclusive categories: true/pseudo (Jones and Knudsen 1928), vibratory/nonvibratory (Fowler 1939), objective/subjective, or extrinsic/intrinsic (Atkinson 1947). In each case, one category involved a tangible signal with a specific physical source. The second category had no identifiable physical or acoustic source (i.e., was pseudo, nonvibratory, subjective or intrinsic).

The dichotomies imposed by these definitions resemble the partition in the contemporary American National Standards Institute (ANSI) definition: Tinnitus is the sensation of sound without external stimulation (ANSI 1969). The ANSI definition hinges on the measurement of external stimulation. If external stimulation is measured, then the sensation is not called tinnitus (unless it originates in the circulatory system in which case a note in the ANSI document includes the resulting sensation as tinnitus). The ANSI definition and its predecessors were widely accepted until a remarkable discovery by Kemp (1978) emphasized a logical difficulty.

Kemp (1978) demonstrated that the ear, which formerly was regarded as a passive receiver, can spontaneously generate sound. These sounds, called spontaneous otoacoustic emissions (SOAEs), can be measured when a sensitive miniature microphone is inserted into the ear canal (see Whitehead, Lonsbury-Martin and McCabe, Chapter 7). Following Kemp's discovery of SOAEs, some of the sensations formerly classified as tinnitus were associated with weak acoustic signals (Penner 1988; Penner 1989a,b) and could no longer be categorized as tinnitus. This conundrum emphasized a shortcoming in the ANSI definition and its predecessors: there are a number of factors involved in detecting external stimulation so the failure to detect it cannot sensibly be taken as evidence that no external stimulation exists.

This reasoning led to the definition of tinnitus as the conscious experience of sound originating in the head (McFadden 1982). This definition of tinnitus, used in this chapter, encompasses perceptions with no apparent basis as well as sounds with a definite source (SOAEs). It is the psychophysical nature of tinnitus that, as yet, has no known physical basis that is now considered.

2.2 Tinnitus Without a Measurable Physical Basis

As a preface to the following sections, it may help to note that the psychophysics of tinnitus that is associated with SOAEs is reminiscent of data that would be obtained if pure tones constituted tinnitus. Conversely, the psychophysics of tinnitus that is not yet associated with physical correlates does not often resemble data obtained if tinnitus were a single pure tone.

2.2.1 Assessment of Tinnitus Pitch

Specification of the tinnitus is of unmistakable scientific interest because the sensation caused by the tinnitus might be linked to the pathology producing it. Although tinnitus may not be characterized as tonal, attempts to define it have focused on its frequency and level while ignoring phase.

In a typical tinnitus pitch matching experiment, a method of adjustment is employed. The frequency of a gated monaural or binaural pure tone is adjusted until the tone is said to match the predominant pitch of the tinnitus. Early reports indicated that matches to the "predominant" pitch of the tinnitus were reliable (Reed 1960; Graham and Newby 1962; Goodwin and Johnson 1980a,b). In these studies, however, some of the matching data were discarded (see Penner 1983 for a critique).

In later reports, the standard deviation, SD, of the matches made to the predominant pitch of the tinnitus exceeded the SD of the matches to a pure tone at the presumed tinnitus frequency (Penner 1983; Burns 1984): the range of all matches to the tinnitus may be as large as one octave (Penner 1983; Tyler and Conrad-Armes 1983; Norton, Schmidt, and Stover 1990). An example of the variability obtained from one subject within one session is presented in Figure 8.1. The data are from a study of pitch matches made to tinnitus that was described as tonal (Penner 1983).

For the data in Figure 8.1, the matches deviate by more than one octave, yet the same subject had no particular difficulty in matching external tones to external tones in the tinnitus region (Penner 1983) during the same session. Among other possibilities, the evanescent nature of the pitch

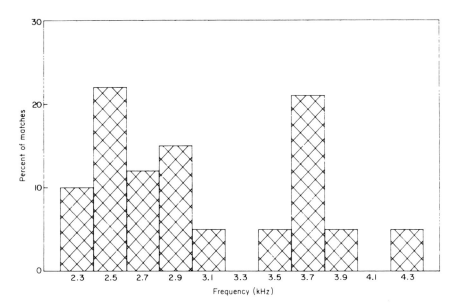

FIGURE 8.1 Distribution of matching responses for one subject. The ordinate displays the percent of 80 matches made to the tinnitus that lay in the 200 Hz region whose midpoint is displayed on the abscissa. (from Penner 1983, with permission from ASHA)

matches could arise because tinnitus is labile, or because it is a multi-component broadband signal and the subject is uncertain about which component of the signal to match, or both. In either case, tinnitus should probably be represented by a range of frequencies rather than by a single frequency. This inference motivates one test, which is discussed later, for exploring the relation of tinnitus and SOAEs.

In order to present the tone in the same ear as the tinnitus, the ear with the tinnitus must be designated. Typically, the designation is based on the subject's statement regarding the source of the tinnitus. However, for normal hearing subjects, binaural sounds that differ in level may be lateralized to one ear. It is therefore possible that supposedly "monaural" tinnitus could be binaural but louder in one ear than in the other.

To test reports of monaural tinnitus, Penner (1983) asked two subjects to place a cross on a line so that the cross represented the intracranial location of the tinnitus. The end points were simply labeled "left ear" and "right ear". Because the tinnitus was said to be monaural, the subjects placed the crosses near the end of the line. When pulsed noise was presented to the ear reported to have the tinnitus, the position of the cross on the line moved toward the contralateral ear. Conversely, when pulsed noise was presented in the ear reported to be contralateral to the tinnitus, the position of the cross on the line moved toward the ipsilateral ear. These data indicate that presumed monaural tinnitus could be binaural. However, the validity of the test for laterality rests on the similarity of the behavior of external tones and tinnitus, and this may be a questionable assumption.

The best strategy for measuring tinnitus pitch and the characteristics of an optimal measurement technique are not yet clear. Typically, in tests and measurements of psychological dimensions, optimal measures are determined to be both reliable and valid. For tinnitus, validity is difficult to assess. If tinnitus fluctuates, then measures of it should also vary, so a valid measure of tinnitus might not appear to be a reliable one.

Penner and Bilger (1992) have recently reported that a two-interval forced-choice adaptive task (2IFC, as in Jesteadt 1980) gave the same within-session SD for matches to the tinnitus as for matches to an external tone, although between-session variability in the matches to tinnitus was comparable to that in previous reports. Their 2IFC procedure thus produced reliable short-term measures of tinnitus pitch and, as discussed in the next section, loudness.

In a 2IFC task, subjects hear sounds in two consecutive temporal intervals and choose between them (Green and Swets 1966). In adapting the paradigm to reveal tinnitus pitch, the tinnitus served as the signal in one of the two intervals and an external sound was presented in the other. The subject was instructed to select, for example, the interval in which the stimulus with the higher pitch occurred. The external sound depended on the subject's responses and changed according to the rules of Levitt (1971) until it was discriminable from the standard, the tinnitus, on 70.7% of the trials. The Penner and Bilger (1992) data suggest either that tinnitus may be stable within a brief time span but fluctuant in the long run or that an unknown artifact produced stable data. Although the forced-choice procedure produced stable data, the general question of whether any single tone can provide a suitable imitation of tinnitus remains open.

2.2.2 Assessment of Tinnitus Loudness

Loudness matching is similar to pitch matching: a gated tone is adjusted until it matches the tinnitus in loudness. The frequency of the matching tone must be chosen and interactions of the tone and the tinnitus are possible.

Despite obvious difficulties, in the majority of studies of tinnitus loudness, tinnitus is matched in loudness to a low-level tone when the matching tone is in the region of the tinnitus frequency. Reed (1960) found that 69% of his subjects matched their tinnitus with a tone of 10 dB sensation level (SL) or less if the frequency of the tone was also matched to the tinnitus pitch. The fact that tinnitus is often matched to an external tone, near it in pitch, with a low sensation level and is nonetheless reported to be "loud" and "annoying" had been viewed as a "paradox" (Fowler 1942, 1943; Reed 1960; Vernon 1976). Why should a low-level sound be loud?

Following Reed, Goodwin and Johnson (1980a), Tyler and Conrad-

Armes (1983) and Penner (1984) have pointed out that in most models of loudness for normal hearing subjects, loudness depends on the threshold of the tone in question. In general, rapid increases in loudness accompany elevated thresholds (Hellman and Zwislocki 1961; Lochner and Burger 1961). Within the context of models of the growth of sensation for normal hearing subjects, recruitment (a supernormal growth of loudness) is a consequence of elevated threshold (Larkin and Penner 1989; Humes and Jesteadt 1991). The existence of recruitment could explain the paradoxical loudness of tinnitus. If recruitment is present, tinnitus can be loud even if the sensation level of a tone with an elevated threshold said to match the tinnitus is low.

To test the idea that recruitment explains the paradoxical loudness of tinnitus, Tyler and Conrad-Armes (1983) and Penner (1984) applied mathematical models for normal-hearing subjects to their data from subjects with tinnitus. Empirically, if the frequency of the matching tone is in a region of normal hearing, the sensation level of the matching tone can be quite large. In particular, Penner (1984) presented equal loudness contours in which tinnitus served as the standard signal. These data are reproduced in Figure 8.2. Note that the greater the absolute threshold of the external comparison tone, the lower the sensation level of the comparison tone said to match the tinnitus.

Penner (1986) fitted the data in Figure 8.2 with the loudness function obtained from normal-hearing observers. For normal-hearing observers, low sensation levels can represent loud tones. The exact relation between loudness and threshold is disputed, but one popular form relates the loudness, L, in sones, of a pure tone at 1 kHz with pressure P to the threshold pressure, p_o, by the equation:

$$L = k(P - p_o)^{0.6} \qquad (1)$$

where k is a constant (Zwicker 1958; Hellman and Zwislocki 1961; Scharf and Stevens 1961). For the data in Figure 8.2, the greater the absolute threshold of the comparison tone, the less the sensation level of the comparison needed to match the tinnitus. Thus Equation 1 is consistent with the general form of the data for matches to the loudness of tinnitus.

In contrast to the fits obtained by Penner (1986), Tyler and Conrad-Armes (1983) have reported that Equation 1 did not describe their data accurately. One possible reason for the difference in the two studies is that Penner's (1986) subjects all had the same etiology whereas Tyler and Conrad-Armes' (1983) subjects did not. Another reason involves Equation 1 itself: the mathematical relation of loudness to threshold is still debated (Larkin and Penner 1989; Humes and Jesteadt 1991).

In order to determine whether the loudness of tinnitus is "paradoxical," it is preferable to bypass the need to quantify the relation between loudness and threshold. To avoid quantitative predictions based on the loudness function, Penner (1986) employed a magnitude estimation task. Ten

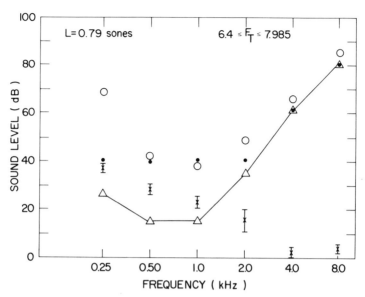

FIGURE 8.2 Isomasking contours (sinusoidal masker level as a function of masker frequency) and audiogram (sinusoidal signal level as a function of frequency). Isomasking contours for pure tones at the labeled frequencies are marked by inverted, filled triangles (forced-choice procedure) and filled circles (adjustment procedure). Tinnitus isomasking contours are marked with open circles. Threshold as a function of frequency is also graphed (open triangles). The average frequency of the tone said to match the predominant pitch of the tinnitus was 3.580 kHz. (from Penner 1987, with permission from ASHA)

subjects with sensorineural hearing loss and tinnitus matched external tones to the tinnitus pitch. These matches were followed by (1) magnitude estimates to measure loudness functions of tones at 1 kHz and at the presumed tinnitus frequency, (2) magnitude estimates of the tinnitus itself, and (3) loudness matches of external tones to the tinnitus.

As seen in Figure 8.3, Penner (1986) found that the slope of the loudness function at 1 kHz was substantially smaller than the slope at the presumed tinnitus frequency. Most importantly, the magnitude estimates of the tinnitus coupled with intensity matches to the tinnitus provided coordinates that typically (for nine of ten subjects) lay near the loudness function of the external tone used in the intensity match, as is also seen in Figure 8.3. This last finding favors the conclusion that the loudness of tinnitus is often simply related to the loudness of external tones at the tinnitus frequency.

2.2.3 Masking of Tinnitus

Having matched the tinnitus with pure tones raises the question of whether tinnitus behaves as a pure tone. One way to approach that issue is to explore the masking of tinnitus. In a classic paper, Wegel and Lane (1924) showed that a pure tone was most effectively masked by tones near it in frequency.

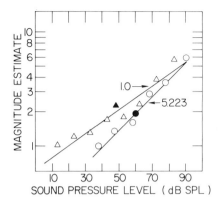

FIGURE 8.3 Magnitude estimates of tones at 1 kHz (open triangles) and at the average frequency matched to the tinnitus (open circles) for Subject 1. The filled triangle represents the mean magnitude estimate and the mean intensity of an external tone at 1 kHz said to match the tinnitus loudness. Similarly, the filled circle represents the same coordinates when the external tone is at the presumed tinnitus frequency. The number next to the arrow is the frequency (in kHz) of the external tone for each loudness function. (from Penner 1986, with permission from ASHA)

In fact, the frequency and level of an unknown masked tone may be deduced from the pattern revealed by its frequency-specific isomasking contour.

For the archetypal subject with tinnitus, however, masking tinnitus does not generally yield the frequency-specific masking contours associated with masking low-level narrow-band signals (Feldmann 1971; Burns 1984; Tyler and Conrad-Armes 1984; Penner 1987). The isomasking pattern (masker level as a function of masker frequency) for tinnitus is occasionally at a constant sensation level (Feldmann 1971). More customarily, it is at a constant dB SPL, even if the isomasking contour for an external tone in the tinnitus region is not constant (Burns 1984; Penner 1987). Typical isomasking contours for tinnitus and external tones are presented in Figure 8.4. The data in Figure 8.4 do not indicate that tinnitus occupies a narrow spectral location.

The lack of frequency-specific masking has led investigators to posit that tinnitus is not likely to be masked in the cochlea. Feldmann (1971) also offered another possibility: the spontaneous activity that may be responsible for tinnitus is not identical to the activity produced by an external stimulus evoking the same sensation.

Additional evidence concerning the locus of the masking of tinnitus involves contralateral masking of tinnitus. For reportedly monaural tinnitus, ipsilateral masker levels sometimes exceed contralateral masker levels (Feldmann 1971), even for masker levels at which no cross-conduction of sound is likely. The difference in masker levels apparently supports the

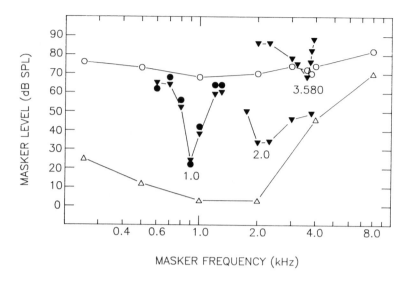

FIGURE 8.4 Threshold and equal loudness contours. The ordinate gives the level of a pure tone, at a frequency specified on the abscissa, that is judged as loud as the tinnitus or is the threshold at that frequency (i.e., the audiogram). The x's represent the sensation level and the open circles the sound pressure level (SPL) of the tone matching the tinnitus loudness. The standard deviation of the measure is also displayed if it exceeded the height of the symbol needed to represent the data point. The open triangles represent the subject's threshold. The dots represent the predicted contour (in dB SPL), based on Equation 1 and the subject's threshold, for the loudness (L), in sones, given in the graph. Matches to the predominant pitch of the tinnitus (F_t) ranged from 6.4 to 7.985 kHz. (from Penner 1984, with permission from ASHA)

inference that tinnitus is masked centrally or that efferent activation inhibits tinnitus (Feldmann 1971).

Tinnitus is sometimes reported to vanish after masking (Feldmann 1971). The noise is said to inhibit the tinnitus. The measured loudness of tinnitus may change substantially when evaluated in an adjustment task, however, and there are data that raise the possibility that some loudness judgments reflect variability in the tinnitus rather than the residual effects of noise on tinnitus (Penner 1988).

An additional masking paradigm of Penner, Brauth, and Hood (1980) showed that for subjects presumed to have sensorineural hearing loss tinnitus became audible in noise that previously masked it, although the noise intensity that masked an external tone remained constant (Penner, Brauth, and Hood 1980). One explanation of this difference is that throughout the presentation of the masking noise and the external tone the firing rates of the primary auditory neurons declined with time (i.e., adapted), but because both the masker and the signal were affected almost equally, the noise intensity required to mask the tone stayed about the same.

That the noise did not remain constant when masking tinnitus could mean that tinnitus originated at a site more central than the adapting primary fibers (i.e., retrocochlearly), or that the neural activity underlying tinnitus is in a chronically adapted state, or adapts less than that underlying noise.

Taken together, the common theme in interpreting the masking data is that tinnitus is masked centrally and therefore does not share the masking characteristics of a pure tone. However, the isomasking patterns do resemble tinnitus isoloudness contours and the possibility that isomasking contours are actually isoloudness contours is being explored. Nonetheless, the generally accepted view is that tinnitus is not masked as is a pure tone.

2.2.4 Two-Tone Suppression in the Tinnitus Region

Another masking paradigm that might shed light on the mechanisms of tinnitus is that of Houtgast (1972). Houtgast (1972) has presented experimental evidence for lateral suppression: the response of the auditory system to a tone of given frequency can be reduced (suppressed) by a tone with a different frequency. In an extension of Houtgast's (1972) paradigm, Shannon (1976) employed a forward-masking task (i.e., the masker preceded the signal in time) in which the masker was either one tone or two. In one of his conditions, the threshold of a signal that followed the masker was tracked as a function of the frequency of the second masker. Shannon's (1976) data showed that two maskers could do less masking than one if the frequency of the second masker was near the first and suppressed it. These data are consistent with the physiological data of Sachs and Kiang (1968), which show that the neural response rate to two tones can be less than to one.

Penner (1980) explored two-tone suppression in tinnitus patients. Her results indicated that tinnitus sufferers did not exhibit two-tone suppression in the tinnitus region. She argued, as Ruggero, Rich, and Freyman (1983) did subsequently to explain SOAEs, that the lack of suppression in regions of hearing loss might cause an imbalance in the suppression and tuning characteristics. Such an imbalance could produce excess spontaneous activity in adjacent regions and the excess activity could be perceived as tinnitus.

2.2.5 Comments on Tinnitus Without a Measured Physical Source

The psychophysics of tinnitus for which no acoustic source has been observed differs from the psychophysics of external tones: pitch matches are generally variable, isomasking contours are often flat, and tinnitus can become audible in noise that previously masked it. The psychophysics of tinnitus caused by externally measured sounds, however, is frequently similar to the psychophysics of external tones. The data relating to this form of tinnitus are now presented.

2.3 Somatosounds

Occasionally, the source of the tinnitus can be heard and recorded by the examiner. Such a sound, called a bruit, is often presumed to be caused by vibrations somewhere in the head or neck that spread to the ear canal. Causes of bruits include palatal myoclonus, hyperpatency of the eustachian tube, temporomandibular joint disease, emissions, and vascular malformations. Several studies of bruits of vascular origin have been conducted (Hentzer 1968; Harris, Brismar, and Conquist 1979). In one recent study by Champlin, Muller, and Mitchell (1990), the patient had tinnitus that fluctuated in a pulsing manner so that it appeared to be of vascular origin. A jugular vein ligation was undertaken and a sensitive miniature microphone inserted into the ear canal to monitor the tinnitus before, during, and after surgery. By using the microphone to monitor the bruit, the effectiveness of the surgery could be immediately evaluated.

The miniature microphone can also be used to detect another source of tinnitus, tinnitus caused by SOAEs. Sounds emitted by the ear may occasionally be heard by an examiner (Glanville, Coles and Sullivan 1971; Wilson and Sutton 1983), although their owner may not hear them. However, most emitted sounds are not perceived by others and are not bruits.

2.4 Tinnitus and SOAEs

2.4.1 Experimental Tests Relating SOAEs and Tinnitus

SOAEs are thought to be of cochlear origin (Brownell 1990), and so some experimental criteria for attributing tinnitus to SOAEs are based on the similarity of the psychophysics of tinnitus and the psychophysics of tones of cochlear origin. In an example of this approach, Wilson (1980) linked SOAEs to tinnitus for one subject whenever the tinnitus percept changed as the ear canal pressure altered the SOAE frequency or whenever an external tone beat with the SOAE.

Wilson (1980) was not the first to note that tinnitus could beat with external sounds. Beats with tinnitus were reported as early as 1931 (Wegel 1931): Wegel's troublesome tinnitus produced "mushy" beats with external sounds. Wegel's (1931) description prompted many subsequent unsuccessful attempts to produce beats with tinnitus. The general failure of external tones to beat with tinnitus might arise because SOAEs underlying tinnitus vanish (Burns and Keefe 1991), or because SOAE caused tinnitus is rare (Penner 1990).

Several additional tests for linking SOAEs and tinnitus have also been reported. Many of these tests are based on perceptual resemblance of the tinnitus and SOAEs. For example, similarity of the tinnitus pitch and a recording of the frequencies of the SOAEs is taken as evidence of their association (Hammel 1981, for five subjects; data of Wilson and Sutton, referred to in Wilson 1986, for one subject). The simultaneous fluctuation of the SOAE spectra and the tinnitus percept may also provide evidence of their relationship (Wilson 1980, for four subjects; Burns and Keefe 1991, for one subject). Finally, for 8 of about 303 subjects, a close correspondence between the SOAE frequency and the frequency of the tone matching the predominant pitch of the tinnitus was used as the criteria for linking SOAEs and tinnitus (Wilson and Sutton 1981; Zurek 1981; Tyler and Conrad-Armes 1982; Hazell 1984; Probst et al. 1986; Rebillard, Abbou, and Lenoir 1987; Zwicker 1987).

The latter test is easily undertaken and, for this reason, is often employed. In the investigations listed above that used this test, only eight subjects with SOAE caused tinnitus were identified (Wilson and Sutton 1981; Hazell 1984; Probst et al. 1986; Rebillard, Abbou, and Lenoir 1987). For four of these, tinnitus was not distressing, and no mention was made of whether tinnitus caused discomfort for one of the remaining subjects. Before accepting a low correspondence between SOAEs and bothersome tinnitus (about 1%), Penner and Burns (1987) suggested that some additional aspects of the empirical connection of the two sounds needed to be considered.

One facet of the problem is that SOAEs are generally stable, objective stimuli (Martin, Probst, and Lonsbury-Martin 1990; see also Whitehead et al., Chapter 7), while the spectra of tinnitus cannot generally be reliably established (Penner 1983; Tyler and Conrad-Armes 1983; Burns 1984). As was seen in Figure 8.1, a range of frequencies corresponds to tinnitus. If so, then SOAEs in a broad region of frequencies bracketing tinnitus could be viewed as corresponding to it.

The use of a broadband measure of tinnitus was proposed by Penner and Burns (1987). In the context of the SOAE-tinnitus link, it was a new approach because previous reports related only the frequency of the "predominant" pitch of tinnitus to the frequency of the SOAE.

Penner and Burns (1987) selected ten subjects known to have both tinnitus and SOAEs. The broad frequency region bracketing tinnitus was taken as a measure of its spectral location. For four of the subjects, this bracketing region contained SOAEs. However, the frequency of a tone matching the predominant pitch of tinnitus did not correspond to any SOAE frequency. The broadband and punctate measures of tinnitus can therefore lead to different implications in assessing the SOAE-tinnitus link. Given the difficulty in specifying tinnitus, it might not be possible to demonstrate a strong correlation between the frequency of the tinnitus and the frequency of the SOAE. Furthermore, even a strong correspondence between the frequency of the SOAE and the region bracketing the tinnitus pitch would not prove that the SOAE is the physical basis of tinnitus. The association or dissociation might be better addressed by exploring whether any SOAE in the tinnitus region or elsewhere causes tinnitus.

Several of the tests mentioned earlier link SOAEs and tinnitus without basing the association on correlations of the frequency matched to the tinnitus pitch and the SOAE frequency. For example, simultaneous fluctuation of the SOAE and the tinnitus seems to link the two sounds (Wilson 1980; Burns and Keefe 1991). However, SOAEs only rarely fluctuate (Frick and Matthies 1988) and might nonetheless be a source of tinnitus. If the tinnitus were to beat with external tones (Wilson 1980), then it is likely due to SOAEs. However, if the SOAE disappeared (as in Burns and Keefe 1991), then beats would not be consistently observed. If pressure changes altered the SOAEs and the tinnitus (Wilson 1980), then either the tinnitus itself fluctuated or it was caused by SOAEs (Wilson 1980). Finally, the judged similarity of the tinnitus to a recording of the SOAEs is a subjective decision and, therefore, might not provide a valid basis for inferring that SOAEs caused tinnitus. Difficulties with these previous tests prompted Penner and Burns (1987) to propose an additional procedure for exploring the link between SOAEs and tinnitus: the masking/suppression demonstration.

Masking occurs when a previously audible signal is made inaudible by the introduction of a masker that conceals the spectrum of the signal. In the present context, tinnitus could be masked without affecting the SOAE. Suppression occurs when a previously present sound, an SOAE, vanishes due to the introduction of a suppressor that is near the SOAE but does not conceal its spectrum. In the present context, an SOAE could be suppressed without affecting tinnitus.

Much data shows that an SOAE may be suppressed by a low-level tone near it in frequency (Wilson 1980; Wilson and Sutton 1981; Zurek 1981) and, as seen in Figure 8.4, tinnitus is often masked by one intense tone (Feldmann 1971; Burns 1984; Penner 1987). If the SOAEs can be suppressed without affecting tinnitus and if tinnitus can be masked without affecting SOAEs, then tinnitus and SOAEs are likely to be independent phenomena.

Figure 8.5 shows the masking/suppression demonstration data of one subject from Penner and Burns (1987). For this subject, the frequency of the tone matching the predominant pitch of the tinnitus does not coincide with either of the two SOAE frequencies. However, the frequency region bracketing tinnitus contains one of the SOAE frequencies, and so it could be argued that the SOAE might be the cause of the tinnitus.

In the top panel of Figure 8.5, the subject's SOAEs are seen as spikes above the noise floor. In the middle panel, the tinnitus is masked and yet the SOAEs are still present. In the bottom panel, the SOAEs are suppressed and the tinnitus was reported to be unchanged. Thus, the SOAEs and the tinnitus are independent phenomena even though the region bracketing the tinnitus pitch contains an SOAE frequency.

Using the masking/suppression demonstration, it is possible to determine whether SOAEs and tinnitus are independent for any particular subject. If

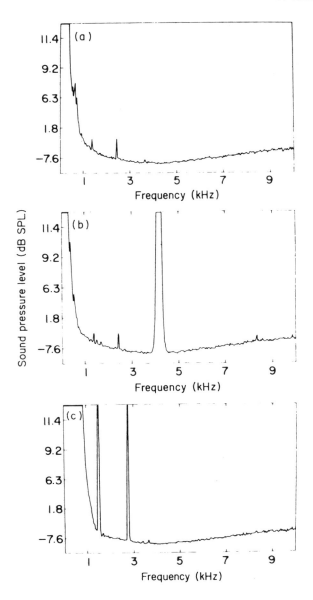

FIGURE 8.5 The SOAEs in silence (a) and with a 4.3 kHz pure tone tinnitus masker at 78 dB SPL (b). In the presence of two tones suppressing each of the subject's SOAEs (c), the tinnitus remained audible and was reported to be unchanged in pitch. (from Penner and Burns 1987 with permission from ASHA)

they are not independent, then some additional psychophysical tests may provide informative results depending on the precise pattern of the audible SOAE(s) and the subject's perceptual capabilities. For example, tinnitus caused by a single SOAE might have the tone-like properties of the SOAE.

If so, the pitch-matches to such a tinnitus might be reliable, tinnitus might beat with external sounds (as Wegel, 1931, reported to be the case for his tinnitus), and the tinnitus isomasking contour (masker level as a function of masker frequency) might be frequency-specific. For tinnitus caused by multiple SOAEs, all but one of the SOAEs could be suppressed and pitch matches made to the remaining component of the tinnitus and the isomasking contour of the remaining component explored.

For the ten subjects in the Penner and Burns (1987) study, tinnitus and SOAEs were judged to be independent. However, the masking/suppression demonstration and some supplementary psychophysical tests were used to infer that SOAEs caused tinnitus in some of the studies that are discussed below.

2.4.2 Psychophysics of SOAE Caused Tinnitus

There are seven studies in which SOAEs were thought to cause disruptive tinnitus (Hammel 1981; Penner 1988; Penner 1989a,b; Plinkert, Gitter and Zenner 1990; Burns and Keefe 1991; Penner and Coles 1992). In some of these studies, tinnitus and SOAEs were not conclusively linked. In the ensuing discussion, data from some of the Penner articles (1988; 1989a,b) are presented so that the reader may see the sort of detailed psychophysical profile that implicates SOAEs as a source of annoying tinnitus.

Penner (1988) studied a twenty-six-year-old woman with binaural tinnitus reported to be louder in her right ear than in her left. The subject described the tinnitus percept as a "changing tone," as "more than one tone," and as a "hum." In her words, her tinnitus was "irritating, miserable, horrible" and "as loud as a police siren." Her primary complaint concerned tinnitus and she had consulted two otolaryngologists about it. She had no other auditory difficulties and had hearing thresholds within 10 dB of normal limits (ANSI 1969).

The masking/suppression demonstration revealed that when the SOAEs were suppressed, the tinnitus was inaudible. Further, the only stimuli which masked her tinnitus coincided with her SOAE suppressors. Thus, the subject had SOAE caused tinnitus.

The SOAE-tinnitus link was also consistent with some aspects of her perception of her tinnitus. For example, SOAEs were found to emanate from both ears with more intense SOAEs in her right ear than in her left. The fact that the right ear had more intense SOAEs agrees with her comment that the tinnitus was louder in her right ear. Furthermore, pitch matches to the lowest frequency of the tinnitus corresponded to the lowest constant frequency of the SOAEs.

The majority of the subject's SOAEs fluctuated in level and, for this reason, might have been audible. Schloth and Zwicker (1983) noted that when the SOAEs of normal subjects are shifted by mechanical forces, they become momentarily audible. However, not all audible and annoying SOAEs are unstable (Wilson and Sutton 1983; Penner 1989a), as a second subject's data demonstrates.

In a second example, Penner (1989a) described a fifty-three-year-old woman with monaural tinnitus characterized as a "steady tone" and a "hum." The subject had four SOAEs in one ear and the tinnitus was audible unless all four SOAEs were simultaneously suppressed. Thus the subject was judged to have tinnitus by the masking/suppression demonstration.

Supplementary tests showed that when all but one SOAE was suppressed the tinnitus isomasking contour resembled the SOAE suppression tuning curve. The SOAEs of this subject did not fluctuate, but may nonetheless have caused discomfort for idiosyncratic reasons: she was an amateur musician and said that the sounds disrupted her musical appreciation.

SOAEs need not be the sole source of tinnitus but may constitute part of tinnitus. For example, Penner (1989b) presented data from a subject whose tinnitus was judged to be due, at least in part, to an SOAE. The subject was a thirty-eight-year-old woman with unilateral tinnitus described as "ringing" and "changing tones." The subject had a sensorineural loss in her left ear, the reported site of her tinnitus. In her right ear, she had normal hearing and an SOAE whose frequency corresponded to the lowest pitch of the tinnitus. Because her tinnitus had two coexisting sources, only one of which was an SOAE, two tones were required to abolish it: a masker in the ear with sensorineural hearing loss and a simultaneous contralateral SOAE suppressor.

Additional circumstantial evidence favoring two coexisting sources of tinnitus was also obtained for this subject. With a continuous suppressor in the right ear, the isomasking contour for tinnitus was quite flat. With a continuous tinnitus masker in the left ear, the isomasking contour for tinnitus was frequency specific. Thus, there appeared to be two coexisting sources of tinnitus.

Treatment of SOAE caused tinnitus may be possible because aspirin generally abolishes SOAEs (McFadden and Plattsmier 1984; Long and Tubis 1988), while in some cases it may induce or enhance preexisting tinnitus. In a recent study (Penner and Coles 1992), a subject with fluctuant SOAEs was judged to have SOAE caused tinnitus based on the masking/suppression demonstration. The tinnitus and SOAEs of this subject disappeared during aspirin intake. Although SOAEs generally disappear during aspirin intake, previous attempts to relieve SOAE caused tinnitus with aspirin had failed (Penner 1989a) because the dose required induced aspirin related tinnitus. Aspirin was, however, apparently effective in abolishing the SOAE caused tinnitus of this one subject.

Several additional studies also employ psychophysical tests to explore the SOAE-tinnitus link. For example, Plinkert, Gitter, and Zenner (1990) noted that the tinnitus of one subject was linked to SOAEs because the tinnitus was not perceived when the SOAE was suppressed. Finally, Burns (1989) and Burns and Keefe (1991) observed the time varying SOAE spectra of one subject and argued that, because the tinnitus was audible only when the SOAEs were observed, the variable SOAEs caused tinnitus.

The reports of subjects with SOAE caused tinnitus raise the question of how frequent is the association. The only estimate of the prevalence of SOAE caused tinnitus among members of a tinnitus self-help group was provided by Penner (1990). About 4.5% of these sufferers had tinnitus that was caused at least in part by SOAEs (Penner 1990).

One interesting question emerging from the data involves identification of the characteristics of subjects who are likely to have SOAE caused tinnitus. At present, the scarcity of such subjects precludes generalizations. The data do, however, provide three hints of the characteristics that future studies might explore.

First, subjects with SOAE caused tinnitus had near normal hearing in the ear with the SOAE. This finding is anticipated because SOAEs are not generally associated with hearing loss exceeding 20 dB HL at the emitted frequency (Probst et al. 1986). A corollary to the connection of SOAE caused tinnitus with normal hearing is that SOAE caused tinnitus is more likely to be found in young people who do not have a hearing loss than in the elderly who generally do. Furthermore, the SOAE caused tinnitus might disappear in a given individual if a hearing loss develops.

Second, the majority of SOAEs are in the 1 to 2 kHz region (Zurek 1981) so SOAE caused tinnitus might be expected to lie in this region, whereas tinnitus unassociated with SOAEs is generally higher in pitch (Reed 1960; Penner 1990). Furthermore, the SD of matches to the pitch of SOAE caused tinnitus is likely to be less than the SD of matches to other forms of tinnitus. The 2.72% SD obtained in Penner (1989a) is nearer to the 1% obtained from highly practiced observers matching external tones to external tones than to the 20% obtained from eight (nonSOAE) tinnitus sufferers matching external tones to tinnitus (Burns 1984). Thus, the pitch matches to components of the SOAE caused tinnitus may lie in the 1 to 3 kHz region and be reliable.

Third, combining the data from surveys of SOAEs in normal hearing people leads to the conclusion that SOAEs are nearly twice as common in women as in men (53% versus 27%; Bilger et al. 1990). Thus, the prevalence of SOAE caused tinnitus would be likely to be greater in women than in men, as was the case in Penner's (1990) survey. It follows that the typical person with SOAE caused tinnitus would be a normal hearing young woman (with thresholds within 10 dB of the 1969 ANSI standards) who matches the tinnitus pitch reliably with tones in the 1 to 3 kHz region.

In summary, available data indicate that about 4% of troublesome tinnitus is caused at least in part by SOAEs. In order to have SOAE-caused tinnitus, however, SOAEs must be measured. Thus, one immediate complication in estimating SOAE caused tinnitus involves the measurement of SOAEs.

2.4.3 Tinnitus and Internal Tones

An ingenious theory that predicted emissions (Gold and Pumphrey 1948) antedated the measurement of SOAEs by thirty-one years (Kemp 1979). Just as SOAEs were only revealed when microphones became sensitive enough to record very faint signals, it is also possible that a more sensitive spectral analysis (increasing the number of spectra averaged to reduce the noise floor) might enable measurement of additional SOAEs. The tympanic membrane must also play a role in transmitting the retrograde wave into the meatus. Indeed, with relatively minor modifications of the tympanic membrane (i.e., by adding mass or by perforating it), the threshold for emissions may be changed by as much as 50 dB (Wiederhold 1990). Measurement of SOAEs is likely to be limited by spectral techniques, the attenuation characteristics of the middle ear (Kemp 1979), and the properties of the tympanic membrane.

Accordingly, it seems possible that there are more cochlear sounds than can be measured as SOAEs. It is consequently not a surprise that there are two reports proposing the existence of internal tones of cochlear origin in the absence of SOAEs. In each report (Formby and Gjerdigen 1981; Bacon and Viemeister 1985), disruptions in pitch perception or changes in masking patterns were consistent with the existence of an internal tone, but SOAEs could not be measured. Additionally, the phenomenon of monaural diplacusis (a condition in which a pure tone does not sound "pure" but elicits sensations of beats, roughness, multiple tones, or an area of silence and possibly an aftertone) led Flottorp (1953) and Ward (1952, 1955) to hypothesize that the ear may generate internal tones called idiotones.

Could such idiotones, which do not produce measurable SOAEs, cause tinnitus? Some data from a subject of Penner (1986) are of special interest in this regard. Thus far, the results of three of this subject's psychophysical tests are consistent with idiotones causing tinnitus.

First, the subject reliably matched the tinnitus pitch with a pure tone. The SD of the matches for this subject, 1.96%, is smaller than Burns (1984) obtained for any of his ten subjects and is obviously markedly different from the SD of the matching data presented in Figure 8.1. The SD of the subject's matches is, in fact, the smallest ever reported for repeated between-session matches to the pitch of tinnitus that is not known to be caused by SOAEs.

The subject's second unusual psychophysical result involved the slope of the psychometric function (the probability of a correct detection as a function of signal level). According to the theory of signal detection (Green and Swets 1966), the slope of the psychometric function flattens as the variability of the background masking noise increases. Thus, if tinnitus served as a source of internal noise, the psychometric function could be flatter in the tinnitus region than above or below it. In a study of the slope of the psychometric functions of seven subjects with tinnitus, only for the subject suspected to have idiotones was the slope the same in the tinnitus region as above and below it.

The third unusual result is that the subject had monaural diplacusis in the neighborhood of 2.463 kHz, the average frequency matched to the pitch of his tinnitus. It was the presence of monaural diplacusis that first prompted speculation that idiotones existed. In a similar vein, monaural diplacusis in the region of tinnitus would seem to imply that idiotones might be irksome.

These three psychophysical results are consistent with the behavior of a cochlear tone. The inference is that if tinnitus behaves like an external tone then it might be generated by an internal tone of cochlear origin even if SOAEs are not present. In other words, psychophysically equivalent signals may have similar origins. This logic has historical precedent: SOAEs that behave like external tones are assumed to be generated in the cochlea (Kemp 1979; Wilson 1980). By analogy, if sounds with similar psychophysics have a common origin, then some cases of tinnitus may be caused by spontaneous internal sounds even when SOAEs cannot be measured. Thus, this last subject's results raise the question of whether tinnitus can be caused by internal tones of cochlear origin even when SOAEs cannot be measured and what sort of psychophysical tests could be used to establish idiotones as a cause of tinnitus.

One important addition to the battery of psychophysical tests linking tinnitus and internal cochlear tones might be the exploration of stimulus-frequency otoacoustic emissions (SFOAEs; Kemp and Chum 1980; Schloth 1982; Lonsbury-Martin et al. 1990). As mentioned in the preceding chapter in this volume, SFOAEs are elicited by the presentation of a low-level pure tone. The maxima and minima of the SFOAEs are thought to correspond to specific tones generated in the cochlea. The interaction of the fixed emission (the internal cochlear tone) and the external stimulus frequency could produce perturbations in the SFOAE measured in the ear canal. Thus, the fluctuations of the SFOAE in the ear canal could reflect the presence of an internal cochlear tone, even if SOAEs cannot be measured.

2.5 *Implications*

Despite marked differences between the psychophysics of SOAE caused tinnitus and the psychophysics of tinnitus without measurable correlates, it is possible that all forms of tinnitus have similar physiological sources. As mentioned earlier, Penner (1980) presented a description of tinnitus in which global discontinuities in the tuning and suppression regions may produce perceptual changes underlying tinnitus. Ruggero, Rich, and Freyman's (1983) model of SOAEs is similar to Penner's (1980) model of tinnitus except that Ruggero, Rich, and Freyman (1983) hypothesized discontinuities that are restricted to a narrow region of the cochlea. It is possible that as the region of damage spreads on the cochlea, the resulting local cochlear disinhibition decreases (so that SOAEs are not measured) while, retrocochlearly, overall disinhibition increases, thereby becoming the basis of tinnitus. In this qualitative model, global changes in cochlear function could prevent tinnitus from behaving as if it consisted of tone sources.

Ultimately, the physiological basis of tinnitus is not solely a psychophysical question: it requires physiological corroboration. Physiological evidence of tinnitus might be present in animals if animals perceive tinnitus. Issues of the production, measurement, and the physiological basis of tinnitus in animals serve as the focus of the second part of this chapter.

3. An Animal Model of Tinnitus

3.1 Introduction

The objective correlates of tinnitus in humans may be possible to determine, even if an acoustic correlate is nonexistent. Two objective measures have recently been explored.

The first measure of tinnitus involved cortical magnetic fields evoked by auditory stimulation. Since tinnitus, by definition, involves the perception of a sound, the expectation was to detect an activity related to tinnitus perception at the cortical level. Although initial results were promising (Pantev et al. 1988; Hari and Lounasmaa 1989; Hoke et al. 1989; Pantev et al. 1989; Jacobson et al. 1991), recent reports from Hoke's group and others did not confirm the original findings (Jacobson et al. 1991; Colding-Jorgensen et al. 1992; Kristeva et al. 1992; Jacobson et al. 1992).

The second measure focused on auditory brainstem evoked responses (ABR) (Berlin and Shearer 1981; Levi and Chisin 1987; Barnea et al. 1990; Cassvan et al. 1990; Jacobson et al. 1991; Sininger 1991; Jastreboff, Ikner, and Hassen 1992). The majority of the work attempted to detect tinnitus related modifications in the transduction properties of the cochlea, as reflected in early waves of ABR, since most tinnitus cases can be related to cochlear damage. The other approach was based on the assumption that the continuous presence of the tinnitus signal causes plastic changes within the auditory subcortical pathways, consequently changing the way the external auditory signals are processed, which is reflected in a part of the ABR occurring at longer latencies (Jastreboff, Ikner, and Hassen 1992). Care must be taken to separate changes in ABR due to tinnitus from changes in ABR due to hearing loss, which so often accompanies tinnitus. Studies of tinnitus using ABR are still in the initial stages of development and do not yet provide confirmation of the existence of tinnitus (Jastreboff, Ikner, and Hassen 1990, 1992; Sininger 1991).

Most attempts to find physical correlates of tinnitus have involved the use of human subjects. Ethical restrictions on research involving humans and the lack of measurable physical correlates of tinnitus in humans have been a stumbling block in understanding the mechanisms of tinnitus generation and perception. Only through increased understanding of these mechanisms will possibilities for new and more effective methods of tinnitus alleviation be available. In addition, understanding the physiological basis of tinnitus is likely to involve measuring and altering the underlying physiology. Such experimentation might involve surgery and the recording of neural signals — these manipulations are not possible in human beings.

3.2 The Need for an Animal Model

The physiological basis of tinnitus might best be understood in an animal model, which could be used for controlled investigations of tinnitus and for testing new treatments. Unfortunately, tinnitus does not have objective correlates, so there is no method for quantifying it or detecting its presence. Therefore, it is often simply assumed that the same factors that produce tinnitus in humans also create tinnitus in animals. The method of choice for evoking tinnitus in animals is the administration of aspirin or other salicylates that reliably induce tinnitus in humans without significant side effects (Mongan et al. 1973; McFadden 1982; McFadden and Wightman 1983; McFadden and Plattsmier 1984; Wier, Pasanen, and McFadden 1988).

3.3 Presumed Salicylate Induced Tinnitus in Animals

Attempts have been made to specify the physiological basis of tinnitus in animals presumed to have tinnitus following injections of salicylate. For example, Evans, Wilson, and Borerwe 1981, described experiments in which high doses of salicylate (400 mg/kg of sodium salicylate) were injected in cats. The spontaneous activity of the auditory nerve single fibers was recorded before and after the injection of salicylate (Evans, Wilson, and Borerwe 1981; Evans and Borerwe 1982). Salicylate increased the rate of spontaneous activity and increased the rate of occurrence of short intervals in the spontaneous activity of the auditory nerve fibers.

Subsequent salicylate studies produced diverse results. Jastreboff and Sasaki (1986) reported increased spontaneous activity and a modified pattern of discharges in the inferior colliculus of guinea pigs two hours after the administration of salicylate. Stypulkowski (1990) presented data showing that salicylate administration slightly altered the spontaneous activity of the auditory nerve in cats. In addition, spectral analysis of the noise recorded from the proximity of the cochlea has indicated that salicylate administration increased noise in the 200 Hz range (Schreiner, Snyder, and Lenarz 1990; Martin et al. 1992).

Although the effects of salicylate on the auditory system are of interest, the relation between salicylate and tinnitus is strongly constrained by a fundamental problem: the lack of proof that animals do experience tinnitus after salicylate administration. Even if physiological changes are observed after salicylate administration, the changes may not be related to tinnitus, but instead might reflect nonspecific drug actions that accompany large doses of salicylate (i.e., pH balance, respiration, temperature control, or vestibular effects as in Flower, Moncada, and Vane 1980).

Knowledge of the time course of the absorption of salicylate in the cochlea is essential in delineating its specific effects on the auditory system. Detailed analysis of the time course of salicylate uptake into the cochlear perilymph, together with careful observation and measurements performed on conscious animals, indicated that, after an i.p. injection, a delay of about 2 to 3 h is needed for the salicylate to achieve a plateau in perilymph and that nonspecific actions of salicylate last up to 1 to 1.5 h (Jastreboff and Sasaki 1986; Jastreboff et al. 1986; Jastreboff, Brennan, and Sasaki 1988e). Therefore, as a rudimentary precaution, it is advisable to consider the effects of salicylate two hours after administration (oral, i.p. or s.c.).

Even allowing adequate time for recovery from the nonspecific action of salicylate and for building up a sufficient perilymphatic level of the drug does not guarantee that tinnitus has developed. Permanent or transient hearing loss caused by salicylate has been shown to increase sensitivity and activity of neurons within auditory pathways (Gerken 1979; Sasaki, Kauer, and Babitz 1980; Sasaki, Babitz, and Kauer 1981; Gerken, Saunders, and Paul 1984; Gerken et al. 1985; Gerken, Simhadri-Sumithra, and Bhat 1986; Salvi et al. 1990; Gerken 1991). These data might reflect the automatic gain control properties of the auditory system, which react to decreased input (hearing loss) by increased neuronal sensitivity. Thus, there is a possibility that the observed changes in single unit activity might result from a salicylate induced increase in the hearing threshold (Flower, Moncada, and Vane 1980; McFadden, Plattsmier, and Pasanen 1984).

3.4 A Behavioral Model of Tinnitus

3.4.1 Justification

In order to prove that an animal has tinnitus, it is necessary to demonstrate that animals are perceiving sound as a result of an experimental manipulation (e.g., administration of salicylate or quinine, sectioning of the auditory nerve, and noise or drug induced damage to the cochlea). By definition, such a model has to be based on an evaluation of the animal's behavior because the animal's response is required to attest to the presence of phantom sound perception.

Behavioral experiments do not directly unveil the electrical and biochemical abnormalities that are responsible for tinnitus. Therefore, the animal model must contain not only a behavioral component to prove that animals do perceive tinnitus, but also other methodologies aimed at investigating the specific physiological mechanisms that are producing tinnitus. The model should permit the use of physiological techniques, while at the same time ensure that tinnitus is actually present.

Physiological analysis might include recording the electrical activity of single neurons and their assembles, investigating the morphological changes in the cochlea and auditory pathways, evaluating the cellular properties of hair cells (such as their mechano-electrical transduction, morphological integrity, motility, and membrane properties), or analyzing neurotransmitter release, biochemical homeostasis of the cochlea, cochlear blood flow, ion channels, expression of genes related to the cochlear function, and the plasticity of the auditory pathways. The physiological and behavioral studies of animals with tinnitus are in progress and relevant data from these studies are presented in the remaining part of the chapter.

3.4.1.1 Description of the Behavioral Approach.

Standard behavioral conditioning is based on creating an association of a brief and well-defined sensory stimulus (labeled the conditioned stimulus – CS) that is stronger than the background, and appetitive or aversive reinforcement (labeled the unconditioned stimulus – US). The CS could be a brief tone, and the US could be food or an electrical shock. In classical Pavlovian conditioning, the delivery of reinforcement does not depend on the animal's action, while for instrumental (operant) conditioning the animal's behavior determines the presence or absence of reinforcement.

The precise temporal relationship of the CS and US, and thus the ability to present the CS for a strictly limited period of time, is an essential requirement for standard behavioral procedures (Estes and Skinner 1941; Annau and Kamin 1961; Borg 1982; Zielinski 1985). Furthermore, this paradigm elicits habituation of any potential response to a continuous background sound. Therefore, the traditional conditioning paradigm could not be applied in an animal model of tinnitus because of one tinnitus feature: it cannot be switched on and off in a controlled manner. Even assuming that tinnitus was induced in animals, it might last for hours or days. In this way, tinnitus could become part of the auditory background and potential reactions to its presence would be extinguished.

The solution to this problem was to transpose a conventional paradigm so that a continuous background was important and its modifications would have consequences for the animal. Specifically, the main idea was to create a paradigm in which the presence of sound is associated with safety, whereas silence is associated with danger (Jastreboff, Brennan, and Sasaki 1988d,e). To achieve these associations, rats were kept in mild, broadband background noise, around-the-clock from the beginning of the procedure. During Pavlovian suppression training the offset of this background noise was terminated with a single footshock. This created an association of silence (offset of noise) with pain, and subsequent presentations of silence (offsets of background noise) induced fear, which can be evaluated by a method proposed by Estes and Skinner (1941). When tinnitus was induced in animals trained to be afraid of silence, even if external background sound was switched off, the animals still perceived sound (tinnitus). Thus the fear evoked by offset of background noise was smaller, resulting only from reaction to new, previously unknown sound of tinnitus and not to the silence to which the animals had been previously conditioned.

3.4.1.2 Practical Implementation of the Animal Model.

The actual experimental procedure has been described in detail (Jastreboff, Brennan, and Sasaki 1988a–e, 1991; Jastreboff 1989, 1990, 1992; Brennan and Jastreboff 1991; Jastreboff and Brennan 1992a,b). In this chapter, only the essentials of the approach are presented.

From the beginning of the procedure, pigmented rats were constantly exposed to a low-level (62 dB SPL), broadband noise in their home and experimental cages. During each daily experimental session the background noise was switched off five times for 1 min each time. The animal's behavior was quantitatively evaluated by calculating the ratio of the number of licks occurring during the 1 min of noise offset divided by the total number of licks occurring 1 min before the offset and during it. This measure is the suppression ratio (R_B), which is widely used in behavioral literature (Estes and Skinner 1941; Annau and Kamin 1961; Church 1969; Brennan and Riccio 1975; Ayres, Haddad, and Albert 1987; van Willigen et al. 1987). The suppression ration equals 0.5 if the licking rate is unchanged, decreases to zero if licking ceases, and could increase to 1.0 if licking markedly increases during noise offset.

During each of the one or two days of Pavlovian suppression training, the offset of noise was terminated by a single footshock. This training associated the offset of CS (background noise) with the US (footshock). Animals quickly learned that the offset of noise indicated incoming shock and, as a result, licking decreased when the background noise was off. This phenomenon is based on the opposite actions of thirst and fear on licking behavior (Estes and Skinner 1941). Because the level of thirst was kept relatively constant by maintaining the same level of water deprivation, modification of the suppression ratio reflects changes in the level of fear (Estes and Skinner 1941).

The final stage, extinction, is of crucial interest for evaluating the presence of tinnitus in animals. During this process noise offset is no longer accompanied by shock. As a result, animals gradually lose the effects of suppression training: fear induced by noise offset gradually diminishes, as reflected in a steady increase of R_B. The results from a control group with daily saline injections are presented in Figure 8.6.

As a result of this suppression training, the absence of sound acquired an aversive property. Any sound, including tinnitus, introduced during the extinction process and present when the background noise is turned off

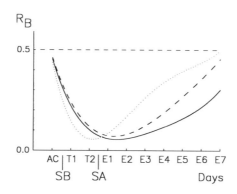

FIGURE 8.6 The behavior of control rats (dashes) and experimental rats injected with sodium salicylate (before suppression training, continuous curve; after suppression training, dots). R_B denotes the suppression ratio of the number of licks during the CS divided by the number of licks before plus during the CS. AC, acclimation day; T1, T2, suppression training; E1 to E7, extinction. SB and SA denote the starting point for daily salicylate administration before or after suppression training. The data have been fitted with a fourth order nonlinear regression. (from Jastreboff and Brennan 1992a with permission from Kugler Publications)

should theoretically cause a decrease of the conditioned fear induced by the offset of the background noise. Any decrease of this conditioned fear should result in a faster process of extinction, so that animals should return to pretraining behavior and disregard noise offset after fewer extinction days.

Notably, the above reasoning is based on the animals' perception of sound. The source of the additional sound is irrelevant (i.e., no distinction should be made between additional sounds that are presented externally or are an internal physiological phenomenon, induced pharmacologically, without any externally measurable auditory counterpart). In order to determine whether the perception of tinnitus has the predicted effect on the animals' behavior, a number of experiments were performed with pharmacologically induced tinnitus or with additional external tones mimicking tinnitus between the suppression training and extinction.

In the main experiment, daily injections of sodium salicylate were started just after the suppression training (group labeled After), with subsequent daily injections two hours before extinction testing (Jastreboff, Brennan, and Sasaki 1988c). The two hour delay allows adequate time for recovery from nonspecific actions of salicylate and for perilymphatic levels of the drug to rise. If salicylate in fact produced tinnitus, animals from this group would exhibit a much faster extinction as compared to the control group given saline injections (Fig. 8.6). Indeed, the group with salicylate administration showed a much faster extinction as compared to the control group. Notably, when 7 or 10 kHz continuous external tones were introduced at the same stage of the experiment as the salicylate injection, virtually identical results were obtained compared with those obtained after the animals had received salicylate injections (Jastreboff, Brennan, and Sasaki 1988c; Jastreboff and Brennan 1992b).

Although these results are in agreement with theoretical predictions that salicylate causes tinnitus in animals, they do not offer incontrovertible evidence. Before accepting the hypothesis that salicylate induces tinnitus in animals, additional matters need to be explored, such as demonstrating that the action of salicylate on animal behavior was indeed restricted to the auditory modality, and that the behavior was not a result of the salicylate induced threshold shift. To eliminate the possibility that the results reflect a different phenomenon rather than tinnitus, a variety of experimental controls were undertaken, four of which are outlined below.

3.4.1.3 Experimental Controls.

The basic control was based on reversing the effect of tinnitus on the animals' behavior by introducing it at a different stage of the paradigm. The idea was that if an extra continuous sound (either tinnitus or an external sound) is introduced before the suppression training (group labeled Before), then during the offset of background noise animals should hear only this additional sound. As a result, this sound could become a CS and the conditioning during suppression training might occur to this particular sound. Because this extra sound is present during the extinction stage and can be perceived between offsets of background noise, it should enhance the general level of fear and thereby prolong the process of extinction.

The experiments with salicylate or external tones confirmed this prediction (Fig. 8.6). Furthermore, for any given experimental situation, the effects of salicylate and continuous external tone introduced before the suppression training were statistically indistinguishable (Jastreboff, Brennan, and Sasaki 1988c; Jastreboff and Brennan 1992b). Therefore, because behavior depends on the stage at which salicylate is introduced, it is unlikely that the nonspecific mechanism of salicylate, rather than the salicylate induced tinnitus, is governing behavior.

The finding that salicylate induced modifications in animal behavior is dependent on the stage of the procedure when salicylate administrations were started argues against attributing these results to other potential effects of salicylate, such as: (1) increase of thirst, (2) pain threshold, (3) hearing threshold, or (4) any other unrelated to auditory system sensation, for example, vestibular disorders (Jastreboff, Brennan, and Sasaki 1988e).

These points were evaluated in a specific series of experiments that have shown that the behavior of animals injected with salicylate is not due to salicylate induced hearing loss (Jastreboff, Brennan, and Sasaki 1988b; Jastreboff 1989), and that the behavior exclusively reflects action of salicylate within the auditory modality (Jastreboff, Brennan, and Sasaki

1988c). All the results are consistent with the effects of perceiving drug induced phantom sound by animals.

3.4.1.4 Replication with Quinine.

Additional support for interpreting these results as an indication that salicylate causes tinnitus in animals has been obtained from work with quinine (Jastreboff, Brennan, and Sasaki 1988d, 1991). One of the standard ways of demonstrating the specificity of a drug is to use a drug from another chemical category. The drug should act on the system being investigated through a dissimilar mechanism, but should share with the first drug the ability to induce the effect studied—tinnitus. Quinine met all these criteria.

The results obtained with quinine in the behavioral paradigm were analogous to those obtained with salicylate. Animals given quinine injections starting after the suppression training showed faster extinction than control animals, while rats given quinine injections starting before suppression training exhibited prolonged extinction (Jastreboff, Brennan, and Sasaki 1988d, 1991). Furthermore, the animals' behavior depended on the drug dose (Jastreboff, Brennan, and Sasaki 1991; Jastreboff and Brennan 1992a). Thus, by using quinine instead of salicylate to induce tinnitus, it can be demonstrated that the results were due to the tinnitus caused by salicylate or quinine rather than any other physiological changes associated with the drugs.

3.4.1.5 Dose-Response Analysis of Salicylate.

Testing of dose-response dependence is a standard pharmacological approach to prove a link between a drug and its effects. Consequently, a series of behavioral experiments were performed using different doses of sodium salicylate. For each dose one group of animals was injected before and another group after suppression training.

Data obtained during previous experiments indicated that the area between the lines showing suppression levels during extinction for Before/After groups could be related to the extent of the perceived tinnitus (see insert in Fig. 8.7) (Jastreboff, Brennan, and Sasaki 1991; Jastreboff and Brennan 1992a). Application of this method to dose-response experiments revealed a highly significant, linear relationship between the area and the salicylate dose, as seen in Figure 8.7.

Interestingly, the levels of salicylate in rat serum and cerebrospinal fluid depend on the dose administered in a linear manner (Jastreboff and Brennan 1992a,b), and it has been shown that, in humans, the loudness of tinnitus is proportional to the serum salicylate level (Day et al. 1989). On the basis of this information, it is reasonable to propose that rats are perceiving tinnitus, the loudness of which is proportionally related to the drug dose. An additional implication from these findings is that this paradigm is able to reflect changes in the intensity of experimentally modified tinnitus.

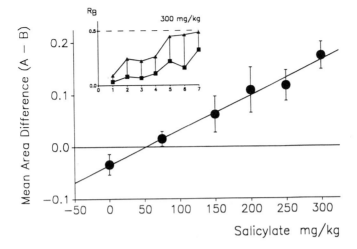

FIGURE 8.7 The dependence of the area difference on salicylate dose. The vertical bars represent SEM. The insert illustrates the method of area difference, using as an example the extinction curves of the 300 mg/kg Salicylate After (triangles) and Salicylate Before (squares) groups and presents the basis for the calculation of all points in this and figure 8.8, showing individual trapezoids used for area calculation. Note the highly significant linear correlation between dose of salicylate and area between extinction curves for Salicylate After and Salicylate Before groups. (from Jastreboff and Brennan 1994 with permission from Karger)

3.4.2 Psychoacoustics of Salicylate Induced Tinnitus in Rats

Both the work on the mechanism of tinnitus and the clinically oriented utilization of the model for testing the effectiveness of the new approaches to alleviate tinnitus would benefit from the measurement of the loudness and the pitch of the perceived tinnitus. As has been indicated, the paradigm described above seemed to be able to detect changes in tinnitus loudness. The next step was to evaluate the loudness and pitch of tinnitus as perceived by animals.

3.4.2.1 Measurement of Tinnitus Loudness.

Measurement of the loudness of tinnitus is based on an approach similar to that used in humans. In humans, the subject is asked to compare the loudness of tinnitus to the loudness of a 1 kHz reference tone. Analogously, by comparing the responses of animal groups given different doses of salicylate and different levels of a 10 kHz external tone, the loudness of tinnitus can be explored. As shown above the area between Before and After groups is linearly related to the salicylate dose and presumably is approximately proportional to the loudness of the perceived tinnitus, since in humans loudness of salicylate induced tinnitus is proportional to salicylate serum level (Day et al. 1989).

To test the hypothesis that the area difference is proportional to the loudness of the perceived sound, a specific series of experiments were performed (Jastreboff and Brennan 1994). In these experiments, in place of salicylate administration, groups of animals were exposed to different intensities (32 to 82 dB SPL) of a 10 kHz tone, around-the-clock, superimposed on the continuous noise background, and introduced to the animals before or after suppression training. At intensities below 72 dB, a linear relationship emerged between group differences based on the time of starting the constant tones (i.e., before or after training) and tonal intensity (Fig. 8.8). This finding supports the postulate that the animal response is proportional to the intensity of the reference sound.

By using the area difference as a common denominator for salicylate dose-response and reference tone intensity data, it is possible to determine the dose of salicylate that induces the same area difference between Before and After groups at a given intensity of the additional reference tone. Because both relationships (area between the lines versus salicylate dose, or versus SPL of the 10 kHz tone) can be described mathematically as linear on a double logarithmic scale, an equation describing the relationship of the intensity of the reference tone to the salicylate dose can be derived as seen in Figure 8.8. This intensity can be used as an approximation of tinnitus loudness, as perceived by rats.

3.4.2.2 Measurement of Tinnitus Pitch.

The approach has been selected based on the generalization principle, which was grounded on the following observation. When a subject (human or

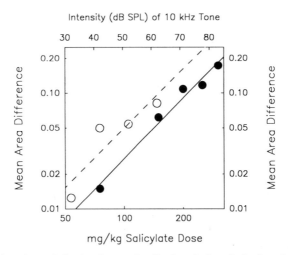

FIGURE 8.8 Estimation of the loudness of salicylate induced tinnitus in rats. Data points show the mean area differences between Before/After groups for various salicylate doses (filled circles) or various intensity levels of the external 10 kHz tone (open circles). Note that for both relationships are linear. (from Jastreboff and Brennan 1992b with permission from Kugler Publications)

animal) is trained to respond to a stimulus and later is tested on a variety of stimuli with similar physical characteristics, the greater the physical similarity of the test stimuli and the original stimulus used for training, the stronger the response (Brennan and Riccio 1975; Brennan and Jastreboff 1991). Therefore, the unknown frequency used during the initial training can be estimated by analyzing the subject's responses to tones of different frequencies.

The animals were injected with salicylate prior to suppression training and thus trained on the tinnitus sound. Test tones ranging from 4 to 11 kHz were presented to rats during extinction, when the background noise was switched off. The average of the suppression ratios from all extinction days was calculated for each given test frequency and is presented in Figure 8.9. The results show an increase in suppression, represented by low R_B values, parallelling the increase of frequency (Brennan and Jastreboff 1991; Jastreboff and Brennan 1992b). Previous data showed that salicylate administered before suppression training resulted in more suppression (smaller R_B) and prolonged duration of extinction (Jastreboff, Brennan, and Sasaki 1988d,e; Jastreboff 1989; Jastreboff and Brennan 1992a,b) than the control groups given saline. Therefore, test tones with frequencies closer to the perceived pitch of tinnitus should evoke stronger suppression (i.e., lower values of R_B). Accordingly, the results in Figure 8.9 indicate that the dominant pitch of a sound induced by salicylate is greater than or equal to 11 kHz.

To test the correctness of the above approach, a series of additional experiments were performed: during suppression training, animals were exposed to a 3 or a 12 kHz tone and tested on tones between 4 and 11 kHz (Brennan and Jastreboff 1991; Jastreboff and Brennan 1992b). The expectations were that opposite gradients for the 3 and 12 kHz groups should be

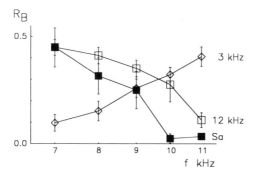

FIGURE 8.9 Estimation of the pitch of salicylate induced tinnitus in rats. Data points represent an average across extinction sessions for a given frequency of testing tone. Groups were: salicylate induced tinnitus (Sa filled squares); 12 kHz external tone (open squares); or 3 kHz external tones (open diamonds). Note the similarity of the salicylate and the 12 kHz groups and the opposite gradient for the 3 kHz group. (from Jastreboff and Brennan 1992b with permission from Kugler Publications)

observed, and that stronger suppression (smaller R_B) would occur for test frequencies closer to the frequency used in training. The results presented in Figure 8.9 confirm this prediction; the group trained with 12 kHz was similar to the group with salicylate administration, while the group trained with 3 kHz demonstrated the most suppression for lower frequencies.

All of the above results indicate that salicylate administration evoked sounds with a dominant pitch greater than or equal to 11 kHz. The upper frequency limit of salicylate induced tinnitus can be found by performing the same series of experiments but with higher test tone frequencies.

3.6 Examples of Application of the Model

Thus far, the rationale motivating the experiment and the success of the behavioral paradigm in demonstrating that animals perceive tinnitus have been discussed. Not only do animals perceive tinnitus, but evaluation of the loudness and pitch of pharmacologically evoked tinnitus is possible. However, the generality of a model utilizing salicylate or quinine induced tinnitus, as well as the clinically relevant mechanisms causing tinnitus, have not yet been addressed.

To properly address this relation, it would be necessary to discuss in detail the various hypotheses of tinnitus generation and the relevance of the salicylate-based model. The interested reader may find the discussion of possible mechanisms of tinnitus in other publications (Jastreboff 1990; Jastreboff et al. 1992). In this part of the chapter, the presentation is restricted to an animal model of tinnitus and examples of application. Nevertheless, some relevant points should be stressed.

First, the paradigm evaluates the perception of phantom sound independently of the method by which it is induced. The behavioral paradigm is independent of the method used to induce tinnitus because it is based on tinnitus perception. It does not matter if tinnitus is induced reversibly by salicylate, by permanent damage to the cochlea induced by loud sound or exposure to ototoxic drugs, or by surgical intervention. Moreover, a modification of this paradigm has been used successfully for detecting visual hallucinations induced by scopolamine, which are also dose-dependent (Jastreboff, Brennan, and Sasaki 1989a; Jastreboff and Brennan 1991a, 1992a). Therefore, the model can be used for the investigation of tinnitus of many origins: it is not restricted to drug induced tinnitus. Presently, work is in progress to include noise induced and surgically induced tinnitus.

3.7 Extent of Appropriateness of the Salicylate-Based Model

The mechanism causing salicylate-based tinnitus might be more universal than initially assumed due to preferential action of salicylate on the outer hair cell system (Jastreboff, Brennan, and Sasaki 1989a; Jastreboff 1990, 1992; Jastreboff and Brennan 1991a). The number of seemingly different mechanisms of tinnitus might be linked by the hypothesis that tinnitus is related to discordant damage of the inner (IHC) and outer (OHC) hair cells (Jastreboff 1990; Stypulkowski 1990). The hypothesis is an extension of Tonndorf's theory (Tonndorf 1987), according to which tinnitus results from unbalanced activity in type I and II afferents, which are innervating IHC and OHC respectively. Discordant damage of hair cells will create such an unbalance. In this view, generation of tinnitus occurs when there is an area on the basilar membrane with damaged or inactivated OHC and relatively intact IHC. This notion is supported by the following reasoning.

Morphological analyses have revealed that factors related most frequently to the appearance of tinnitus (e.g., exposure to a loud sound or various families of ototoxic drugs) have a greater impact on the OHC than on the IHC system (Liberman and Kiang 1978; Bohne and Clark 1982; Liberman and Mulroy 1982; Bohne, Yohman, and Gruner 1987; Liberman 1987; Liberman and Dodds 1987). Furthermore, the damage is more pronounced at the basal, high-frequency region of the basilar membrane. As a result, if the basilar membrane (from the base to the apex) is analyzed in a subject with a partially damaged cochlea, an area is encountered with both hair systems damaged, followed by an area with OHC predominantly damaged and IHC relatively intact, followed by an area with both systems unaltered.

If tinnitus can be related to this intermediate area of discordant damage of OHC and IHC, then the predominant pitch of tinnitus should be localized near abrupt changes in the audiogram. Clinical data indicate that this indeed is frequently the case (Hazell et al. 1985). Furthermore, because it is possible to have normal thresholds with profound but diffuse damage of OHC (Bohne and Clark 1982), this hypothesis might explain the presence of tinnitus in patients without hearing loss.

Extensive electrophysiological results of Stypulkowski (Stypulkowski 1990), substantiated by data published by Bobbin and colleagues (Puel et al. 1989; Puel, Bobbin, and Fallon 1990), strongly indicate that salicylate acts predominantly on OHC. Furthermore, McFadden and Plattsmier (1984) showed that salicylate abolishes SOAEs in humans, a finding that Penner and Coles (1992) replicated by eliminating SOAE related tinnitus with aspirin. Finally, Brownell (1990) demonstrated that exposure of OHC to the same concentration of salicylate, as measured in the perilymph of rats injected with tinnitus inducing doses of salicylate (Jastreboff et al. 1986; Jastreboff, Brennan, and Sasaki 1988e) resulted in reversible disintegration of their internal skeleton (Dieler, Shehata, and Brownell 1991a,b; Shehata, Brownell, and Dieler 1991). All these data support the hypothesis that salicylate administration results in temporary and reversible discordant deactivation of the OHC and IHC systems.

It follows that salicylate may provide an advantageous drug for changing cochlear transduction properties and neuronal activity resulting in tinnitus emergence by producing mild and reversible variations in the cochlea. Irreversible and more extensive versions of these dysfunctions occur in cases of permanent tinnitus (Jastreboff 1990; Jastreboff 1992).

There is a wide variety of potential mechanisms of tinnitus (Jastreboff 1990). In the following sections two examples of the work on tinnitus mechanisms using the animal model are outlined.

3.8 Calcium and Tinnitus

One application of the model involves the evaluation of the hypothesis that cochlear calcium homeostasis and calcium channels might be involved in tinnitus. The calcium hypothesis predicted the possibility of diminishing salicylate or quinine induced tinnitus by increasing the exogenous Ca^{2+} uptake, or by blocking Ca^{2+} influx to hair cells with nimodipine, a highly specific blocker of voltage dependent L-type Ca^{2+} channels (Jastreboff et al. 1985; Jastreboff, Brennan, and Sasaki 1987; Jastreboff and Brennan 1988; Bobbin et al. 1990; Jastreboff 1990; Jastreboff, Brennan, and Sasaki 1991; Jastreboff, Nguyen, and Sasaki 1991).

To evaluate the calcium hypothesis, several series of behavioral experiments were performed in which six classes of situations were analyzed: (1) control animals given saline injections; (2) animals given a tinnitus inducing drug (salicylate or quinine); (3) rats injected with these drugs and nimodipine; (4) rats injected with salicylate or quinine and given Ca^{2+} supplement by providing them with 50 mM $CaCl_2$ in place of tap water throughout the duration of the experiment; (5) rats with nimodipine injections; and (6) rats kept on an exogenous calcium supplement. For each of these conditions, two groups of rats were used. One group was administered daily salicylate or quinine, and eventually nimodipine, starting before training. For the other group, administrations followed training.

The area difference (see Fig. 8.7) between these two groups has been used to assess the loudness of tinnitus. The results have been described in detail (Jastreboff, Hansen, and Sasaki 1989; Jastreboff, Brennan, and Sasaki 1991; Jastreboff and Brennan 1992a; Jastreboff et al. 1992) and are summarized in Figure 8.10. Comparisons of the results of salicylate or quinine alone with those obtained when additional experimental manipulations were introduced revealed the attenuating effects of Ca^{2+} supplement or nimodipine on drug induced tinnitus. Control groups with saline, nimodipine, and Ca^{2+} supplement alone did not show any noticeable difference between the Before and After groups. These results support the hypothesis that calcium blocks the perception of tinnitus and supports the postulate that extra- and intra-cellular calcium homeostasis might play a role in tinnitus development and its alleviation.

To investigate the physiological mechanism involving modulation of

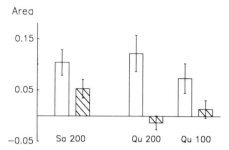

FIGURE 8.10 Effects of exogenous calcium supplement on salicylate or quinine induced tinnitus. The mean area differences between groups with salicylate/quinine administered before or after suppression training are plotted for groups with drug alone (open bars) and for combined treatment of a drug with exogenous calcium supplement (50 mM $CaCl_2$ substituted for tap water - hatched bars). Sa, salicylate; Qu, quinine. Doses in mg/kg s.c. as marked. Note the dose dependence of quinine induced effects and attenuation produced by calcium supplement. Bars denotes SEM calculated from data for individual days. (from Jastreboff, Nguyen, Brennan, and Sasaki 1992 with permission from Kugler Publications.)

calcium levels, it was necessary to prove that: (1) the level of free calcium in perilymph decreases after salicylate, and (2) that providing animals with calcium supplements attenuates this decrease. Therefore, a series of experiments were performed during which animals were exposed to exactly the same doses of salicylate and parameters of calcium supplements as during the behavioral experiments. Free calcium was measured continuously by a calcium selective microelectrode inserted through the cochlear round window (Jastreboff and Brennan 1991b; Jastreboff, Nguyen, and Sasaki 1991; Jastreboff and Brennan 1992a; Jastreboff et al. 1992).

The main results presented in Figure 8.11 show that salicylate does indeed cause a decrease in free perilymphatic calcium, which is dose dependent (Jastreboff and Brennan 1991b; Jastreboff, Nguyen, and Sasaki 1991; Jastreboff et al. 1992). Interestingly, the calcium supplement did not change the initial baseline before salicylate administration, but did attenuate a salicylate induced calcium decrease. This indicates that calcium supplements enhance cochlear calcium homeostasis, rather than increasing the free calcium level. An important aspect is that analogous and consistent results were obtained with both behavioral methodology and free calcium measurements.

3.9 Neuronal Activity Related to Tinnitus

The consensus that tinnitus results from abnormal neuronal activity has limited experimental support. Characterization of this neuronal activity is crucial for understanding the contribution of the neuronal pathways in the emergence of tinnitus and in the search for methods of tinnitus alleviation.

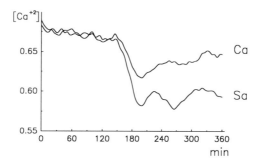

FIGURE 8.11 Effect of exogenous calcium supplement on salicylate induced decrease of perilymphatic free calcium level. Salicylate was injected at a dose level of 233 mg/kg. Sa, group with the drug alone; Ca, group with additional calcium supplement. Note the same initial level of free calcium and the decrease of salicylate induced effects. (from Jastreboff, Nguyen, Brennan, and Sasaki 1992 with permission from Kugler Publications.)

In continuation of our efforts with recordings performed on guinea pigs (Jastreboff and Sasaki 1986), we performed single unit recordings from the external nucleus of the inferior colliculus of pigmented rats before and after: (1) salicylate injection, (2) saline injection, (3) salicylate injection on animals receiving calcium supplement (Jastreboff and Chen 1993; Chen and Jastreboff 1995). The spontaneous activity of all units recorded in the control situations, i.e., before salicylate and before or after saline, exhibited stable and similar levels. Salicylate administration resulted in a significant increase of the activity as compared to activity of units recorded before salicylate and from activity recorded during the control experiments before or after saline administration. Units recorded from animals pretreated with calcium did not exhibit any increase of the spontaneous activity after salicylate.

Human reports (McFadden 1982; McFadden, Plattsmier, and Pasanen 1984) show that salicylate induced tinnitus is perceived as similar to tonal or narrow band noise, and behavioral experiments indicate that salicylate induces perception of a sound with a dominant frequency of about 11 kHz (Jastreboff and Brennan 1992b). To evaluate the relationship of salicylate induced increase of spontaneous activity and the characteristic frequency of a unit, the spontaneous activity of all sharply tuned units were plotted as a function of their characteristic frequency. The average spontaneous activity did not show any clear difference between the before/after salicylate recordings for frequencies below 10 kHz. However, for units in the 10 to 16 kHz range, administration of salicylate resulted in a high and statistically significant increase of the spontaneous activity. Calcium pretreatment totally abolished this effect. Notably, this range of characteristic frequency corresponds closely to our data from evaluating the pitch of salicylate induced tinnitus in rats, which indicated the pitch of tinnitus to be close to 11 kHz (Jastreboff and Brennan 1992).

Temporal patterns of spontaneous activity were affected as well and showed abnormal patterns of epileptic-like activity, characterized by short bursts of multiple discharges. The emergence of this epileptic-like activity displayed similar dependence on the characteristic frequency as the spontaneous activity increase. While for characteristic frequencies below 10 kHz the ratio of units exhibiting high frequency activity was practically the same before and after salicylate administration, it increased 2.4 times for units with characteristic frequencies of 10 to 16 kHz, and for characteristic frequencies above 16 kHz, the ratio decreased after salicylate. Notably, calcium pretreatment abolished this high frequency activity both before and after salicylate.

These results for the first time show changes in neuronal patterns of activity in animals experiencing tinnitus as demonstrated at the behavioral level and support the possibility of epileptic like activity as being related to tinnitus. Notably, both a salicylate induced increase of the spontaneous activity and the emergence of bursting patterns of spike discharges were predominant in units with characteristic frequencies corresponding closely to the behaviorally determined pitch of tinnitus.

The above outline of the work on the calcium hypothesis and neuronal activity demonstrates the approach and possible applications of the animal model for tinnitus. Other experiments were also performed to assess the effectiveness of various drugs for attenuating tinnitus and successfully identified effective substances. It is worth noting that there are a variety of different types of tinnitus and a number of other factors that might be involved in tinnitus generation, perception, and evaluation (Jastreboff 1990). Other works involving analysis of the cochlear potentials, single unit spontaneous activity recorded from auditory pathways, and biochemical and pharmacological methodologies are in progress in Jastreboff's laboratory and are performed in close conjunction with behavioral evaluation of the perception of experimentally induced tinnitus.

In this part of the chapter, an animal model of tinnitus has been presented and validated. The model includes a behavioral component for detecting the presence of tinnitus and measuring its pitch and loudness. By combining the behavioral approach with physiological measures, a paradigm permitting examination of the mechanisms underlying tinnitus has been developed.

4. Summary

The psychophysical measures of tinnitus in humans seem to indicate that tinnitus only rarely has a measurable acoustical concomitant (Penner 1990).

The lack of an acoustical concomitant and the unreliability of psychophysical data make models based on human observations highly tenuous. Human data do, however, indicate that tinnitus is associated with a disruption of suppression in the periphery. Future work by the first author and her colleagues will extend quantitative models of the periphery (Patterson and Holdsworth 1992) to the case in which suppression fails selectively at high frequencies. Studying the pattern of spontaneous firings when the suppression region is distorted may unveil a relation between the audiogram and suppression regions and the spectrum of the tinnitus produced.

Because uncovering the specific physiological basis of tinnitus will likely require experimental manipulations that cannot be made on intact human beings, animal models of tinnitus are essential. Using an animal model, it is possible to show that animals perceive high-pitched tinnitus following injections of salicylate. Future work will be directed at uncovering the physiological basis of drug induced tinnitus and extending the animal model by using noise and surgery to induce tinnitus in animals.

Acknowledgments. This work was supported by the National Institute of Health, National Institute on Deafness and Other Communication Disorders grants (RO1 DC00068 to M. J. Penner and RO1 DC00299 and RO1 DC00445 to P. J. Jastreboff), by a grant from Deafness Research Foundation (P. J. Jastreboff) and by a grant from the University of Maryland (M. J. Penner). Portions of this article were reprinted with permission of the British Journal of Audiology and the American Speech, Language, Hearing Association.

References

American National Standards Institute (1969) Specifications for Audiometers ANSI, New York, p. S3.6.

Annau Z, Kamin LJ (1961) The conditioned emotional response as a function of intensity of the US. J Comp Physiol Psych 54:428–432.

Atkinson M (1947) Tinnitus aurium: some considerations on its origin and treatment. Arch Otolaryngol 45:68–76.

Ayres JJB, Haddad C, Albert M (1987) One-trial excitatory backward conditioning as assessed by conditioned suppression of licking in rats: Concurrent observations of lick suppression and defensive behaviors. Anim Learn Behav 15:21–217.

Bacon SP, Viemiester NF (1985) A case study of monaural diplacusis. Hear Res 19:49–56.

Barnea G, Attias J, Gold S, Shahar A (1990) Tinnitus with normal hearing sensitivity: Extended high-frequency audiometry and auditory-nerve brainstem-evoked responses. Audiol 29:3645.

Berlin CI, Shearer PD (1981) Electrophysiological simulation of tinnitus. In: CIBA Foundation Symposium 85, Tinnitus. London: Pitman Books Limited, pp. 139-150.

Bilger RC, Matthies ML, Hammel DR, Demorest M (1990) Genetic implications of gender differences in the prevalence of spontaneous otoacoustic emissions. J Speech Hear Res 33:418-432.

Bobbin RP, Jastreboff PJ, Fallon M, Littman T (1990) Nimodipine, an L-channel Ca^{+2} antagonist, abolishes the negative summating potential recorded from the guinea pig cochlea. Hear Res 46:277-288.

Bohne BA, Clark WW (1982) Growth of hearing loss and cochlear lesion with increasing duration of noise exposure. In: Hamernik RP, Henderson D, Salvi R (eds) New Perspectives on Noise-Induced Hearing Loss. New York: Raven Press pp. 283-302.

Bohne BA, Yohman L, Gruner MM (1987) Cochlear damage following interrupted exposure to high-frequency noise. Hear Res 29:251-264.

Borg E (1982) Auditory thresholds in rats of different age and strain. A behavioral and electrophysiological study. Hear Res 8:101-115.

Brennan JF, Jastreboff PJ (1991) Generalization of conditioned suppression during salicylate-induced phantom auditory perception in rats. Acta Neurobiol Exp 51:15-27.

Brennan JF, Riccio DC (1975) Stimulus generalization of suppression in rats following aversively motivated instrumental or Pavlovian training. J Comp Physiol Psychol 88:570-579.

Brownell WE (1990) Outer hair cell electromotility and otoacoustic emissions. Ear and Hear 11:82-92.

Brusis T, Loennecken I (1985) Treatment of tinnitus with iontophoresis and local anesthesia. Laryngol-Rhinol-Otol Stuttg 64(7):355-358.

Burns EM (1984) A comparison of variability among measurements of subjective tinnitus and objective stimuli. Audiol 23:426-440.

Burns EM (1989) Alternate-state OAEs as a basis for episodic intermittent tinnitus. J Acoust Soc Amer 85:S35.

Burns EM, Keefe DH (1991) Unstable spontaneous otoacoustic emission as a cause of tinnitus. Abstracts of the Fourteenth Midwinter Meeting of the Association for Research in Otolaryngology, p. 239.

Carmen R, Svihovec D (1984) Relaxation-biofeedback in the treatment of tinnitus. Amer J Otol 5(5):376-381.

Cassvan A, Ralescu S, Moshkovski FG, Shapiro E (1990) Brainstem auditory evoked potential studies in patients with tinnitus and/or vertigo. Arch Physical Med Rehabil 71:583-586.

Champlin CA, Muller SP, Mitchell SA (1990) Acoustic measurements of objective tinnitus. J Speech Hear Res 33:816-821.

Chen G-D, Jastreboff PJ (1995) Salicylate-induced abnormal activity in the inferior colliculus in rats. Hear Res (In press).

Chouard CH, Meyer B, Maridat D (1981) Transcutaneous electrotherapy for severe tinnitus. Acta Otolaryngol 91:415-422.

Church RM (1969) Response suppression. In: Campbell BA, Church RM (eds) Punishment and Aversive Behavior. New York: Appleton-Century-Crofts, pp. 111-156.

Colding-Jorgensen E, Lauritzen M, Johnsen NJ, Mikkelsen KB, Saermak K (1992) On the value of the auditory evoked magnetic P200 as an objective measure of tinnitus. In: Aran, J-M, Dauman R (eds) Tinnitus 91, Proceedings IV. International Tinnitus Seminar, Bordeaux 1991. Amsterdam: Kugler Publications, pp. 323–326.

Day RO, Graham GG, Bieri D, Brown M, Cairns D, Harris G, Hounsell J, Platt-Hepworth S, Reeve R, Sambrook PN, Smith J (1989) Concentration-response relationships for salicylate-induced ototoxicity in normal volunteers. Br J Clin Pharmacol 28:695–702.

Dieler R, Shehata-Dieler WE, Brownell WE (1991a) Concomitant salicylate-induced alterations of outer hair cell subsurface cisternae and electromotility. J Neurocytol 20:637–653.

Dieler R, Shehata WE, Brownell WE (1991b) Salicylates reversibly alter subsurface cisternae and rapid electromotile responses in isolated outer hair cells. Abstracts of the Fourteenth Midwinter Meeting of the Association for Research in Otolaryngology, p. 13.

Dobie RA, Hoberg KE, Rees TS (1986) Electrical tinnitus suppression: A double-blind study. Arch Otolaryngol Head Neck Surg 95:319–323.

Emmett JR, Shea JJ (1984) Medical treatment of tinnitus. J Laryngol Otol, Suppl 9:264–270.

Estes WK, Skinner BF (1941) Some quantitative properties of anxiety. J Exp Psychol 29:390–400.

Evans EF, Borerwe TA (1982) Ototoxic effects of salicylates on the responses of single cochlear nerve fibers and on cochlear potentials. Br J Audiol 16:101–108.

Evans EF, Wilson JP, Borerwe TA (1981) Animal models of tinnitus. In: CIBA Foundation Symposium 85, Tinnitus. London: Pitman Books Limited, pp. 108–138.

Feldmann H (1971) Homolateral and contralateral masking of tinnitus by noise-bands and by pure tones. Audiol 10:138–144.

Flottorp G (1953) Pure-tone tinnitus evoked by acoustic stimulation: The idiophonic effect. Acta Otolaryngol 43:395–415.

Flower RJ, Moncada S, Vane JR (1980) Analgesic-antipyretics and anti-inflammatory agents. In: Gilman AG, Goodman LS, Gilman A (eds) The Pharmacological Basis of Therapeutics. New York: Macmillan, pp. 682–728.

Formby C, Gjerdingen DB (1981) Some systematic observations on monaural diplacusis. Audiol 20:219–233.

Fowler EP (1939) The use of threshold and louder sounds in clinical diagnosis and the prescribing of hearing aids. New methods for accurately determining the threshold for the bone conduction and for measuring the tinnitus and its effects on obstructive and neural deafness. Laryngoscope 48:572–578.

Fowler EP (1942) The "illusion of loudness" in tinnitus: Its etiology and treatment. Laryngoscope 52:275–285.

Fowler EP (1943) Control of head noises: Their illusions of loudness and timbre. Arch Otolaryngol 37:391.

Fowler EP (1944) Head noises in normal and disordered ears: significance, measurement, differentiation and treatment. Arch Otolaryngol 39:498–503.

Frick LR, Matthies ML (1988) Effects of external stimuli on spontaneous otoacoustic emissions. Ear and Hear 9:190–197.

Gerken GM (1979) Central denervation hypersensitivity in the auditory system of the cat. J Acoust Soc Amer 66:721–727.

Gerken GM, Saunders SS, Paul RE (1984) Hypersensitivity to electrical stimulation of auditory nuclei follows hearing loss in cats. Hear Res 13:249–259.

Gerken GM, Saunders SS, Simhadri-Sumithra R, Bhat KHV (1985) Behavioral thresholds for electrical stimulation applied to auditory brainstem nuclei in cat are altered by injurious and noninjurious sound. Hear Res 20:221–231.

Gerken GM, Simhadri-Sumithra R, Bhat KHV (1986) Increase in central auditory responsiveness during continuous tone stimulation or following hearing loss. In: Salvi RJ, Henderson D, Hamernik RP, Colletti V (eds) Basic and Applied Aspects of Noise-Induced Hearing Loss. New York: Plenum Publishing Corporation, pp. 195–211.

Glanville JD, Coles RRA, Sullivan BM (1971) A family with high tonal tinnitus. J Laryngol Otol, 85:1–10.

Gold T, Pumphrey RJ (1948) Hearing. I. The cochlea as a frequency analyzer. Proc Roy Soc, London Serial B 135:462–491.

Goodwin PE, Johnson RM (1980a) The loudness of tinnitus. Acta Otolaryngol 90:353–359.

Goodwin PE, Johnson RM (1980b) A comparison of reaction times to tinnitus and nontinnitus frequencies. Ear and Hear 1(3):148–155.

Graham JT, Newby HA (1962) Acoustical characteristics of tinnitus. Arch Otolaryngol 75:82–87.

Green DM, Swets J (1966) Signal Detection Theory and Psychophysics. John Wiley and Sons, Inc., New York, NY.

Hammel DR (1981) Spontaneous otoacoustic emissions. [MSC thesis] Urbana-Champaign (IL): University of Illinois.

Hari R, Lounasmaa OV (1989) Recording and interpretation of cerebral magnetic fields. Science 244:432–436.

Harris S, Brismar J, Cronquist S (1979) Pulsatile tinnitus and therapeutic embolization. Acta Otolaryngol 88:220–226.

Hazell JWP (1984) Spontaneous cochlear acoustic emissions and tinnitus. Clinical experience in the tinnitus patient. J Laryngol Otol 9:106–110.

Hazell JWP (1987) A cochlear model for tinnitus. In: Feldmann H (ed) Proceedings III International Tinnitus Seminar, Muenster 1987. Karlsruhe: Harsch Verlag, pp. 121–128.

Hazell JWP, Jastreboff PJ (1990) Tinnitus I. Auditory mechanisms: A model for tinnitus and hearing impairment. J Otolaryngol 19:1–5.

Hazell JW, Wood SM, Cooper HR, Stephens SD, Corcoran AL, Coles RR, Baskill JL, Sheldrake JB (1985) A clinical study of tinnitus maskers. Br J Audiol 19:65–146.

Hazell JWP, Meerton LJ, Conway MJ (1989) Electrical tinnitus suppression (ETS) with a single channel cochlear implant. J Laryngol Otol, Suppl 18:39–44.

Heller MF, Bergman M (1953) Tinnitus aurium in normally hearing persons. Ann Otol Rhinol Laryngol 62:73–83.

Hellman RP, Zwislocki JJ (1961) Some factors affecting the estimation of loudness. J Acoust Soc Amer 33:687–694.

Hentzer E (1968) Objective tinnitus of the vascular type. Acta Otolaryngol 66:273–281.

Hoke M, Feldmann H, Pantev C, Lutkenhoner B, Lehnertz K (1989) Objective evidence of tinnitus in auditory evoked magnetic fields. Hear Res 37:281-286.

Houtgast T (1972) Psychophysical evidence for lateral inhibition in hearing. J Acoust Soc Amer 51:1885-1894.

Humes LE, Jesteadt W (1991) Models of the effects of threshold on loudness growth and summation. J Acoust Soc Amer 90:1933-1943.

Jacobson GP, Ahmad BK, Morgan J, Newman CW, Tepley N, Wharton J (1991) Auditory evoked cortical magnetic field (M100 - M200) measurements in tinnitus and normal groups. Hear Res 56:44-52.

Jacobson GP, Ahmad BK, Moran J, Newman CW, Wharton JA, Tepley N (1992) Auditory evoked cortical field (M100/M200) measurements in tinnitus and normal groups. In: Aran J-M, Dauman R (eds) Tinnitus 91, Proceedings IV International Tinnitus Seminar, Bordeaux 1991. Amsterdam: Kugler Publications, pp. 317-322.

Jakes SC, Hallam RS, Rachman S, and Hinchcliffe R (1985) The effects of reassurance, relaxation training and distraction on chronic tinnitus sufferers. Behav Res Ther 24:497-507.

Jastreboff PJ (1989) An animal model of tinnitus: development, present status, perspectives. Hear J 42:58-63.

Jastreboff PJ (1990) Phantom auditory perception (tinnitus): mechanisms of generation and perception. Neurosci Res 8:221-254.

Jastreboff PJ (1992) Appropriateness of salicylate-based models of tinnitus. In: Aran J-M, Dauman R (eds) Tinnitus 91, Proceedings IV International Tinnitus Seminar, Bordeaux 1991. Amsterdam: Kugler Publications, pp. 309-313.

Jastreboff PJ, Brennan JF (1988) Specific effects of nimodipine on the auditory system. Ann NY Acad Sci 522:716-718.

Jastreboff PJ, Brennan JF (1991a) Animal model of tinnitus. Abstracts of the Fourteenth Midwinter Meeting of the Association for Research in Otolaryngology, p. 3.

Jastreboff PJ, Brennan JF (1991b) An animal behavioral model of hallucinations in rats. Abstracts of the Annual Meeting of the Society for Neuroscience 17:1484.

Jastreboff PJ, Brennan JF (1992a) Animal model of tinnitus: Recent developments. In: Aran J-M, Dauman R (eds) Tinnitus 91, Proceedings IV International Tinnitus Seminar, Bordeaux 1991. Amsterdam: Kugler Publications, pp. 283-292.

Jastreboff PJ, Brennan JF (1992b) The psychoacoustical characteristics of tinnitus in rats. In: Aran J-M, Dauman R (eds) Tinnitus 91, Proceedings IV International Tinnitus Seminar, Bordeaux 1991. Amsterdam: Kugler Publications, pp. 305-308.

Jastreboff PJ, Brennan JF (1994) Evaluating the loudness of phantom auditory perception (tinnitus) in rats. Audiol 33:202-217.

Jastreboff PJ, Chen G-D (1991) Salicylate induces a decrease in free calcium in rat cerebrospinal fluid. Abstracts of the Fourteenth Midwinter Meeting of the Association for Research in Otolaryngology, p. 145.

Jastreboff PJ, Chen G (1993) Salicylate-induced modification of the spontaneous activity in the inferior colliculus in rats. Abstracts of the Sixteenth Midwinter Meeting of the Association for Research in Otolaryngology, p. 40.

Jastreboff PJ, Sasaki CT (1986) Salicylate-induced changes in spontaneous activity of single units in the inferior colliculus of the guinea pig. J Acoust Soc Amer 80:1384–1391.

Jastreboff PJ, Jastreboff MM, Hansen R, Sasaki CT (1985) Salicylate-related changes in intracochlear calcium as a cause of auditory dysfunction. Abstracts of the Annual Meeting of the Society for Neuroscience 11:244.

Jastreboff PJ, Brennan JF, Sasaki CT (1987) Behavioral and electrophysiological animal model of tinnitus. In: Feldmann H (ed) Proceedings of the III International Tinnitus Seminar, Muenster. Karlsruhe: Harsch Verlag, pp. 95–99.

Jastreboff PJ, Brennan JF, Sasaki CT (1988a) Animal model of tinnitus: Quinine effects. Abstracts of the Eleventh Midwinter Meeting of the Association for Research in Otolaryngology, p.260.

Jastreboff PJ, Brennan JF, Sasaki CT (1988b) Pigmentation, anesthesia, behavioral factors and salicylate uptake. Arch Otolaryngol Head Neck Surg 114:186–191.

Jastreboff PJ, Brennan JF, Sasaki CT (1988c) Animal model of tinnitus: Specificity of the paradigm. Abstracts of the Annual Meeting of the Society for Neuroscience 14:1099.

Jastreboff PJ, Brennan JF, Sasaki CT (1988d) An animal model for tinnitus. Laryngoscope 98:280–286.

Jastreboff PJ, Brennan JF, Sasaki CT (1988e) Phantom auditory sensation in rats: An animal model for tinnitus. Behav Neurosc 102:811–822.

Jastreboff PJ, Brennan JF, Sasaki CT (1989) An animal model of tinnitus: Perceptual complexity. Abstracts of the Twelfth Midwinter Meeting of the Association for Research in Otolaryngology, pp. 191–192.

Jastreboff PJ, Brennan JF, Sasaki CT (1991) Quinine-induced tinnitus in rats. Arch Otolaryngol Head Neck Surg 117:1162–1166.

Jastreboff PJ, Hansen R, Sasaki CT (1989) Dose-dependent serum calcium decrease after salicylate. Abstracts of the Annual Meeting of the Society for Neuroscience 15:210.

Jastreboff PJ, Hansen R, Sasaki PG, Sasaki CT (1986) Differential uptake of salicylate in serum, CSF and perilymph. Arch Otolaryngol Head Neck Surg 112:1050–1083.

Jastreboff PJ, Ikner CL, Hassen A (1990) A mathematical evaluation of brainstem evoked response aimed at detection of tinnitus in humans. Abstracts of the Thirteenth Midwinter Meeting of the Association for Research in Otolaryngology, p. 214.

Jastreboff PJ, Ikner CL, Hassen A (1992) An approach to objective evaluation of tinnitus in humans. In: Aran J-M, Dauman R (eds) Tinnitus 91, Proceedings IV International Tinnitus Seminar, Bordeaux 1991. Amsterdam: Kugler Publications, pp. 331–390.

Jastreboff PJ, Nguyen Q, Sasaki CT (1991) Perilymphatic free calcium and pH changes induced by salicylate in rats. Abstracts of the Fourteenth Midwinter Meeting of the Association for Research in Otolaryngology, p. 75.

Jastreboff PJ, Nguyen Q, Brennan JF, Sasaki CT (1992) Calcium and calcium channel involvement in tinnitus. In: Aran J-M, Dauman R (eds) Tinnitus 91, Proceedings IV International Tinnitus Seminar, Bordeaux 1991. Amsterdam: Kugler Publications, pp. 109–119.

Jesteadt W (1980) Effects of masker level and signal delay on forward masking in normal and impaired listeners. J Acoust Soc Amer 68:S1.

Jones IH, Knudsen VO (1928) Certain aspects of tinnitus, particularly treatment. Laryngoscope 38:597-611.

Kemp DT (1978) Stimulated acoustic emissions from within the human auditory system. J Acoust Soc Amer 64:1386-1391.

Kemp DT (1979) Evidence of mechanical nonlinearity and frequency selective wave amplification in the cochlea. Arch Oto Rhino Laryngol 224:37-45.

Kemp DT, Chum RA (1980) Observations on the generator mechanism of stimulus frequency emissions—Two tone suppression. In: van den Brink G, Bilsen FA (eds) Psychophysical, Physiological, and Behavioral Studies in Hearing. Delft: Delft University, pp. 34-42.

Kiang NYS, Moxon EC, Levine RA (1970) Auditory-nerve activity in cats with normal and abnormal cochleas. In: Wolstenholme GEW, Knight J (eds) Ciba Foundation Symposium on Sensorineural Hearing Loss. London: Churchill Livingstone, pp. 241-273.

Kristeva R, Lütkenhöner B, Ross B, Elbert T, Zowalik S, Hampson S, Hoke M, Feldmann H (1992) The amplitude ratio M200/M100 of the auditory evoked magnetic field in normal hearing subjects and tinnitus patients. In: Aran J-M, Dauman R (eds) Tinnitus 91, Proceedings IV International Tinnitus Seminar, Bordeaux 1991. Amsterdam: Kugler Publications, pp. 327-329.

Larkin WD, Penner MJ (1989) Partial masking in electrocutaneous sensation: A model for sensation matching, with applications to recruitment. Percept Psychophys 46:207-219.

Letowski TR, Thompson MV (1985) Interrupted noise as a tinnitus masker: an annoyance study. Ear and Hearing 6:65-70.

Levi H, Chisin R (1987) Can tinnitus mask hearing? A comparison between subjective audiometric and objective electrophysiological threshold in patients with tinnitus. Audiol 26:153-157.

Levitt H (1971) Transformed up-down methods in psychoacoustics. J Acoust Soc Amer 49:467-477.

Liberman MC (1987) Chronic ultrastructural changes in acoustic trauma: Serial-section reconstruction of stereocilia and cuticular plates. Hear Res 26:65-88.

Liberman MC, Dodds LW (1987) Acute ultrastructural changes in acoustic trauma: Serial-section reconstruction of stereocilia and cuticular plates. Hear Res 26:45-64.

Liberman MC, Kiang NYS (1978) Acoustic trauma in cats. Acta Otolaryngol, Supplement 358:1-63.

Liberman MC, Mulroy MJ (1982) Acute and chronic effects of acoustic trauma: Cochlear pathology and auditory nerve pathophysiology. In: Hamernik RP, Henderson D, Salvi R (eds) New Perspectives on Noise-Induced Hearing Loss. New York: Raven Press, pp. 105-136.

Lochner JPA, Burger JF (1961) Form of the loudness function in the presence of masking noise. J Acoust Soc Amer 33:1705-1707.

Long GR, Tubis A (1988) Modification of spontaneous and evoked otoacoustic emissions and associated psychoacoustic microstructure by aspirin consumption. J Acoust Soc Amer 84:1343-1353.

Lonsbury-Martin BL, Harris FP, Stanger BB, Hawkins MD, Martin GK (1990) Distortion-product emissions in humans: II. Sites of origin revealed by suppression contours and pure-tone exposures. Hear Res 28:191-208.

Marlowe FI (1973) Effective treatment of tinnitus through hypnotherapy. Am J Clin Hypnosis 15:162–165.
Martin GK, Probst R, Lonsbury-Martin BL (1990) Otoacoustic emissions in human ears: Normative findings. Ear and Hear 11:106–120.
Martin WH, Schwegler JW, Pratt H, Rosenberg S, Rosenwasser RH, Flamm E (1992) Spectral analysis of human auditory nerve activity: Tinnitus vs. non-tinnitus subjects. Abstracts of the Fifteenth Midwinter Meeting of the Association for Research in Otolaryngology, p. 134.
McFadden D (1982) Tinnitus: Facts, Theories and Treatments. Washington, D.C.: National Academy Press.
McFadden D, Plattsmier HS (1984) Aspirin abolishes spontaneous oto-acoustic emissions. J Acoust Soc Amer 76:443–448.
McFadden D, Wightman FL (1983) Audition: Some relations between normal and pathological hearing. Annu Rev Psychol 34:95–128.
McFadden D, Plattsmier HS, Pasanen EG (1984) Aspirin-induced hearing loss as a model of sensorineural hearing loss. Hear Res 16:251–260.
Mongan E, Kelly P, Nies K, Porter WW, Paulus HE (1973) Tinnitus as an indication of therapeutic serum salicylate levels. JAMA 226:141–145.
National Center for Health Statistics (1968) Hearing status and ear examination: findings among adults, United States, 1960-1962. Vital and Health Statistics Series 11(32) Washington, D.C.: U.S. Department of Health, Education and Welfare.
Norton SJ, Schmidt AR, Stover LJ (1990) Tinnitus and otoacoustic emissions: Is there a link? Ear Hear 11:159–166.
Pantev C, Hoke M, Lehnertz K, Lutkenhoner B, Anogianakis G, Wittkowski W (1988) Tonotopic organization of the human auditory cortex revealed by transient auditory evoked magnetic fields. Electroencephal Clin Neurophysiol 69:160–170.
Pantev C, Hoke M, Lutkenhoner B, Lehnertz K (1989) Tonotopic organization of the auditory cortex: Pitch versus frequency representation. Science 246:486–488.
Patterson RD, Holdsworth A (1992) A functional model of neural activity patterns and auditory images. Adv Speech Hear Lang Proc (In press).
Penner MJ (1980) Two-tone forward masking patterns and tinnitus. J Speech Hear Res 23:779–786.
Penner MJ (1983) Variability in matches to subjective tinnitus. J Speech Hear Res 26:263–267.
Penner MJ (1984) Equal loudness contours using subjective tinnitus as the standard. J Speech Hear Res 27:267–274.
Penner MJ (1986) Tinnitus as a source of internal noise. J Speech Hear Res 29:400–406.
Penner MJ (1987) The masking of tinnitus and central masking. J Speech Hear Res 30:147–152.
Penner MJ (1988) Audible and annoying spontaneous otoacoustic emissions. Arch Otolaryngol Head Neck Surg 114:150–153.
Penner MJ (1989a) Aspirin abolishes tinnitus caused by spontaneous otoacoustic emissions. Arch Otolaryngol Head Neck Surg 115:871–875.
Penner MJ (1989b) Empirical tests demonstrating two coexisting sources of tinnitus: A case study. J Speech Hear Res 32:458–462.
Penner MJ (1990) An estimate of the prevalence of tinnitus caused by spontaneous

otoacoustic emissions. Arch Otolaryngol Head Neck Surg 116:418–423.

Penner MJ, Bilger RC (1992) Consistent within-session measures of tinnitus, J Speech Hear Res 35:694–700.

Penner MJ, Burns EM (1987) The dissociation of SOAEs and tinnitus. J Speech Hear Res 30:396–403.

Penner MJ, Coles RRA (1992) Indication for aspirin as a palliative for SOAE-caused tinnitus. Br J Audiol 26:91–96.

Penner MJ, Brauth S, Hood L (1980) The temporal course of the masking of tinnitus as a basis for inferring its origin. J Speech Hear Res 24:257–261.

Plinkert PK, Gitter AH, Zenner H (1990) Tinnitus associated spontaneous otoacoustic emissions. Acta Otolaryngol 110:342–347.

Probst R, Coats A, Lonsbury-Martin B, Martin G (1986) Otoacoustic emissions from ears with hearing loss. Am J Otolaryngol 8:73–81.

Puel J-L, Bledsoe Jr SC, Bobbin RP, Ceasar G, Fallon M (1989) Comparative actions of salicylate on the amphibian lateral line and guinea pig cochlea. Comp Biochem Physiol 93:73–80.

Puel J-L, Bobbin RP, Fallon M (1990) Salicylate, mefenamate, meclogenamate, and quinine on cochlear potentials. Arch Otolaryngol Head Neck Surg 102:66–73.

Rebillard G, Abbou S, Lenoir M (1987) Les Oto-Emissions Acoustiques II. Les oto-emissions spontanees:resultats chez des sujets normaux ou presentant des acouphenes. Ann Otolaryngol 104:363–368.

Reed GF (1960) An audiometric study of two hundred cases of subjective tinnitus. Arch Otolaryngol 71:94–104.

Ruggero MA, Rich NC, Freyman R (1983) Spontaneous and impulsively evoked otoacoustic emissions: indicators of cochlear pathology? Hear Res 10:283–300.

Sachs MB, Kiang NYS (1968) Two-tone inhibition in auditory-nerve fibers. J Acoust Soc Amer 68:858–875.

Saito H, Yokoyama A, Takeno S, Sakai T, Ueno K, Masumura H, Kitagawa H (1982) Fetal toxicity and hypocalcemia induced by acetylsalicylic acid analogues. Res Commun Chem Pathol Pharmacol 38:209–220.

Salvi RJ, Ahroon WA (1983) Tinnitus and neural activity. J Speech Hear Res 26:629–632.

Salvi RJ, Saunders SS, Gratton MA, Arehole S, Powers N (1990) Enhanced evoked response amplitudes in the inferior colliculus of the chinchilla following acoustic trauma. Hear Res 50:245–258.

Sasaki CT, Kauer JS, Babitz L (1980) Differential [14C]2-deoxyglucose uptake after deafferentation of the mammalian auditory pathway – a model for examining tinnitus. Brain Res 194:511–516.

Sasaki CT, Babitz L, Kauer JS (1981) Tinnitus: Development of a neurophysiologic correlate. Laryngoscope 91:2018–2024.

Scharf B, Stevens JC (1961) The form of the loudness function near threshold. Proceedings of the Third International Congress of Acoustics I. Amsterdam: Elsevier.

Schloth E (1982) Akustische Aussendungen des menschlichen Ohres (oto-akustiche Emissionen) [dissertation]. Munchen: Technische Universitat.

Schloth E, Zwicker E (1983) Mechanical and acoustical influences on spontaneous oto-acoustic emissions. Hear Res 11:285–293.

Schreiner CE, Snyder RL, Lenarz TH (1990) Spectral and temporal characteristics of abnormal ensemble spontaneous activity of cat auditory nerve. Abstracts of the Thirteenth Midwinter Meeting of the Association for Research in Otolaryngology, pp. 197-198.

Scott B, Lindberg P, Lyttkens L, and Melin L (1985) Psychological treatment of tinnitus. An experimental group study. Scand Audiol 14:223-230.

Shannon RV (1976) Two-tone unmasking and suppression in a forward-masking situation. J Acoust Soc Amer 2:275-287.

Shehata WE, Brownell WE, Dieler R (1991) Effects of salicylate on shape, electromotility and membrane characteristics of isolated outer hair cells from guinea pig cochlea. Acta Otolaryngol 111:707-718.

Sininger YS (1991) Electrophysiologic measurements of tinnitus. Abstracts of the Fourteenth Midwinter Meeting of the Association for Research in Otolaryngology, p. 3.

Smith P, Coles R (1987) Epidemiology of tinnitus: An Update. In: Feldmann H (ed) Proceedings of the III International Tinnitus Seminar. Karlsruhe: Harsch Verlag, pp. 147-153.

Stypulkowski PH (1990) Mechanisms of salicylate ototoxicity. Hear Res 46:113-145.

Tonndorf J (1987) The analogy between tinnitus and pain: A suggestion for a physiological basis of chronic tinnitus. Hear Res 28:271-275.

Tyler RS, Baker LJ (1983) Difficulties experienced by tinnitus sufferers. J Speech Hear Disord 48:150-154.

Tyler RS, Conrad-Armes D (1982) Spontaneous acoustic cochlear emissions and sensorineural tinnitus. Br J Audiol 16:193-194.

Tyler RS, Conrad-Armes D (1983) Tinnitus pitch: A comparison of three measurement methods. Br J Audiol 17:101-107.

Tyler RS, Conrad-Armes D (1984) Masking of tinnitus compared to masking of pure tones. J Speech Hear Res 27:106-111.

Ueno K, Shimoto Y, Yokoyama A, Kitagawa H, Takeno S, Sakai T, Saito H (1983) Alleviation of acetylsalicylic acid-induced fetal toxicity. Res Commun Chem Pathol Pharmacol 39:179-188.

van Willigen F, Emmett J, Cote D, Ayres JJB (1987) CS modality effects in one-trial backward and forward excitatory conditioning as assessed by conditioned suppression of licking in rats. Anim Learn Behav 15:201-211.

Vernon J (1976) The loudness(?) of tinnitus. Hear Speech Action 44:17.

Vernon J (1977) Attempts to relieve tinnitus. J Amer Audiolog Soc 2:124.

Ward WD (1955) Tonal monaural diplacusis. J Acoust Soc Amer 29:365-372.

Wegel RL (1931) A study of tinnitus. Arch Otolaryngol Head Neck Surg 14:160-165.

Wegel RL, Lane CE (1924) The auditory masking of one pure tone by another and its probable relation to the dynamics of the inner ear. Physics Rev 23:266-285.

Wiederhold ML (1990) Effects of tympanic membrane modification on distortion product otoacoustic emissions in the cat ear canal. In: Dallos P, Geisler CD, Matthews JW, Ruggero M, Steele CR (eds) Mechanics and Biophysics of Hearing. New York: Springer-Verlag, pp. 251-258.

Wier CC, Pasanen EG, McFadden D (1988) Partial dissociation of spontaneous otoacoustic emissions and distortion products during aspirin use in humans. J Acoust Soc Amer 84:230-237.

Wilson JP (1980) Evidence for cochlear origin for acoustic re-emissions, threshold for fine structure, and tonal tinnitus. Hear Res 2:233-252.

Wilson JP (1986) Otoacoustic emissions and tinnitus. Scand Audiol Suppl 25:109-118.

Wilson JP, Sutton GJ (1981) Acoustic correlates of tonal tinnitus. In: Evered D, Lawrensen G (eds) Tinnitus. London: Pitman, pp. 82-107.

Wilson JP, Sutton GJ (1983) A family with high tonal objective tinnitus—An update. In: Klinke R, Hartman R (eds) Hearing—Physiological Bases and Psychophysics. Berlin: Springer-Verlag, pp. 97-103.

Zielinski K (1985) Jerzy Konorski's theory of conditioned reflexes. Acta Neurobiol Exp 45:173-186.

Zurek P (1981) Spontaneous narrowband acoustic signals emitted by human ears. J Acoust Soc Amer 69:514-522.

Zwicker E (1958) Ueber psychologishe und methodische Grunderlagen der Lautheit. Acoustica 13.194-211.

Zwicker E (1987) Objective otoacoustic emissions and their uncorrelation to tinnitus. In: Feldmann H (ed) III International Tinnitus Seminar, Karlsruhe: Harsch Verlag, pp. 75-81.

Index

Different species cited in the text are included in the general index. In most cases species are indexed by their scientific names. Common names are included with a note to the scientific name. The only exceptions are a few of the most common species such as humans, cat, chicken, etc. that are indexed by the common name.

α-Difluromethylornithine, ototoxicity, 139
ABR, see Auditory Brainstem Response
Acoustic distortion, effects of ototoxic drugs, 128
Acoustic neuroma, OAEs, 233
Acoustic overpressure
  chinchilla, 225–226
  effects on DPOAEs, 222–225
  effects on OAEs, 222ff
  guinea pig, 222
  human effects, 222–225
  rabbit, 225–226
Acoustic reflex, OAEs, 235
Acoustic trauma, see Acoustic Overpressure
Adenovirus, hearing loss, 157, 178
African clawed frog, see *Xenopus laevis*
Age, effects on types of OAEs, 232
AIDS, cytomegalovirus, 164
  hearing loss, 179ff
  temporal bone histopathology, 181
AIED, see Autoimmune Inner Ear Disease
Albinism-deafness syndrome, 31
Alport syndrome, 31–32

Altricial species, development, 89
Aminoglycosides (see also Ototoxic drugs), 125ff
  OAEs, 217–219
Amphibian, regeneration of eighth nerve, 42–44
Animal model, Meniere's disease, 228–230
Animal models, OAEs, 223ff
Animal perception, tinnitus, 282
Anoxia, effects on OAEs, 221–222
Anthelminic oil of Chenopodium, ototoxicity, 117
Antineoplastic agents, see Cisplatin
Arenavirus, hearing loss, 181
Ascaridole, ototoxicity, 117
Aspirin, see Salicylates
Audiology, OAEs, 200
Audiometric profiles, inheritance, 20
Auditory brainstem response (ABR)
  effects of deprivation, 94–95
  cytomegalovirus, 171
  otitis media, 106ff
  tinnitus, 277
Auditory deprivation, 3–4, 86ff
  anatomical consequences, 96ff
  behavioral consequences, 90ff
  binaural effects in rats, 91

305

Auditory deprivation (cont.)
cell size, 99–100
chicken, 92
cochlear removal, 98
communication, 86ff
defined, 87–88
duck, 92
effects on brainstem nuclei, 97ff
effects on cochlear nucleus, 94
effects on inferior colliculus, 94–95
effects on sound localization, 93, 95
environmental, 97–98
hearing aids, 103
humans, 100ff
in owls, 93
noise effects, 92
physiological consequences, 93ff
research questions, 108–109
territorial behavior, 91
Auditory nerve, See Eighth nerve
Auditory neuron survival, role of neurotrophins, 62ff
Auditory neurons
development and survival, 65
neurotrophin effects, 62–63
regeneration, 62ff
response to NGF, 62ff
Auditory pattern deprivation, rat, 92
Auditory system, development of innervation, 47ff
Autoimmune disease, 5–6
*IH7*Autoimmune inner ear disease (AIED), and viral infection, 182ff
IH0 animal models, 185ff
diagnosis, 182ff
viral etiology, 184–185
Autoimmune reaction, hearing loss, 159
Autosomal recessive nonsyndromic deafness, 32

Barn owl, see *Tyto alba*
Basilar papilla, ototoxicity, 127
BDNF (Brain derived neurotrophic factor), 54–55
binding, 57
cochleovestibular ganglion, 63
effects on auditory neurons, 62–63
role in CNS development, 65
spiral ganglion, 63
Binaural hearing
auditory deprivation effects, 91
deprivation and inferior colliculus, 95
Bird lagena, ototoxicity model, 119–120
Brain derived neurotrophic factor, see BDNF
Bronx waltzer mutant, 231
Bullfrog, see *Rana catesbiana*

Calcium channels, ototoxic drugs, 129
Calcium, tinnitus, 290ff
Candidate gene, defined, 21–22
Carrier drugs, for neurotrophin delivery, 69–70
Cat
DPOAEs, 222
ototoxicity, 127
ototoxicity model, 119
regeneration of eighth nerve, 45
TEOAEs, 203
Cavia procellus, regeneration of eighth nerve, 45
cDNA, inner ear, 12ff, 22, 29
cDNA libraries, 2
Cell size, auditory deprivation, 99–100
Cell transplantation, for neurotrophin delivery, 70
Charcot-Marie Tooth disease, genes, 33
Chemoaffinity theory, 44, 46–47
Chemoattractive fields, 49, 51
Chicken
auditory deprivation, 92
effects of auditory deprivation, 99–100
monaural cochlear deprivation, 98
otocyst, 49
Rubella, 174
Chinchilla langier
acoustic overpressure, 225–226
CN in auditory deprivation, 96–97
cochlear effects of deprivation, 100
DPOAEs, 225–226

effects of acoustic overpressure, 222
ototoxicity model, 119
regeneration of eighth nerve, 45
SOAEs, 209
Chinchilla, see *Chinchilla langier*
Chromosome 5q, 19
Ciliary ganglion, 57
Ciliary neurotrophic factor, see CNTF
Cis-diamminedichloroplatinum II, see Cisplatin
Cisplatin, 131–133
  cochlear microphonics, 132
  effects on DPOAEs, 214–216
  effects on guinea pigs, 215
  effects on hearing, 132
  effects on OAEs, 214–215
  effects on TEOAEs, 214–215
  evoked potentials, 132
  guinea pig, 132
  hearing loss, 214
  organ of Corti, 132
  ototoxic potential, 131
  outer hair cells, 132
  Pryer reflex, 132
Clone, yeast artificial chromosome (YAC), 25ff, 32
CNS development, roles of BDNF and NT-3, 65
CNTF, 57–58
  binding protein (CNTF R), 58
  history, 57
  role in auditory neuron development and survival, 66
  structure, 57–58
Cochlear deprivation, 98
Cochlear deprivation effects, monaural, 98
Cochlear hearing loss, 88
Cochlear homeostasis, effects of ototoxic drugs, 123
Cochlear implant, communication development, 101–102
Cochlear microphonics
  cisplatin, 132
  effects of ototoxic drugs, 123
  effects of salicylates, 134–135
Cochlear nucleus
  auditory deprivation in chinchilla, 96–97

auditory deprivation in ferret, 97,99
auditory deprivation in mouse, 96ff
effects of conductive hearing loss on, 99–100
physiological effects of deprivation, 94
Cochleovestibular ganglion, BDNF, 63
Cockayne syndrome, 32, 33
COL2A1, 32
COL2A2, 32
Columbia SK virus, hearing loss, 157, 181–182
Communication deficits, hearing deprivation, 101–102
Communication development, cochlear implant, 101–102
Communication, otitis media, 104–105
Compound action potential (CAP), 222
  OAEs, 232
  DPOAEs, 228–230
Conductive hearing loss, 87–88
  early onset, 104–105
  effects on neuron size, 99–100
  late onset, 107–108
Congenital deafness, pedigree, 17
Congenital hearing loss, statistics, 156
Coxsackievirus, hearing loss, 157
Critical period, 4, 90ff
  hearing development in owls, 93, 95
Cryptococcus neoformans, HIV, 180–181
Cytomegalovirus
  animal models, 166ff
  auditory brainstem response, 171
  diagnosis, 163–164
  hearing loss, 157ff
  symptoms, 164–165, 167–168
  temporal bone histopathology, 165–166

Deafness genes, 15
  human, 18, 30ff
Deafness
  mitochondrial inheritance, 31
  molecular genetics, 10ff
  nonsyndromic, 18
  regressively inherited, 16

Degeneration, 44ff
    auditory nerve, 49
    neurotrophic factors, 52
Deletion map, X chromosome, 26
Denervation, lateral line, 51
Deprivation, see Auditory Deprivation
Development
    efferent system, 49
    hair cells, 47, 49
    hearing, 88–89
    innervation, 47–50
    innervation of auditory system, 47ff
Diabetes mellitus, genes, 33
Dinucleotide repeats, deafness, 15
Diplacusis, tinnitus, 276
Distortion Product Otoacoustic Emissions, see DPOAEs
Diuretics, 136–138
    administration with ototoxic drugs, 217
    clinical effects, 136
    effects on OAEs, 215
    effects on primate TEOAEs, 215
    endolymphatic potential, 215
    humans OAEs, 215
    mechanism of action, 137–138
    ototoxic effects, 136–138
    pathology, 136
    stria vascularis, 136–138, 215
DNF, delivery to ear, 69
DPOAEs (see also OAEs, SOAEs, SFOAEs, TEOAEs)
    Bronx waltzer mutant, 231
    cat, 222
    chinchilla, 225–226
    compound action potential, 228–230
    definition, 210–213
    differences with other OAE types, 213
    diuretics in primates, 215
    effects of acoustic overpressure, 222–225
    effects of age, 232
    effects of anoxia and hypoxia, 222
    effects of cisplatin, 214–216
    effects of gentamicin, 218–219
    effects of ototoxic drugs, 220–221
    effects of quinine, 220
    effects of salicylate, 219–220
    effects of toluene, 220
    generators, 221
    guinea pig, 215, 222, 231
    hair cell loss, 231
    hereditary hearing loss, 230–231
    human, 210–213, 222–225
    influence of efferent system, 233
    Macaca sp., 206, 212
    Meniere's disease, 227–228
    mice, 230–231
    primate, 206, 212
    rabbit, 212, 220, 225–226
    rodent, 212, 220
Duck, auditory deprivation, 92

Ear canal, OAEs, 199
Efferent system, development, 49
Efferent system, influence on OAEs, 233
Eighth nerve, degeneration, 49
    regeneration, 42ff
Encephalomyocarditis, hearing loss, 157
Endolymphatic hydrops, animal model for Meniere's, 228–230
Endolymphatic hydrops, OAEs, 227–230
Endolymphatic potential
    diuretics, 215
    effects of ototoxic drugs, 128
Environmental deprivation, 92
Epstein-Barr virus, hearing loss, 157, 181–182
Erythromycin, ototoxicity, 139
Ethacynic acid, see Diuretics
Evoked potential, cisplatin, 132
    effects of ototoxic drugs, 128, 217–218
Exon trapping, 26–27
Extinction, tinnitus, 281

Felis domestica, see cat
Ferret, see *Mustellla putorius*
FGFs 1–7, 60–62
    physiological role, 62
    role in auditory neuron development and survival, 67
Fibroblast Growth Factors, see FGFs
Furosemide, see Diuretics

Index

Gallus domesticus, see Chicken
Genes, human deafness, 30ff
Genetic isolates, 19
Genetics, 2
  experimental designs, 10ff
Gentamicin
  effects on DPOAEs, 218-219
  effects on OAEs, 218-219
  hair cells, 130
  metabolites, 130
  ototoxic drugs, also see Kanamycin
Gentamicin, time course of effects, 130
Gerbil, SFOAEs, 208
Glutathione, effects on ototoxic drugs, 130
Glycerol
  effects on Meniere's disease, 230
  effects on OAEs, 230
Gorilla, herpes zoster, 176
Grieg cephalopolysyndactyly syndrome, gene, 33
Growth factors, history
Growth factors, see Neurotrophic Factors, Neurotrophins
Guinea pig
  autoimmune inner ear disease, 186
  cisplatin, 132
  cytomegalovirus, 166ff
  DPOAEs, 215, 222, 231
  effects of cisplatin, 215
  hereditary hearing loss, 231
  hair cells, salicylate, 289
  herpes zoster, 176
  influenza, 177
  middle ear effects on OAEs, 235
  ototoxic drugs, 126-127
  ototoxicity model, 119
  SFOAEs, 208
  SOAEs, 209
  TEOAEs, 203, 208

Hair cell
  calcium, 290-291
  development, 47, 49
  DPOAEs, 231
  effect of ototoxic drugs, 124, 126-129
  effects of salicylate, 219-220
  effects of toluene, 220
  gentamicin, 130
Hamster
  influenza, 177
  mumps, 161
  rubeola, 162-163
  variola virus, 179
Haplotype analysis, deafness, 23
Hearing aids, auditory deprivation, 103
Hearing deprivation, communication deficits, 101-102
Hearing development, otitis media, 104-105
Hearing
  effects of cisplatin, 132
  effects of salicylates, 133-134
Hearing impairment
  early identification, 102
  early onset, 101-102
Hearing impairment, late onset, 102-103
Hearing loss
  adenovirus, 178
  AIDS, 179ff
  arenavirus, 181
  autoimmune disease, 182ff
  cisplatin, 214
  conductive, 87-88
  cytomegalovirus, 163ff
  deprivation, 87ff
  eighth nerve, 99-100
  herpes simplex, 178-179
  herpes zoster oticus, 174ff
  HIV, 179ff
  humans, 232
  mumps, 160ff
  parainfluenza, 178
  rubella, 172ff
  rubeola, 161ff
  tinnitus, 279
  variola virus, 179
  viral infection, 155ff
Hearing, newborns, 7
Hearing, onset, 88-89
Hepatitis virus, hearing loss, 157
Hereditary hearing loss, 230-231
  DPOAEs, 230-231
  guinea pig, 231
  mice, 230-231

Herpes viruses, hearing loss, 157, 178–179
Herpes zoster oticus, hearing loss, 174ff
HIV, hearing loss, 157, 179ff
Human
  deafness genes, 30ff
  DPOAEs, 210–213, 222–225
  effects of acoustic overpressure, 222–225
  genes, 33
  hearing loss, 232
  OAEs, 199ff
  ototoxic drugs, 217
  ototoxicity model, 117–118
  SFOAEs, 205–208
  SOAEs, 208, 209
  TEOAEs, 202–203
  use of OAEs in newborns, 240–241
Human immunodeficiency virus (HIV), hearing loss, 157, 179ff
Hunter syndrome, 31, 33
HuP2, human gene, 21
Hurler syndrome, gene, 32–33
Hyla squirelia, regeneration of eighth nerve, 42–43
Hypoxia, effects on OAEs, 221–222

IGF (Insulin-like growth factors), 58–60
  role in auditory neuron development and survival, 66–67
Implantable polymers, for neurotrophin delivery, 70
In vitro systems, ototoxicity model, 120
Inferior colliculus
  auditory deprivation effects, 97
  effects of auditory deprivation, 99–100
  physiological effects of deprivation, 94–95
  tinnitus, 292
Influenza, hearing loss, 177
Influenza virus, hearing loss, 157
Inheritance, deafness, 10ff
Inner ear
  application of neurotrophins, 68–70

  cisplatin effects, 132
  development in mouse, 48–49
  infection, virus, 158ff
Inner hair cell
  Bronx waltzer mutant, 231
  effect of ototoxic drugs, 126–129
  NGF, 63
Innervation, auditory system, 47ff
Insulin, 58–60
Insulin-Like Growth Factors I and II, see IGF
Intense sound, also see Acoustic Overpressure
  effects on OAEs, 222ff

Kanamycin, effect on hearing, 128
KIT gene, 32
Koch's postulates, viral infection, 157

Labyrinthitis, 5–6
Language development, otitis media, 104ff
Late-onset auditory deprivation, 103
Lateral line
  chemoattraction, 49
  denervation, 51
  ototoxicity model, 119–120
  Rana clamitans, 51
  reversibility of ototoxic drugs, 128
Lateral superior olive, auditory deprivation effects, 97–98
Leopard frog, see *Rana pipiens*
Linguistic ability and hearing, 4
Lizard, ototoxicity, 127
Loop diuretics, see Diuretics
Loudness, tinnitus, 262ff

Macaca sp., DPOAEs, 206, 212
Magnitude estimation, tinnitus, 265
Mammal, regeneration of eighth nerve, 44ff
Masking level difference, otitis media, 106–107
Masking, tinnitus, 264ff
Measles, see Rubeola
Melanogenesis, 29

Meniere's disease
  animal model, 228–230
  DPOAEs, 227–228
  effects of glycerol, 230
  endolymphatic hydrops, 228–230
  TEOAEs, 227–228
  OAEs, 227–230
Middle ear disorders
  effects on OAEs, 233–237
  effects on TEOAEs, 235
Mitochondrial inheritance, deafness, 31
Models of genetic deafness, mouse, 20ff
Molecular genetics, deafness, 10ff
Monaural hearing loss, sequelae, 103–104
Monge's disease, 31
Monkey (also see Primate)
  herpes zoster, 176
  mumps, 161
Mononucleosis virus, hearing loss, 157
Mouse
  cochlear nucleus in auditory deprivation, 96ff
  cytomegalovirus, 166ff
  development of inner ear, 48–49
  DPOAEs, 230–231
  effects of auditory deprivation, 99–100
  hereditary hearing loss, 230–231
  models of genetic deafness, 20ff
  monaural cochlear deprivation, 98
  mutants, human syndromes, 33
  mRNA, inner ear, 12ff
  otic placode, 47
  ototoxicity model, 119
  NGF, 63
Mumps
  animal models, 161
  hearing loss, 157ff
  temporal bone histopathology, 160–161
  vestibular effects, 160ff
Mustellla putorius (ferret)
  auditory deprivation effects 96–97, 99–100
  influenza, 177
  monaural cochlear deprivation, 98–99

Mycoplasma pneumoniae, hearing loss, 178

Neomycin
  effects on hearing, 128
  ototoxicity, 127
Nephrotoxicity, 117, 125
Nerve growth factor, see NGF
Neurofibromatosis, gene, 33
Neuronal repair, history, 41
Neurotrophic factors, 2–3, 41ff; also see Neurotrophins
  CNTF, 57–58
  definition, 52
  FGF, 60–62
  history, 52–53
  IGF, 58–60
  insulin, 58–60
Neurotrophins, also see Neurotrophic factors
Neurotrophins, see also BDNF, CNTF, FGF, IGF, NGF, TGF, NT3, NT4, NT5, NT-4/5, NT6
  BDNF, 54–55
  clinical applications, 72–73
  delivery systems, 68–70
  NGF, 52–53
  role in auditory neuron survival, 62ff
NGF (Nerve Growth Factor), 52–53
NGF binding, 57
  effects on auditory neurons, 62
  hair cells, 63
  high-affinity receptor, 55–56
  low-affinity receptor, 55
  mRNA expression, 63
  organ of Corti, 63
  receptors, 55–57
Noise, auditory deprivation, 92
Noise damage, see Acoustic Overpressure
Norrie's disease, gene, 31ff
Northern blot, 28
NT-3, 51, 54, 56
  delivery to ear, 69
  effects on auditory neurons, 62–63
  role in CNS development, 65
  spiral ganglion, 63, 64
NT-4/5, 56–57

NT-4, 54–55, 56–57
NT-5, 54–55
NT-6, 55
Nucleus magnocellularis, auditory deprivation in chicken, 99–100

OAEs (see also DPOAEs, SOAEs, SFOAEs, TEOAEs), 6–7 199ff
  acoustic neuroma, 233
  acoustic reflex, 235
  animal models, 223ff
  clinical applications, 237ff
  compound action potential, 232
  definition, 199–200
  differences between, 213
  effects of acoustic overpressure, 222ff
  effects of aminoglycoside antibiotics, 217–219
  effects of anoxia, 221–222
  effects of cisplatin, 214–215
  effects of diuretics, 215
  effects of diuretics on rabbits, 215
  effects of gentamicin, 218–219
  effects of hypoxia, 221–222
  effects of intense sound, 222ff
  effects of middle ear disorders, 233–237
  effects of ototoxic drugs, 123
  effects of quinine, 220
  effects of salicylate, 219–220
  effects of toluene, 220
  endolymphatic hydrops, 227–230
  glycerol effects, 230
  guinea pig, 235
  hereditary hearing loss, 230–231
  human, 199ff, 215
  influence of efferent system, 233
  measurement, 200–202, 204–205
  Meniere's disease, 227–230
  neural involvement, 232
  origin, 232
  ototoxic drugs, 213ff, 217–219
  primates, 221
  retrocochlear hearing loss, 232–233
  suppression, 209
  types, 201
  use with human newborns, 240–241
Oncogenes, 55

Operant conditioning, tinnitus, 281ff
Organ cultures, ototoxicity model, 120
Organ of Corti
  autoimmune inner ear disease, 185ff
  cisplatin, 132
  cytomegalovirus, 166
  herpes 179
  mumps, 161
  NGF, 63
  parainfluenza, 178
  regeneration of eighth nerve, 44ff
  rubeola, 162
Osteogenesis imperfecta, gene, 33
Osteopetrosis, gene, 33
Otic placode, mouse, 47
Otitis media
  auditory brainstem response, 106–107
  communication sequelae, 104–105
  masking level difference, 106–107
  sound localization, 106
Oto-palatal-digital syndrome, 31
Otoacoustic emissions, see also DPOAEs, OAEs, SOAEs, SFOAEs, TEOAEs
Otocyst, 14, 51
  chicken, 49
  ototoxicity model, 120
Otolith organs, rubeola, 162
Ototoxic drugs (also see α-Difluromethylornithine, Cisplatin, Diuretics, Erythromycin, Gentamicin, Kanamycin, Neomycin, Polypeptide antibiotics, Quinine, Salicylates, Streptomycin), 124ff
  administration with diuretics, 217–218
  adverse effects, 117
  aminoglycosides, 125ff
  basilar papilla, 127
  calcium channels, 129
  cisplatin, 131–133
  classification, 125
  effects on auditory function, 116, 123–124, 126–129
  effects on cochlear homeostasis, 123
  effects on cochlear microphonics, 123

effects on DPOAEs, 220-221
effects on hair cells, 124, 126-129
effects on hearing, 128
effects on OAEs, 123, 217-219
effects on Pryer reflex, 123
effects on vestibular function, 124
evaluation, 122-124
guinea pig, 126-127
incidence of ototoxicity, 126
mechanisms of action, 129-130
metabolites, 130
nephrotoxicity, 125
OAEs, 200, 213ff
outer hair cells, 130, 217
pathology, 126-129
pharmacokinetcs, 120-121
routes of administration, 121-122
salicylates, 133-136
time course of effects, 130
tissue culture, 127
vestibular disorders, 129
Ototoxicity, also see Ototoxic Drugs
α-difluromethylornithine, 139, 116ff
amelioration of adverse effects, 141-142
animal models, 117ff
anthelminic oil of chenopodiumm, 117
antioxidants, 142
cat, 127
definition, 116
diuretics, 136-138
drug-drug interactions, 140-141
effects on vestibular system, 126
erythromycin, 139
evaluation, 122-124
factors influencing, 140-142
lizards, 127
molecular structure, 116-117
noise trauma, 142
polypeptide antibiotics, 139
quinine, 138-139
rabbit, 127
reversibility, 128
squirrel monkey, 127
symptoms, 117
transduction channels in hair cells, 128-129
Ototoxicity model
bird lagena, 119-120

cat, 119
chinchilla, 119
guinea pig, 119
humans, 117-118
in vitro systems, 120
lateral line, 119-120
mouse, 119
primates, 118-119
rabbit, 119
rat, 119
Outer hair cell
active processing, 199-200
Bronx waltzer mutant, 231
cisplatin, 132
effect of ototoxic drugs, 126-129
NGF, 63
ototoxic drugs, 130, 217
salicylates, 136
Oxygen intake, effects on OAEs, 221-222

Parainfluenza, hearing loss, 157, 178
organ of Corti, 178
Pathology, ototoxic drugs, 126-129
PAX3 gene, 21-22, 30, 32
Perception by animals, tinnitus, 282
Permanent threshold shift (PTS), and ototoxicity, 116, 117
Pharmacokinetics, ototoxic drugs, 120-121
Phosphoinositide lipids, ototoxic drugs, 129
Piebald trait, gene, 33
Pigtail macaque, see *Macaca nemestrina*
Pitch, animal tinnitus, 286ff
tinnitus, 260ff, 275
Platyfish, see *Xiphophorus maculatus*
PMP-22 gene, 32
Polio virus, hearing loss, 157, 181-182
Polypeptide antibiotics, ototoxicity, 139
Positional cloning, 23ff
Precocial species, development, 89
Presbycusis, 102-103
Primate
DPOAEs, 206, 212
effects of diuretics on TEOAEs, 215
effects of salicylate on OAEs, 221

Primate (cont.)
    ototoxicity model, 118–119
    SFOAEs, 207–208
    SOAEs, 209
    TEOAEs, 203, 206
Programmed cell death, 63
Prostaglandins, salicylates, 135
Proto-oncogenes, 55
Pryer reflex
    cisplatin, 132
    effects of ototoxic drugs, 123
Psychometric function, tinnitus, 275

Quinine
    effect on OAEs, 220
    effects on DPOAEs, 220
    effects on SOAEs, 220
    effects on TEOAEs, 220
    ototoxicity, 138–139
    tinnitus, 284

Rabbit
    acoustic overpressure, 225–226
    DPOAEs, 220, 212s, 225–226
    effects of diuretics on OAEs, 215
    herpes simplex, 179
    monaural cochlear deprivation, 98
    ototoxicity, 127
    ototoxicity model, 119
Ramon y Cajal, 41, 52
Ramsy Hunt syndrome, herpes, 175
Rana catesbiana, regeneration of
        eighth nerve, 44
Rana clamitans, lateral line
        development, 49
Rana clamitans, lateral line
        regeneration, 51
Rana pipiens, regeneration of eighth
        nerve, 43–44
Rana temporaria, regeneration of
        eighth nerve, 43
Rat
    animal model for tinnitus, 277ff
    auditory deprivation effects on
        cochlear nucleus, 94
    auditory deprivation effects on
        inferior colliculus, 94–95
    auditory pattern deprivation, 92

behavioral model for tinnitus, 280ff
binaural auditory deprivation, 91
effects of auditory deprivation,
    99–100
herpes zoster, 176
ototoxicity model, 119
Receptor binding, 57
Receptors, NGF, 55–57
Regeneration, 41ff
    amphibian, 42–44
    eighth nerve, 2–3, 42ff
    mammals, 44ff
    of frequency response, 43–44
    spiral ganglion, 44ff
    vestibular system, 43, 44
Retinitis pigmentosa, deafness, 17, 22
Retrocochlear hearing loss, 232–233
Rhesus macaque, see *Macaca mulatta*
RNA amplification, hair cells, 13
Rodent
    DPOAEs, 212, 220
    ototoxicity model, 119
Rubella
    animal models, 174
    audiogram, 173
    clinical findings, 173
    hearing loss, 172ff
    temporal bone histopathology,
        173–174
    teratology, 172
Rubella virus, hearing loss, 157
Rubeola
    animal models, 162–163
    audiogram, 162
    hearing loss, 157ff
    otolith organs, 162
    temporal bone histopathology, 162
    Warthin-Finkeldy giant cells, 162

Salicylates, 133–136
    animal tinnitus, 282ff
    calcium, 291
    cochlea, 279
    cochlear microphonics, 134–135
    dose-response, 284
    effect on TEOAEs, 219–220
    effects on DPOAEs, 219–220
    effects on hair cells, 219–220

effects on hearing, 133–134
effects on OAEs, 219–220
hair cells, 289
hearing loss, 279
mechanism of action, 135–136
outer hair cells, 136
pathology, 134–135
prostaglandins, 135
SOAE, 273
spontaneous neural activity, 278–279
tinnitus, 133, 135, 278ff
tinnitus loudness, 285–286
tinnitus pitch, 287–288
Sarcoma 180, 53
Sensitive period, development, 90ff
Sensorineural hearing loss, 17, 88, 199–200
Sensorineural hearing loss, TEOAEs, 232–233
SFOAEs, see also OAEs, DPOAEs, SOAEs, TEOAEs)
    gerbil, 208
    guinea pig, 208
    human, 205–208
    influence of efferent system, 233
    measurement,, 207
    primates, 207–208
    suppression, 209
    tinnitus, 276
SOAE-caused tinnitus, psychophysics, 272ff
SOAEs
    aspirin, 273
    chinchilla, 209
    cochlear damage, 276–277
    definition, 208
    effects of age, 232
    effects of quinine, 220
    fluctuation, 270, 272
    frequency range, 274
    gender, 274
    guinea pig, 209
    human, 208
    incidence, 274
    masking, 270ff
    measurement techniques, 204–205, 275
    mechanism of production, 208–209
    predominant pitch, 269

primate, 209
suppression, 209, 270ff
tinnitus, 7, 259ff, 268ff
tinnitus in animals, 289
Somatosounds, tinnitus, 268
Sound localization
    effects of auditory deprivation, 93, 95
    otitis media, 106
Speech quality, hearing deprivation, 101–102
Sperry, Roger, 41–44
Spiral ganglion
    BDNF, 63, 64
    degeneration, 44ff
    development, 47
    explant, 64
    NT-3, 63
Splotch, mouse chromosome, 21, 31
Spontaneous activity, tinnitus, 292–293
Spontaneous Otoacoustic Emissions, see SOAE
Squirrel monkey, ototoxicity, 127
St. Louis encephalitis, hearing loss, 157
Stickler syndrome, 32
Stimulus Frequency Otoacoustic Emissions, see SFOAEs
Stimulus generalization gradients, auditory deprivation, 92
Streptomycin, 125
Streptomycin, blocking transduction channels, 128–129
Stria vascularis, effects of diuretics,136–138, 215
Superior olivary complex, auditory deprivation effects, 97ff
Suppression, OAE, 209
Synaptogenesis, 47
Syndromal deafness, human, 31
Systemic groups, 21

Temporal pattern discrimination in rats, auditory deprivation, 91
Temporary threshold shift (TTS), and ototoxicity, 116, 117
    cat, 203
    diuretics in primates, 215
    effects of age, 232

Temporary threshold shift (TTS), and
    ototoxicity (cont.)
  effects of cisplatin, 214–215
  effects of middle ear disorders,
      235–235
  effects of quinine, 220
  effects of salicylates, 219–220
  guinea pig, 203, 207, 208
  human, 202–203
  human hearing loss, 232
  measurement, 201–205
  Meniere's disease, 227–228
  primates, 203, 206
  sensorineural hearing loss, 232–233
  suppression, 209
Territoriality in rats, auditory
    deprivation, 91
TGF, 62
  role in auditory neuron development
      and survival, 67–68
Tickborne encephalitis, hearing loss,
    157
Tinnitus, 7, 258ff
  ABR, 277
  Amer. Nat. Stand. Inst. (ANSI)
      definition, 259
  and SOAEs, 7
  animal behavioral model, 288
  animal model, 277ff
  behavioral animal model, 279ff
  behavioral animal model paradigm,
      280ff
  calcium, 290–291
  conditioning and learning, 280ff
  controls in animal model, 283–284
  diplacusis, 276
  hair cells, 290–291
  hearing loss, 279
  human, 259–260
  inferior colliculus, 292
  loudness, 262ff
  loudness in animal, 285–286
  magnitude estimation, 265
  masking, 264ff
  mechanisms, 289ff
  neuronal activity, 291ff
  operant conditioning, 281ff
  ototoxicity, 116, 117
  pitch, 260ff, 275, 286ff

  psychoacoustical measurement,
      261–262
  quinine, 284
  rubeola, 162
  salicylate dose-response, 284
  salicylate-induced, 278ff
  salicylate model, 288ff
  salicylates, 133, 135, 282ff
  SOAEs, 259, 289
  somatosounds, 268
  spontaneous activity, 292–293
  two-tone suppression, 267
  without physical basis, 260ff
Tissue culture, ototoxic drugs, 127
Toluene
  effects on DPOAEs, 220
  effects on hair cells, 220
  effects on OAEs, 220
Transforming Growth Factor Beta
    Subfamily, see TGF
Transiently Evoked Otoacoustic
    Emissions, see TEOAE
Trapezoid body, auditory deprivation
    effects, 97ff
Tree frog, see *Hyla squirelia*
Trembler mutants, 32 trk
    proto-oncogene, 55–56 trk-A,
    55–56, trk-B, 56–57 trk-C,
    56–57
Trp-I gene, 29
Two-tone suppression, tinnitus, 267
Tyto alba (barn owl), critical period,
    93, 95
Tyto alba (barn owl), monaural
    auditory deprivation, 93, 95

Unilateral hearing loss, 4
Usher syndrome, 2, 17, 31

Varicella-zoster, hearing loss, 157
Variola virus, hearing loss, 157, 179
Vertigo, 156
Vertigo, rubeola, 162
Vestibular disorders, AIDS and HIV,
    180
Vestibular effects, influenza, 177
Vestibular effects, mumps, 160ff

Vestibular function
  deafness, 20
  effects of ototoxic drugs, 124
Vestibular ganglion, development and survival, 66
Vestibular malfunction, deafness, 17
Vestibular system
  ototoxic drugs, 129
  regeneration, 43, 44
Viral infection, 5-6
  autoimmune disease, 182ff
  consequences, 158-159
  diagnosis, 155ff
  otopathology, 155ff
  routes to ear, 158
Viral vector gene therapy, 71-72

Wardenburg syndrome, 2, 21-22, 25, 30-31

Warthin-Finkeldy giant cells, rubeola, 162
Western equine encephalitis, hearing loss, 157
Whirler, chromosomes, 24
Wolfram syndrome, gene, 33

X chromosome, deletion map, 26
X-linked genes, deafness, 10ff
Xenopis laevis, neurotrophins, 54-55
Xiphophorus maculatus, neurotrophin-6, 55

Yeast artificial chromosome (YAC), 25ff, 32
Yellow fever, hearing loss, 157, 181-182